Interstellar Magnetic Fields

Interstellar Magnetic Fields

Observation and Theory

Proceedings of a Workshop,
Held at Schloß Ringberg, Tegernsee,
September 8–12, 1986

Editors: R. Beck and R. Gräve

With 125 Figures

Springer-Verlag Berlin Heidelberg New York
London Paris Tokyo

Dr. Rainer Beck
Dr. Roland Gräve

Max-Planck-Institut für Radioastronomie, Auf dem Hügel 69,
D-5300 Bonn 1, Fed. Rep. of Germany

ISBN-13:978-3-642-72623-1   e-ISBN-13:978-3-642-72621-7
DOI: 10.1007/978-3-642-72621-7

2153/3150-543210

Preface

The subject of interstellar magnetic fields is of increasing interest to both observational and theoretical astronomers. Consequently, it was opportune for both groups to convene and discuss the current state of research and future prospects. For this purpose, the Max-Planck-Gesellschaft zur Förderung der Wissenschaften e.V. generously provided its unique conference centre Schloß Ringberg.

The scientific organising committee consisted of Profs. W. Hillebrandt (Max-Planck-Institut für Physik und Astrophysik, München) and R. Wielebinski (Max-Planck-Institut für Radioastronomie, Bonn). The local organising committee comprised Gabriele Breuer (conference secretary) and the Editors.

These proceedings summarise the lectures and short contributions presented at the workshop. Although extremely productive, the discussions could not, unfortunately, be included.

We thank the staff of the castle, and especially its administrator, Mr. Hörmann, for providing the excellent environment conducive to smooth running of the conference. We acknowledge the Max-Planck-Gesellschaft for its support in financing the workshop. Thanks are also due to all participants for their enthusiasm and (generally!) prompt submission of manuscripts. Last but not least, we thank Gabriele Breuer for her invaluable help before, during and after the days at Schloß Ringberg. She skillfully typed and retyped many of the manuscripts for these proceedings and unstintingly performed last minute corrections.

We hope that some of the ideas presented here will stimulate further discussion and research in this area.

Bonn, December 1986

R. Beck
R. Gräve

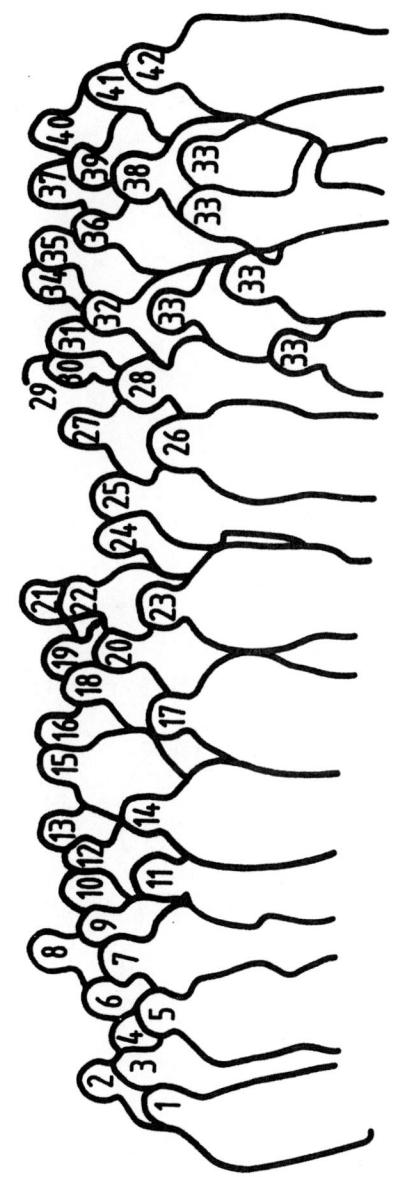

1 Arp, H.
2 Gräve, R.
3 Beck, R.
4 Meyer, F.
5 Klein, U.
6 Crusius, A.
7 Wielebinski, R.
8 Hillebrandt, W.
9 Fujimoto, M.
10 Meyer-Hofmeister, E.
11 Sofue, Y.
12 Feitzinger, J.V.
13 Cugnon, P.
14 Kronberg, P.P.

15 Kundt, W.
16 Aly, J.J.
17 Ruzmaikin, A.
18 Scarrott, S.M.
19 Anzer, U.
20 Verschuur, G.L.
21 Thiemann, H.
22 Lesch, H.
23 Nelson, A.H.
24 Sanchez-Saavedra, M.L.
25 Spicker, J.
26 Loiseau, N.
27 Krause, F.
28 Asseo, E.

29 Fürst, E.
30 Buczilowski, U.R.
31 Reif, K.
32 Krause, M.
33 Tosa, M.
34 Völk, H.
35 Junkes, N.
36 Hummel, E.
37 Breuer, G.
38 Reich, W.
39 Browne, P.F.
40 Schmidt-Voigt, M.
41 Kössl, D.
42 Sieber, W.

Summary of Part I

Contents

Part II   Large-Scale Magnetic Fields in Our Galaxy

Large-Scale Magnetic Fields in Galaxies

Interstellar Magnetic Fields –
Past, Present and Future

R. Beck

Max-Planck-Institut für Radioastronomie, Auf dem Hügel 69,
D-5300 Bonn 1, Fed. Rep. of Germany

Interstellar magnetic fields have become popular in astrophysics in recent years, stimulated by a significant improvement of the observational techniques in the radio and optical range and a refinement of theoretical models. It now seems promising to attack the question of the origin of magnetic fields and their influence on the dynamics of the interstellar medium.

1. Past

Until the middle of this century cosmical magnetic fields were known to exist only in our solar system. The importance of magnetic effects on the sun was widely discussed at that time. In 1963 IAU Symposium No. 22 about "Stellar and Solar Magnetic Fields" was held only a few kilometers away from the place of this workshop. The presence of uniform interstellar magnetic fields had been inferred from the observation of optical polarization in our Galaxy [1] as well as from the detection of linearly polarized radio emission from our Galaxy [2-4]. Optical polarization had been explained by grain alignment perpendicular to the field lines [5], while radio polarization had been attributed to synchrotron emission. By 1971 the dynamo theory of magnetic field amplification originally developed for the sun [6] had been modified for galactic disks [8-10]. The zero mode of the "thin disk dynamo" produces an axisymmetric, mainly toroidal field.

In the sixties the influence of magnetic fields on the evolution of spiral galaxies was lively discussed. Several astronomers tried to explain the Hubble sequence by the angle between the rotation axis and magnetic field orientation in the protogalaxy [11,12]. The estimates of the field strength in our Galaxy, however, indicated that the magnetic energy is far too small to influence the rotation of the gaseous disk significantly (however compare NELSON, this volume). The following decline of interest in interstellar fields is marked by WOLTJER'S summary remarks at various conferences:

> *"The role of magnetic fields in interstellar gas dynamics and spiral structure remains very uncertain."* (Brussels, 1964)

> *"The larger one's ignorance, the stronger the magnetic field."* (Noordwijk, 1966)

> *"The magnetic field cannot be the main cause of spiral structure".* (Basel, 1969)

In the next decade there was little interest in interstellar magnetic fields. Theoreticians then started to consider fields in the process of star formation [13,14] and for the stabilization of jets [15]. The dynamo theory was applied to real galaxies [16]. An alternative theory of a large—scale, bisymmetric magnetic field compressed from an intergalactic field was developed [17]. A dynamo operating in the first mode was invoked to achieve a steady—state solution [18].

The advance of radio observation techniques enabled the study of the field structure in our Galaxy by means of rotation measures of extragalactic sources. The interpretation is, however, still under debate [19,20]. Maps of the polarized radio emission from nearby spiral galaxies could be confronted with axisymmetric and bisymmetric field models [21,22].

2. Present

Figure 1 draws the present picture of the role of magnetic fields in interstellar space. Magnetic fields interact with all other components of the interstellar medium. The various interaction processes may lead to a distribution of energy which — averaged over the whole galaxy — is observed as energy equipartition between e.g. cosmic rays and magnetic fields [23].

The role of magnetic fields may be even more important than is suggested in Fig. 1. Magnetic effects in the interstellar medium have been invoked for almost every phenomenon (Table 1). It now seems that galaxies would appear much different without magnetic fields, maybe they would not exist at all.

The origin of magnetic fields, however, is still dubious. Theoreticians agree that some sort of dynamo action is needed to maintain the observed field strengths. The

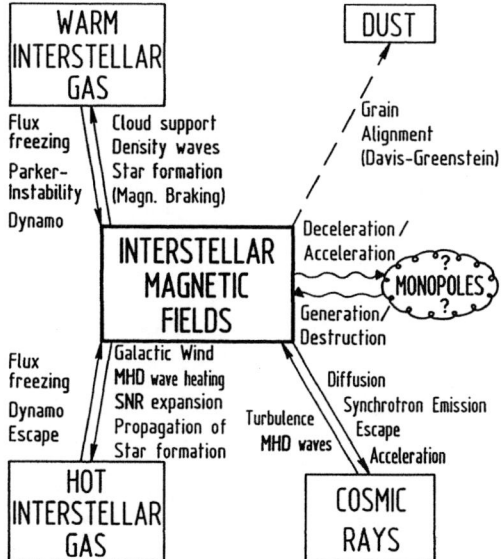

Figure 1: Interaction processes in the interstellar medium

Table 1. Effects of interstellar magnetic fields on:

	possible	probable	almost certain
gas cloud support			+
star formation			+
stellar evolution	+		
stellar activity			+
supernova explosion	+		
SNR expansion		+	
interstellar gas dynamics			+
interstellar dust dynamics		+	
cosmic ray dynamics			+
spiral structure		+	
density waves		+	
galactic wind		+	
galaxy interaction	+		
nuclear activity		+	
jet formation			+
galaxy rotation	+		
galaxy evolution	+		
galaxy formation	+		

kind of dynamo (α^2, $\alpha\omega$, local turbulent), the mode of operation (axisymmetric, bi-symmetric, or higher) and the origin of seed fields are controversal.

The dynamo model is supported by the observational result that the average field strength in a galaxy and the average surface mass density (determined from the rotation curve in the same radius interval) are correlated (Fig. 2). In the dynamo theory the maximum field strength is reached as soon as magnetic stresses balance field amplification which leads to a condition [24] consistent with observations:

$$B_p\ B_t\ \propto\ R\ \rho\ v\ \Omega$$

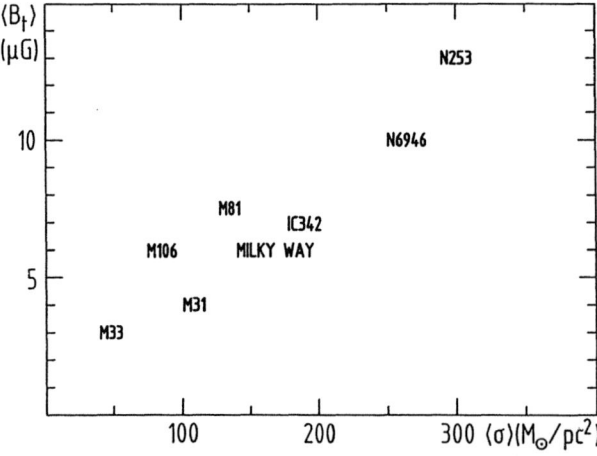

Figure 2: Relation between the average effective strength of the total magnetic field $\langle B_t \rangle$ and the average surface mass density $\langle \sigma \rangle$. $\langle B_t \rangle$ has been computed from radio observations; $\langle \sigma \rangle$ from HI rotation curves in the same radius range (see [22] for details)

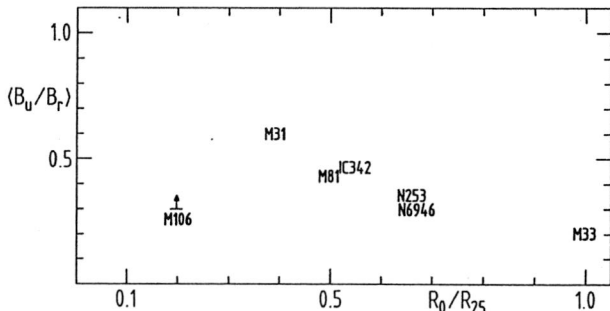

Figure 3: Relation between the average degree of uniformity of the magnetic field $\langle B_u/B_r\rangle$ and the shape of the rotation curve characterized by the ratio between the radius R_0 where the rotation curve becomes flat and the radius R_{25} at optical surface brightness level $\mu_B = 25$ mag arcsec^{-2}. $\langle B_u/B_r\rangle$ depends on the spatial resolution of the observations which is comparable in all galaxies except for M106 (lower resolution)

where B_p and B_t are the poloidal and toroidal field components, R is the radius of the source, ρ the gas density, v the turbulent velocity and Ω the angular velocity.

Galactic dynamos produce large-scale field structures of axisymmetric, bisymmetric or higher modes. The degree of uniformity of the field (expressed by the ratio of uniform to random field strengths B_u/B_r) may be used as a measure of the effectivity of the $\alpha\omega$-dynamo; its average value depends on the form of the rotation curve (Fig. 3). The question arises whether the rotation curve determines the basic mode of dynamo action. Tidal interaction with companion galaxies has also been quoted to explain the different magnetic field configurations [25].

Not all spiral galaxies show a "grand design" spiral structure. These are believed to be candidates of the SSPSF theory (stochastic, self-propagating star formation). Magnetic fields have been introduced as guiding lines for propagation of star formation and support the formation of spiral segments (FEITZINGER, this volume).

3. Future

The present sensitivity of optical and radio polarization observations enables the study of the large-scale structure of the field for a large sample of galaxies. The relative frequencies of axisymmetric and bisymmetric field structures could then be determined and compared with the theories of field origin. Large-scale field reversals between neighbouring spiral arms would indicate the occurrence of higher-order dynamo modes. The effect of tidal interaction and peculiarities in the rotation curve on the field structure should be investigated.

Nothing is known yet about the field structure in the hot coronal medium of galaxies (observed as a "thick disk" in radio continuum) because the polarized radio emission is extremely weak. The bisymmetric field model predicts a field perpendicular to the plane which is strongest in the interarm region [18]. The toroidal dynamo field is accompanied by a weaker poloidal field perpendicular to the disk [9].

5

The Westerbork synthesis telescope was the first to detect linearly polarized radio emission from nearby galaxies [26]. It has been demonstrated that the Very Large Array is also a useful instrument to detect polarized radio emission with high angular resolution and sensitivity (HUMMEL et al., KRAUSE et al., LOISEAU et al., this volume). This makes it possible to study the fine structure of the field on different scales. The structure of the field in star-forming regions and around supernova remnants is of special interest.

New input of observational data will stimulate theoretical studies. More work is needed on the timescale and effectivity of galactic dynamos and the origin of seed fields. The interaction between magnetic fields and galactic density waves has been neglected during the past 16 years [27 and TOSA, this volume]. The effect of interstellar fields on the star-forming clouds [13,14,28] and on the expansion of supernova remnants [29 and FÜRST, this volume] should be investigated in more detail.

4. Conclusions

A new era of research on interstellar magnetic fields has just begun. The significant improvement of observational techniques to detect radio and optical polarization during the last decade opens new ways to study the role of magnetic fields in the interstellar medium.

References

1. W.A. Hiltner: Astrophys. J. 114, 241 (1951)
2. V.A. Razin: Astron. Zh. AJ 235, 241 (1958)
3. G. Westerhout, Ch.L. Seeger, W.N. Brouw, J. Tinbergen: Bull. Astron. Inst. Netherlands 16, 187 (1962)
4. R. Wielebinski, J.R. Shakeshaft: Nature 195, 982 (1962)
5. L. Davis, J.L. Greenstein: Astrophys. J. 114, 206 (1951)
6. M. Steenbeck, F. Krause: Zeitschr. Naturforsch. 21a, 1285 (1966)
7. E.N. Parker: Astrophys. J. 163, 255 (1971)
8. S.I. Vainshtein, A.A. Ruzmaikin: Astron. Zh. AJ 48, 902 (1971)
9. M. Stix: Astron. Astrophys. 42, 85 (1975)
10. M.P. White: Astron. Nachr. 299, 209 (1978)
11. J.H. Piddington: Monthly Notices Roy. Astron. Soc. 128, 345 (1964)
12. R.N. Henriksen, M. Reinhardt: Astrophysics and Space Science 49, 3 (1977)
13. L. Mestel, R.B. Paris: Astron. Astrophys. 136, 98 (1984)
14. T.Ch. Mouschovias: Adv. Space Res. 2, no. 12, 71 (1983)
15. G. Benford: Monthly Notices Roy. Astron. Soc. 183, 29 (1979)
16. A.A. Ruzmaikin, D.D. Sokoloff, A.M. Shukurov: Astron. Astrophys. 148, 335 (1985)
17. T. Sawa, M. Fujimoto: Publ. Astron. Soc. Japan 32, 551 (1980)
18. T. Sawa, M. Fujimoto: Publ. Astron. Soc. Japan 38, 133 (1986)

19. Y. Sofue, M. Fujimoto: Astrophys. J. <u>265</u>, 722 (1983)

20. J.P. Vallée: Astrophys. Lett. <u>23</u>, 85 (1983)

21. Y. Sofue, M. Fujimoto, R. Wielebinski: Ann. Rev. Astron. Astrophys. <u>24</u>, 459 (1986)

22. R. Beck: IEEE Trans. on Plasma Science, in press (Dec. 1986)

23. E. Hummel: Astron. Astrophys. <u>160</u>, L4 (1986)

24. E.H. Levy, W.K. Rose: Astrophys. J. <u>193</u>, 419 (1974)

25. J.P. Vallée: Astron. J. <u>91</u>, 541 (1986)

26. A. Segalovitz, W.W. Shane, A.G. de Bruyn: Nature <u>264</u>, 222 (1976)

27. W.W. Roberts, C. Yuan: Astrophys. J. <u>161</u>, 887 (1970)

28. E. Dorfi: Astron. Astrophys. <u>138</u>, 378 (1984)

29. J.E. Borovsky, M.B. Pongratz, R.A. Roussel-Dupré, T.H. Tan: Astrophys. J. <u>280</u>, 802 (1984)

Dynamo Excitation in Very Large Scales

F. Krause

Zentralinstitut für Astrophysik der Akademie der Wissenschaften der DDR, Rosa-Luxemburg-Str. 17a, DDR-1502 Potsdam-Babelsberg, GDR

1. Introduction

When Ludwig BIERMANN in 1950 published his paper with the title "Über den Ursprung der Magnetfelder auf Sternen und im interstellaren Raum" [1], no direct information about the interstellar magnetic field was at hand, however, its existence was inferred from the observed properties of the cosmic radiation, especially their isotropy, by Alfvén, Richtmyer and Teller, and Fermi. Interestingly, the direct observations, which started 10 years later, confirmed not only the postulated existence of an interstellar magnetic field, but also the estimated order of magnitude.

In order to remind the facts: The isotropy of the cosmic radiation can only be understood in case the radiation is often deflected, and this can hardly be explained other than by a magnetic field. From the maximal observed particle energy of 10^{16} eV it was concluded that the magnetic field has to be stronger than 10^{-8} G in order to prevent escaping from the Galactic disc. Another estimation is based on the idea that the magnetic energy density cannot be smaller than the observed energy density of the cosmic radiation, otherwise the magnetic field is uncapable to resist the pressure of cosmic ray particles. Here a minimal field strength of some 10^{-6} G is found, that is the value which until now is representative for the interstellar field.

The abovementioned investigation by Biermann concerns the origin of the interstellar magnetic field. A mechanism of the following kind is proposed:

- Firstly, the difference of the masses of the ions and the electrons causes a different behaviour with respect to mass independent forces. Consequently, ions and electrons move differently with the result of a net current, which is accompanied by a magnetic field. Based on the data of the interstellar plasma Biermann estimated a field of the order of 10^{-16} G.

- Secondly, the turbulent motion amplifies this weak background field up to the value of equipartition of kinetic energy of this motion and the magnetic energy.

Here is taken into account that, due to the very large scales of the turbulence elements, the magnetic field does not decay within a time of the order of the age of the universe.

This proposal provides for a plausible explanation of a magnetic field which is of turbulent character, with a length scale equal to that of the turbulence elements; i.e. of about 100 pc. However, the direct observations of the Galactic field by Faraday rotation and polarization of starlight, which started in the early sixties, provided for results which indicate a large-scale field. The development of better observational means confirmed this for our own Galaxy, moreover, the magnetic fields of a number of nearby galaxies were also investigated and likewise a large-scale structure was found. Consequently, an explanation different to that of Biermann has to be found. There is at hand the concept of the self-excited dynamo, which already proved to be applicable to the planets and the Sun. In this case the concerted action of different turbulence elements provides for growth and maintenance of a large-scale magnetic field.

2. The self-excited dynamo

The self-excited dynamo is in the modern language of non-linear physics a process of self-organization. It was discovered and first constructed by Werner v. Siemens in 1867.

In Fig. 1 a simple model of a self-excited dynamo is presented. It consists of a circular copper disc with an axis (also of copper) about of which it can rotate. An electrically conducting wire connects the rim of the disc with the axis, thereby entangling this axis forming a coil. The disc is assumed to rotate about the axis with the angular velocity ω. Now the behaviour of an initial magnetic field B is considered in dependence on the angular velocity. For small ω the field will simply decay. However, in case ω is beyond a critical angular velocity, the B=0 field is no more stable: Small fluctuations will exponentially grow and a large-scale magnetic field is formed and maintained as long as the rotational motion is maintained.

The bifurcation diagram to the right reflects this behaviour. Such diagrams are representative for all self-organization processes, the processes, which play the dominant role in the evolution of the universe. The qualitative difference left and right of the critical point is important: The system is a mechanical one for $\omega < \omega_{crit}$, and an electromagnetic power station for $\omega > \omega_{crit}$.

In cosmical objects this dynamo looks a bit different although being basically the same. Since the dynamo process takes place in a homogeneously conducting body – no isolated wires can be arranged in a proper way as in Fig. 1 – it needs small-

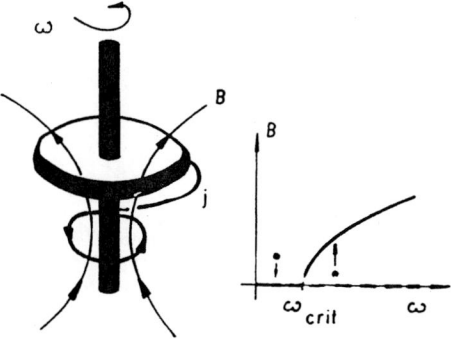

Figure 1: Simple model of a self-excited dynamo

scale motions, which in cosmical objects are present in form of convection or turbulence. In stratified rotating turbulent (or convective) layers a mean electromotive force parallel to the magnetic field appears. This is the so-called α-effect.

Due to the action of Coriolis forces the turbulence elements show internal helical motions, and all elements in the same hemisphere show the same kind of helical motion: either according to a right-handed screw or a left-handed one. A magnetic field line undergoing the influence of a local screw motion is deformed to a twisted Ω, where because of the twist a current parallel to B appears (Fig. 2). In this way, because of the overall rotation, all turbulence elements in one hemisphere of a cosmical object drive the same current parallel or anti-parallel to the mean magnetic field.

This effect is the key to understand dynamo excitation. Assume (Fig. 3) a special conducting body with α ≠ 0. Assume a poloidal magnetic field. Because of the α-effect we have a poloidal current which is combined with a toroidal magnetic field. This toroidal magnetic field drives a toroidal current which is combined with the poloidal field. This is a so-called α²-dynamo. The production of the toroidal field from the poloidal can also be caused by differential rotation. The field lines are tracted with the motion thus forming toroidal field belts. This is the so-called αω-dynamo. Generally, if there is differential rotation, this is more efficient. The Sun forms an αω-dynamo.

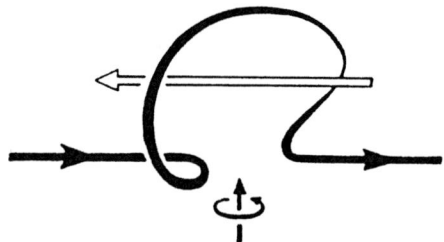

Figure 2: A magnetic flux rope undergoing the influence of a helical motion is shaped into a twisted Ω. The loop is accompanied by a current which, in case of right-handed helical motions, has a component anti-parallel to the magnetic field.

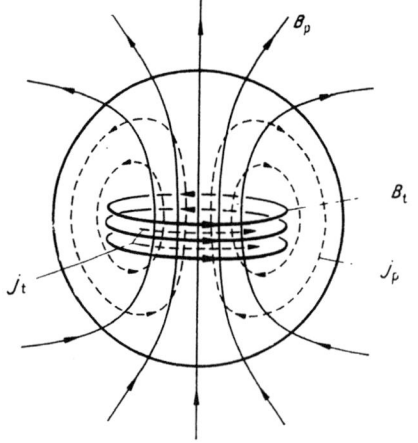

Figure 3: A self-maintaining magnetic field configuration in an electrically conducting sphere with α-effect. The sphere is embedded in the empty, insulating space

3. Galactic dynamo models

A great number of spherical dynamo models have been calculated, especially for the Sun and the Earth [2,3]. As it concerns the Galaxy the situation is different: The Galaxy forms a flat disc which has been approximated either by plane layer models [4,5,6] or by oblate spheroids as was done by STIX [7] and WHITE [8]. Worth to mention is also a paper by SOWARD [9], who developed an asymptotic formula for the limit of small axis ratio, in this way linking both kinds of models.

We consider an electrically conducting oblate spheroid embedded in the insulating space. Let R denote the large axis and b the small one. The spheroid shows differential rotation about its short axis. This rotational shear is the first induction action, which provides for the generation of an azimuthal field B_ϕ in case there is a radial field B_r. The characteristic parameter is

$$C_\omega = \frac{R^2}{\eta_T} \left(r \, \frac{d\omega}{dr} \right) , \qquad (1)$$

where η_T denotes the turbulent magnetic diffusivity and r the radial coordinate.

The α-effect drives a current parallel to the azimuthal field and with this current a poloidal magnetic field appears, i.e. one with its field lines in the meridional planes. Here the characteristic parameter is

$$C_\alpha = \frac{R\alpha}{\eta_T} . \qquad (2)$$

In order to assess whether or not the Galaxy is above the critical value of self-excitation we need an estimation of the order of these two parameters. Here the most uncertain quantities are α and η_T.

Mean-field magnetohydrodynamics provide for the relations [2]

$$\alpha \approx \overline{u \cdot curl\ u}\ \tau_{cor} , \qquad \eta_T \approx \overline{u^2}\ \tau_{cor} , \qquad (3)$$

where u is the turbulent velocity, τ_{cor} the correlation time (life time of turbulence elements), the bar denotes the average. It has to be noted that these relations (3) are derived on a low approximation level. Therefore, the following results should not be overestimated, rather looked upon as a means of orientation.

From observations we know the large axis R $\approx$ 15 kpc, the rotational shear in the neighbourhood of the Sun (Oort's constant), i.e. (r dω/dr) $\approx$ −30 km/s kpc, the turbulent velocity u $\approx$ 10 km/s and the correlation length (size of turbulence elements) λ_{cor} $\approx$ 100 pc.

The corrrelation time is not observed. Generally it is inferred from mixing length theory that τ_{cor} $\approx$ λ_{cor}/u. Hence we find τ_{cor} $\approx$ 10^7 a.

Only due to Coriolis forces the average $\overline{u \cdot curl\ u}$ is unequal zero, therefore we estimate it to be

$$\overline{u \cdot curl\ u} \approx \frac{\overline{u^2}}{L} (\omega\ \tau_{cor}) , \qquad (4)$$

where L denotes the scale height. Obviously $\lambda_{cor} \lesssim L$ has to be assumed.

11

The above considerations result in

$$C_\alpha \approx 50 \frac{\lambda_{cor}}{L}, \quad C_\omega \approx -7000 . \tag{5}$$

We clearly see that $C_\alpha \ll |C_\omega|$, i.e. that the induction action of the differential rotation dominates. Therefore, the α-effect is negligible for the production of the azimuthal field, the toroidal field. This is the situation of an $\alpha\omega$-dynamo which is characterized by one parameter only, the product

$$C_1 = C_\alpha C_\omega . \tag{6}$$

For our Galaxy we find according to (5)

$$C_1 \approx -10^5 . \tag{7}$$

The negative sign indicates that the angular velocity decreases with the distance from the axis of rotation.

The investigations of the dynamo problem always run along the following path: The induction equation for a model considered here possesses a countable set of eigensolutions, we will speak of B-modes. Each B-mode has a certain growth rate. For small values of C_1 all growth rates are negative, i.e. any initial field will decay. In case C_1 is beyond the critical value (see Fig. 1) for some B-modes, at least for one, the growth rates will be positive. It is generally assumed that the B-mode, for which first the growth rate becomes positive, will most rapidly grow and take all the available energy. The dynamo problem is therefore often formulated as the question for the critical values, i.e. the smallest value of C_1, where one B-mode has the growth rate zero. Subsequently the structure of this B-mode is determined and compared with the observations. If C_1 exceeds the critical value by a small amount, this B-mode will grow exponentially with time unless the growth is stopped by the backreaction to the motions. In this way it is clear that the excited field completely forgets its history, its structure is only determined by the structure of the whole system, which acts as a dynamo.

The system under consideration consists of an electrically conducting oblate spheroid with internal motions, the induction actions of which are characterized by the two constants C_α and C_ω in (5). This system shows reflection symmetry with respect to the equatorial plane and axisymmetry with respect to the axis of rotation. Therefore, the excited B-mode will also reflect these symmetries. In more detail, the excited B-mode will either be symmetric (S) or antisymmetric (A) with respect to the equatorial plane. In addition, it will depend on the azimuth ϕ according to $e^{im\phi}$, where m is an integer.

m=0 corresponds to the axisymmetric case. Here fields showing symmetry with respect to the equatorial plane are of interest. For the poloidal field part the leading term in the multipole expansion is a quadrupol. The toroidal field part is a ring field in and around the equatorial plane encircling the axis of rotation. A B-mode of this kind is denoted by S0.

m=1 describes fields depending on $e^{i\phi}$, which are generally called as being of bisymmetric structure (BSS fields). A typical example of this kind is a dipole with its moment in the equatorial plane. In our denotation this is a field of type S1.

We are now in a position to illuminate the results of STIX [7], WHITE [8] and SOWARD [9]. In Table 1 the critical values of C_1 in dependence on the axis ratio $\varepsilon = b/R$ are represented. We see that these values, where first a B-mode with positive growth rate appears, are of the order of 10^5 or more. For positive values of C_1, which are not realistic in galaxies, a field of type A0 is first excited, where the poloidal part is of dipolar structure. For negative values of C_1 the most easily excited B-mode is of type S0. Here the poloidal field part is of quadrupolar structure. The toroidal part is a ring field, and, since $C_\alpha \ll |C_\omega|$, this part is much stronger than the poloidal part. Consequently, the appearance of this B-mode is mainly that of a ring field. Such fields are observed in a small number of galaxies, e.g. M31.

These investigations are restricted to the axisymmetric case, i.e. no answer is given to the question whether or not a field of type S1, which is of bisymmetric structure, is excited more easily, or to the question for the conditions on which BSS fields compete with the axisymmetric ones. Here further investigations are needed.

The excitation of BSS modes was considered for plane layer models in the so-called "local approximation". We renounce to comment on these investigations and refer to other papers in this volume.

Table 1: *Critical values of C_1 in dependence on the axis ratio, resp. on the small diameter b, where R = 15 kpc [10]*

axis ratio $\varepsilon = b/R$	b [kpc]	dipolar A0 $10^{-3} C_1$	quadrupolar S0 $10^{-3} C_1$	asymptotic S0 $10^{-3} C_1$
0.13	2.0	—	− 64	− 51
0.10	1.5	1090	− 134	− 115
0.07	1.0	2790	− 399	− 363
0.03	0.5	17340	−2805	−2678

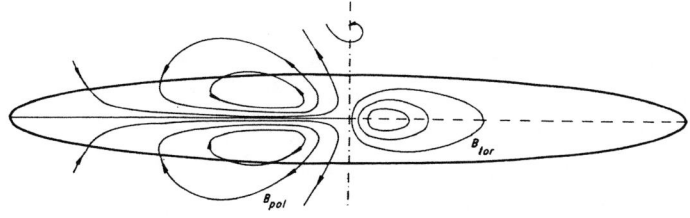

Figure 4: Representation of the magnetic field excited by an $\alpha\omega$-dynamo working in an oblate spheroid. The field is axisymmetric with respect to the axis of rotation. At the left-hand side the field lines of the poloidal field part are depicted. It is of quadrupolar type. At the right-hand side the lines of constant field strength of the toroidal part are drawn (after WHITE [8])

One point, however, should be mentioned here. As a general result of dynamo theory [11,2] B-modes with $m \neq 0$, e.g. the BSS fields S1, show a dependence on the time according to $e^{i(m\phi - \Omega t)}$, where the frequency of this drift Ω is given by

$$\Omega = \frac{c_\Omega}{b^2/\eta_T} \quad , \tag{8}$$

c_Ω is the imaginary part of the eigenvalue, its growth rate is the real part.

In spherical objects this drift motion is very slow compared with the rotational motion, e.g. for the Earth it reflects the westward drift of the geomagnetic field.

For the Galaxy we find with the foregoing data

$$\Omega \approx c_\Omega \, 10^{-9} \, a^{-1} \, , \tag{9}$$

i.e. a value of about the same order or even larger than the rotational frequency $\omega \approx 5 \, 10^{-9} \, a^{-1}$. Consequently, it may happen that the drift motion of the magnetic field is significantly different from the rotational motion of the Galactic matter.

4. The non-linear aspect

The magnetic fields which we observe in the universe are highly influenced by non-linear interactions. Dynamo theory, as far as it is well elaborated, is placed at and near the border to the non-linear region. We have no well-developed non-linear theory, but we try to understand what will happen there.

This is presumably easy when C_1 exceeds the critical value by a small amount so that only one B-mode has a positive growth rate. The estimation for our Galaxy (5) in comparison with the results of the calculations (Table 1) let this case appear realistic. We expect one well developed B-mode of low order, i.e. of type S0 or S1. The observations, which revealed a ring field or a BSS field in some galaxies, apparently support this view.

For larger C_1 more than one B-mode may have positive growth rates. In the competition between them one expects that the one with the largest growth rate will dominate. It will take the available energy and suppress the others.

Independent of this, however, it is necessary to analyse which of the growing B-modes will reach a stable state. This is a question of non-linear stability. The fastest growing mode is not necessarily that one which reaches a stable state in the non-linear region. There may also exist stable solutions which are not found by the kinematic analysis. In all these cases we may expect one stable field, whereas the other nonstable modes will be found in the fluctuating part.

Furthermore, as we know from hydrodynamic turbulence, it may happen that none of the B-modes with positive growth rates is stable. Then we expect a turbulent mixture of these B-modes, a case which apparently is observed among the solar-type stars. There is a number of these stars which show one dominating B-mode with a clear period like the Sun. Some of these stars show also magnetic activity but with a random behaviour in time. Higher rotational rates indicate larger values of C_1 here.

Finally, even in the case of galaxies we have to take into account the very large life times of the fluctuations. Therefore, the dynamo excited field may be highly veiled by long living fluctuations of the same order of magnitude as the average field or even larger. Apparently we find these conditions in the northern hemisphere of our Galaxy.

References

1.   L. Biermann: Z. Naturforschg. <u>5a</u>, 65 (1950)

2.   F. Krause, K.-H. Rädler: <u>Mean-Field Magnetohydrodynamics and Dynamo Theory</u> (Akademie-Verlag, Berlin 1980)

3.   Ya.B. Zeldovich, A.A. Ruzmaikin, D.D. Sokolov: <u>Magnetic Fields in Astro-physics</u> (Gordon and Breach Science Publishers, New York, London, Paris, Montreux, Tokyo 1983)

4.   E.N. Parker: Astrophys. J. <u>163</u>, 255 (1971)

5.   S.I. Vainstain, A.A. Ruzmaikin: Astron. J. (USSR) <u>49</u>, 449 (1972) (Sov. Astron. <u>16</u>, 365 (1972))

6.   A.A. Ruzmaikin, D.D. Sokolov, V.I. Turchaninoff: Astron. J. (USSR) <u>57</u>, 311 (1980) (Sov. Astron. <u>24</u>, 182 (1980))

7.   M. Stix: Astron. Astrophys. <u>42</u>, 85 (1975)

8.   M.P. White: Astron. Nachr. <u>299</u>, 209 (1978)

9.   A.M. Soward: Astron. Nachr. <u>299</u>, 25 (1978)

10. The values published by Stix in [7] are erroneous by a factor $1/\varepsilon$. We present here the corrected values.

11. F. Krause: Astron. Nachr. <u>293</u>, 187 (1971)

Magnetic Fields of Galaxies

A. Ruzmaikin

Keldysh Institute of Applied Mathematics, USSR Academy of Sciences,
Miusskaya sq. 4, 125047 Moscow, USSR

Intensity of galactic magnetic field is minute. It is measured in
microgausses. However, the field has a record scale measured in ki-
loparsecs. (Unfortunately, the Guinness book does not mention this
giant magnet.) To create $B = 2$ μGauss in a disk of $R = 15$ kpc
in radius and $h = 0.4$ kpc in half-thickness one needs an electromotive force of about $BRh/ct = 10^{10}$ Volts acting for $t = 10^{10}$ years.
F.Hoyle, who first noted the fact, did not see any other possibility except the idea of the relic origin of the galactic magnetic
field.

In reality, the largescale galactic magnetic field can be amplified and supported by hydrodynamical motions of ionised gas. Under
this dynamo action a weak initial magnetic field, which probably
can be a product of star bursts (see the last section), grows exponentially. A characteristic time of the growth is approximately
$5 \cdot 10^8$ years, i.e. much less as compared with the galactic life-time.
Then the field is stabilized due to its back-action on the motions.
The electromotive force created by the motions, $|v \times B| \cdot h/c$, is en-
tirely sufficient to explain 10^{10} Volts provided that even 0.1 of
random velocity intensity, 10 km/sec, is acting effectively.

The conception of the mean-field galactic dynamo acting in the
gaseous disk due to the differential rotation, and the mean helicity
of gas motion was introduced by Parker /1/, and Vainshtein and
Ruzmaikin /2/. In these and some subsequent publications, see for
instance /3/, the conditions for a local magnetic field generation
(at given radial distance) were formulated. The mode which is the
easiest to excite in the thin galactic disk has an even azimuthal
field B_φ under the reflection in the central plane and a corresponding quadrupole component (B_r, B_z) in the cylindrical coordinates r, φ, z (Fig. 1).

Now the observations and the theory allow us to clarify a global
structure of the mean magnetic field in spiral galaxies. The most
exciting result is the discovery of two-spiral, or bisymmetric, con-

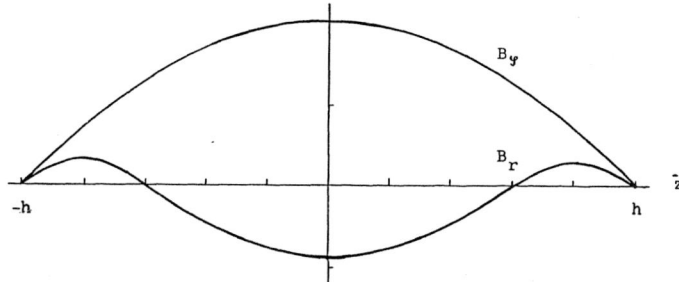

Fig. 1. The dynamo generated magnetic field distribution
across the disk at a given radius

figurations in M51, M81, M33, NGC 6946, see review /4/. The magnetic
lines corresponding to this configuration are directed inward in one
part of the galaxy and outward in the other part. Along with the
bisymmetric configurations the ring-like structures first pointed
out in the theory /5/ are also observed (M31, the Galaxy?). Note
that magnetic lines of axisymmetric fields are spirals as well. However the field intensity and the inclination angle do not depend on
azimuth φ . In both cases this angle decreases with r.

The bisymmetric magnetic structure can be interpreted as the lowest non-axisymmetric mode excited by the dynamo action /6-8/.

1. Degree of Non-axisymmetry

Let us assume that the velocity of gas motions in the galactic disk
is distributed on the average axisymmetrically. Specifically, the
angular velocity and the mean helicity do not depend on φ -coordi-
nate. A local rate of exponential growth of the mean magnetic field,
γ , does not depend on the azimuthal number   m /1-3/:

$$\gamma = \gamma_0 \frac{\beta}{h^2}(R_\alpha \cdot R_\omega)^{1/2},$$

where β is a turbulent diffusivity, $\gamma_0 \simeq$ 0.4, the dimensionless
numbers

$$R_\omega = \frac{\omega h^2}{\beta}(=10), \qquad R_\alpha = \frac{\alpha h}{\beta} (=1)$$

characterize the intensity of the differential rotation and the mean
helicity. It gives a possibility to estimate· m /6/.

Consider an initial magnetic field having a component in the disk
plane. The differential rotation ω (r) twists the field into spi-

ral. A distance between the coils with opposite directions of the field is decreasing in time as $\Delta r_m = r/m \omega t$ at given r. A characteristic time for the diffusion of the magnetic mode m is proportional to

$$\tau_d = \frac{(\Delta r_m)^2}{\beta} = \frac{r^2}{\beta \, m^2 \omega^2 t^2}$$

The mode will evidently be excited, when $\gamma^{-1} < \tau_d$. It gives the required estimate

$$m \leq \frac{r}{h} \gamma_o^{2/3} R_\omega^{-1/4}$$

At $r = 3$ kpc the right-hand side is of about two. Thus, the modes $m = 1,2$ can be excited in the axisymmetric disk. Note that the mode $m = 1$ corresponding to the bisymmetric structure is more preferable because it has two times larger τ_d than the $m = 2$ mode.

2. Dynamo Solutions

The dynamo equation for the mean magnetic field has the form:

$$\frac{\partial B}{\partial t} = \text{rot} \, (\boldsymbol{\omega} \times r \times + \, \alpha - \beta \, \text{rot}) \, B \qquad (1)$$

Sawa and Fujimoto /7/ following to Parker /1/ developed a method of solution of (1) in the local rectangular coordinates whose origin moves with the galactic rotation. It appears more natural to use the cylindrical coordinates (r, φ, z) and some approximations specific for the thin disk /8/. There are a small parameter $\lambda = h/R$ (=0.01) and a large parameter R_ω or $R_\alpha R_\omega$ (=0.1) so that $\partial/\partial z = 0_\lambda$ (1), $\partial/\partial r = 0(\lambda^{1/2})$ and

$$\frac{B_r}{B_\varphi} = 0_\lambda (1) \doteq 0 \quad (R_\alpha R_\omega)^{-1/2} \quad , \quad \frac{B_z}{B_\varphi} = 0 \, (\lambda^{1/2}).$$

The solution of (1) is sought for in the form:

$$\begin{pmatrix} B_r \\ B_\varphi \end{pmatrix} = Q(r) \begin{pmatrix} b_r(r,z) \\ b(r,z) \end{pmatrix} \exp (\Gamma t + im \varphi)$$

The B_z component is separated and can be found afterwards. The functions b_r and b_φ are obtained in the local $\lambda = 0$ approximation from the earlier derived equations /1-3/ depending on z with r as a parameter. These equations do not include m!

The radial distribution of the magnetic field is determined by the Schrödinger-type equation

$$\lambda^2 \frac{d}{dr} \frac{1}{r} \frac{d}{dr} rQ \ + \ U(r)Q = \ \Gamma Q \qquad (2)$$

with a complex potential

$$U = \gamma(r) \ - im \ R_\omega \cdot \omega(r),$$

where $\gamma(r)$ is the local growth rate (Fig. 2). The boundary conditions can be used in a simple form, say, $Q(0) = Q(R) = 0$. The global growth rate Γ is complex, $\operatorname{Im} \Gamma$ is proportional to $m\omega R_\omega$. Then after a discretization this eigen-value problem may be solved numerically by the QR - algorithm. The solutions for some spiral galaxies of interest are presented in /8/, for the case $m = 0$ see /9/. Here only main results are briefly discussed.

The axisymmetric mode $m = 0$ is preferable in all cases. In the case of M31 all nonaxisymmetric modes are decaying. The field has a ring-like distribution, it is excited in the regions $r < 2$ kpc and $r > 7$ kpc. This distribution is qualitatively in agreement with observations /10, 11/. The growth rate of the mode $m = 1$ has a significant value $\operatorname{Re} \Gamma = 2.5$ in the case of M51 (Fig. 3), where in fact a bisymmetric structure is observed. The mode $m = 2$ here is decaying.

There are two regions of the axisymmetric field generation in the Galaxy: $r < 5$ kpc and a ring, $5 < r < 15$ kpc, with a gap at $r = 5$ kpc. The possibility of the mode $m = 1$ excitation depends crucially on the disk thickness. For $0.5 < h < 0.7$ kpc in the solar vicinity it may be excited.

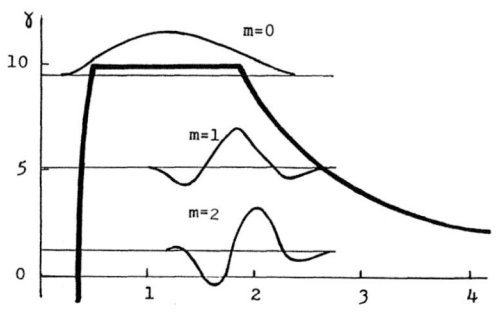

Fig. 2. A local growth rate $\gamma(r)$ and radial eigenfunctions for the modes m = 0, 1, 2

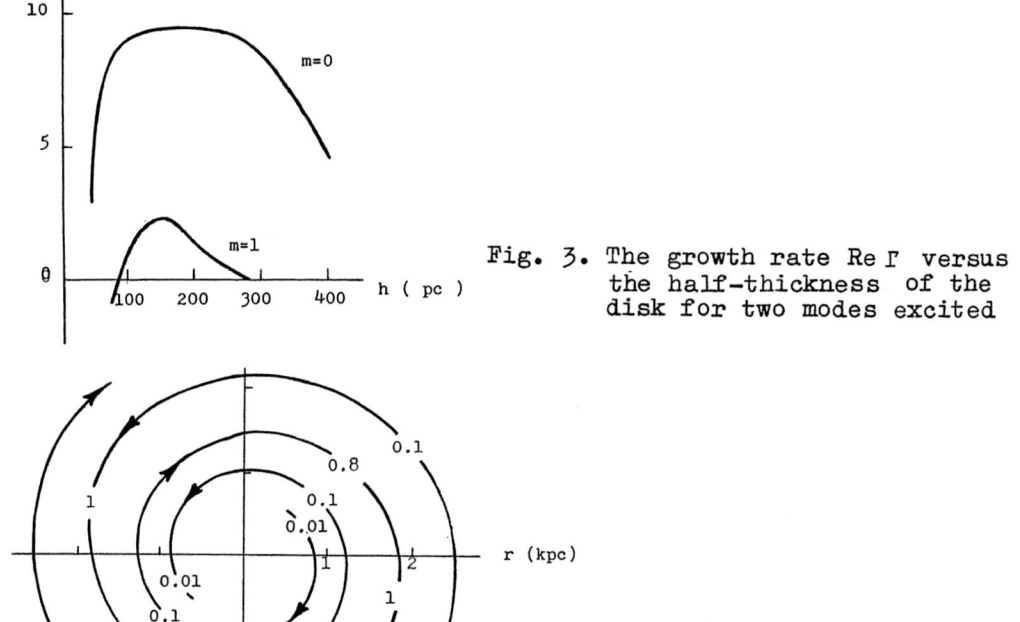

Fig. 3. The growth rate Re Γ versus the half-thickness of the disk for two modes excited

Fig. 4. The structure of maximal magnetic field intensity for a galaxy like M51

Thus, the non-axisymmetric magnetic field in the axisymmetric disk can be generated only in common with the axisymmetric field. However, non-axisymmetric modes have a wider region of localization, see Fig. 2. It means that they must be dominant in the outer parts of galaxies (Fig. 4). This fact can be tested by observations, say, in M51. The reason for the axisymmetric mode to be concentrated in the inner part of a galaxy is a decrease of the differential rotation (the main factor for symmetrization!) in outer parts and outward expulsion of non-axisymmetric modes.

It is worth to note that if the observations reject the presence of the axisymmetric mode in the inner parts of a galaxy it will not mean the failure of the dynamo. Simply one should consider a non-axisymmetric gaseous disk by taking into account the spiral structure, etc.

3. Seed Magnetic Field

To put dynamo into action one needs an initial seed field. Cosmological magnetic fields generated due to the interaction between the

electron-proton plasma and a relict radiation are very weak /3/.
Besides, they are dominated by the dipole mode which can hardly be
amplified by dynamo action in the galactic disk. The field ejections
from supernovae and, perhaps, the other stars are more promising.
In the infinite space, a mean value from the sum of random ejections
is certainly zero. However, in the finite galactic disk the mean
value does not vanish /6/.

A naive evaluation $B_0 = bN^{-1/2}$, where b is a mean square
field in a correlation cell, N is the number of cells, overestima-
tes the seed field because it assumes that the field in a cell is
homogeneous, i.e. considers a sum of δ - functions. Really, the
field changes its sign inside the cell as a divergenceless loop. It
is better to consider a model in the form of a sum of δ - function
derivatives. It gives an additional small factor

$$B_0 = \frac{b}{N^{1/2}} \frac{1}{\triangle r} ,$$

where l (=100 pc) is the size of correlation cell, $\triangle r$ (=3 kpc) is
the range of localization for a main mode of the mean field excited
by dynamo action. Putting b = 1 μG, N = 300 in the region $\triangle r$
one finds $B_0 = 10^{-3} \mu G$.

The fluctuating magnetic fields come not only from the star ejections. They are produced by the dynamo action as well. The first
observations of a correlation function for the intensity of the Galactic background synchrotron radiation (due to the magnetic field)
were made by Dagkesamansky and Shutenkov /12/. Weak nonthermal sources near the Galactic plane may possibly be explained as fluctua-

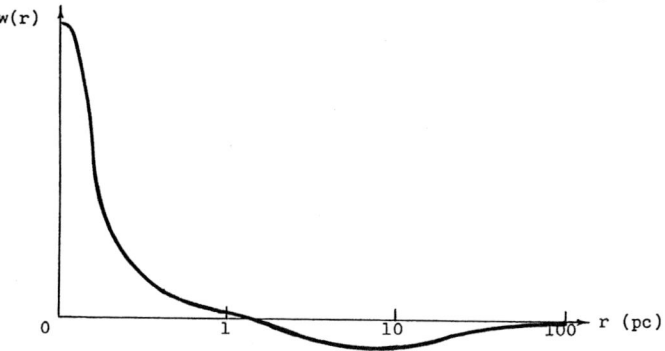

Fig. 5. The correlation function for fluctuations of the
magnetic field,
$w(r) = \langle (H - B)_{r_1} (H - B)_{r_2} \rangle$, $r = |r_1 - r_2|$,
calculated in the idealized case of shortcorrelated
turbulence

tions of the magnetic field /13/. The correlation function of the
field excited by dynamo action has been found recently /14/.
The general dynamo theory predicts /15/ that the magnetic field is
intermittent. It means, in particular, that the mean magnetic field
generally considered by theorists does not resemble a typical field
realization.

4. Literature

1. E.N. Parker: Astrophys. J. 163, 255 (1971)
2. S.I. Vainshtein, A.A. Ruzmaikin: Astron. J. (USSR), 48, 902
    (1971)
3. Ya.B. Zeldovich, A.A. Ruzmaikin, D.D. Sokoloff: Magnetic Fields
    in Astrophysics Gordon-Breach, New York, London, Paris 1983)
4. Y.Sofue, M.Fujimoto, R.Wielebinski: Ann. Rev. Astron. Astrophys.
    Vol. 24 (1986), p.459
5. A.A. Ruzmaikin, A.M. Shukurov: Astron. J.(USSR), 58, 969 (1981)
6. A.A. Ruzmaikin, D.D. Sokoloff, A.M. Shukurov: In Plasma Astrop-
    hysics, Proc. Varenna-Abastumani Workshop, Sukhumi (ESA Publ.
    1986)
7. T.Sawa, M.Fujimoto: Publ. Astron. Soc. Japan, 38, 133,(1986)
8. Yu.S. Baryshnikova, A.A. Ruzmaikin, D.D. Sokoloff, A.M. Shuku-
    rov: Generation of Large-scale Magnetic Fields in Spiral Gala-
    xies, Preprint (Space Research Institute, Moscow, 1986)
9. A.A. Ruzmaikin, D.D. Sokoloff, A.M. Shukurov: Astron. Astro-
    phys. 148, 135 (1985).
10. R.Beck: In Proc. IAU Symp.No.100, p.159 (Reidel, Dordrecnt
    1983)
11. Y.Sofue, U.Klein, R.Beck, R.Wielebinski: Astron. Astrophys.
    144, 257 (1985)
12. R.D. Dagkesamanski, V.R. Shutenkov: Preprint (Lebedev Physical
    Institute, Moscow 1985), submitted to Pisma Astron. J.
13. Y.Sofue, H.Hirabayashi, K.Akabane, M.Inoue, T.Handa, N.Nakai:
    Preprint (NRO, Report No.31, 1984), submitted to Publ. Astron.
    Soc. Japan
14. N.I. Kleeorin, A.A. Ruzmaikin, D.D. Sokoloff: In Plasma Astro-
    physics, Proc. Varenna-Abastumani Workshop, Sukhumi (ESA Publ.
    1986)
15. S.A. Molchanov, A.A. Ruzmaikin, D.D. Sokoloff: Sov. Phys.
    Uspeki, 145, 593 (1985)

Bisymmetric Spiral Magnetic Fields in Spiral Galaxies

M. Fujimoto

Department of Physics, Nagoya University, Nagoya 464, Japan

1. Introduction to the "Bisymmetric" Spiral (BSS) Magnetic Fields

A radio polarization analysis was presented in 1978 by TOSA and FUJIMOTO [1] to determine a large-scale configuration of magnetic fields in the spiral galaxy M51. On the basis of the distributions of planes of polarization at 6 cm and 21 cm [2], they determined the rotation measures (RMs) of the Faraday effect and the intrinsic polarization angles. The former gives the line-of-sight directions of magnetic fields and the latter the orientations projected onto the plane of sky. Figure 1 shows the latter result for M51. We are impressed that the projected field lines are not closed but spiral and parallel to the luminous arms. Another method was proposed for distinguishing between the circular and the bisymmetric spiral magnetic fields by

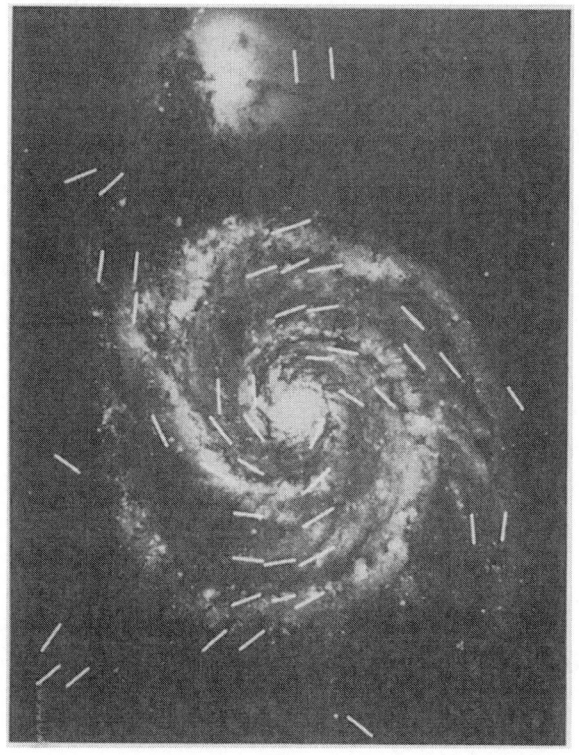

Fig. 1: Magnetic field orientations in M51 superimposed on a photograph (TOSA and FUJIMOTO [1]; photo by the Carnegie Institution of Washington [3])

measuring the variation of RM or the angle between the polarization planes at 6 and
21 cm along a concentric circle in the M51 disk. If the magnetic field is circular, we
have one maximum and one minimum in RM along the circle from 0 to 2π (Fig. 2a),
and if it is bisymmetric spiral, we have two minima and two maxima (Fig. 2b). Figure
3 suggests that M51 is the latter case rather than the former. See [4,5] and papers
presented at this workshop for the BSS fields observed in more detail and extensively in other galaxies.

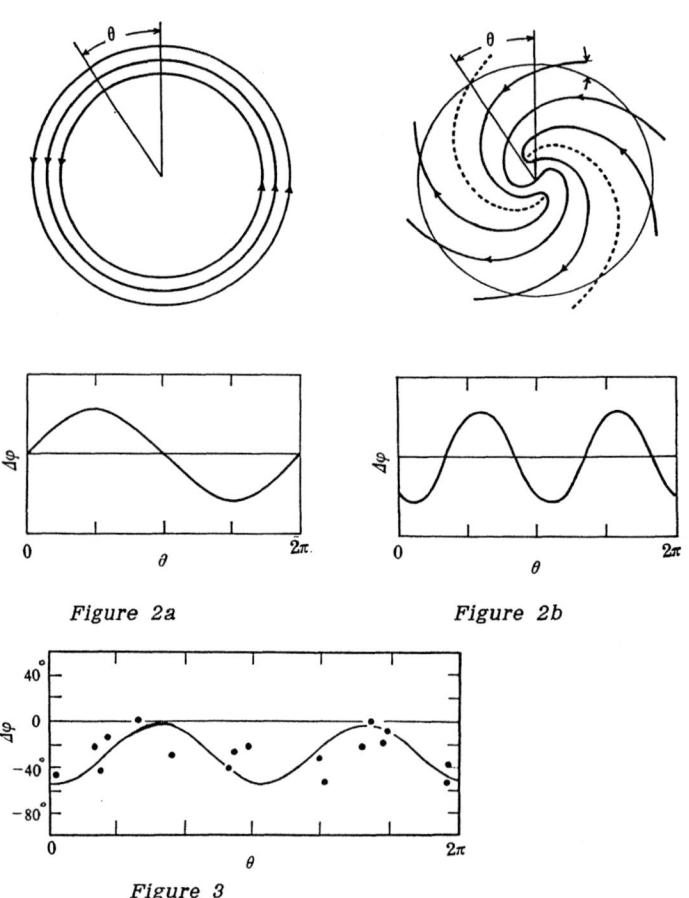

Figure 2a   Figure 2b

Figure 3

2. Turbulent Diffusion of Magnetic Fields and Dynamo Action

It is natural to consider that the BSS field is a pattern of the magnetic field distribution rotating rigidly. It resembles the spiral arm phenomenon in the density
wave theory. The turbulent diffusion must be operative, therefore, for enhancing the
decoupling of interstellar gas from the BSS mean fields. At the same time, the
dynamo action due to the cyclonic turbulence must work, otherwise the magnetic
fields in the disk would be dissipated away quickly, because the thickness of the
hydrogen gas layer is as small as 100 pc or so [6].

2.1 Kinematic Dynamo Equation

We consider the following induction equation governing the evolution of the mean magnetic field in a rotating gaseous disk locally in turbulent motion,

$$\frac{\partial \underset{\sim}{B}}{\partial t} = \text{rot } (\underset{\sim}{U} \times \underset{\sim}{B}) + \kappa \, \nabla^2 \underset{\sim}{B} + \text{rot } (\alpha \underset{\sim}{B}), \tag{1}$$

and

$$\text{div } \underset{\sim}{B} = 0, \tag{2}$$

where κ and U denotes respectively the diffusion constant and the mean circular velocity of gas, $|U| = V(r)$. The last term on the right-hand side of equation (1) is the mean magnetic field generated per unit time by the local cyclonic turbulence [6], in which we call α the dynamo strength. It is to be noted that the sign of α is opposite above and below the galactic plane, though $|\alpha|$ is the same.

We write down equations (1) and (2) in the cylindrical coordinates (r,ϕ,z) and search for asymptotic solutions in the following bisymmetric spiral form rotating rigidly [7,8],

$$\underset{\sim}{B}(r,\phi,z;t) = \underset{\sim}{b}(r) \, \exp \, i\left(-\omega t + \phi + mz + \frac{1}{\varepsilon} \ln r\right), \tag{3}$$

where $\phi + (1/\varepsilon) \ln r = \text{const}$ with a small number ε ($\simeq 0.1$) represents a logarithmic spiral for a constant phase of the field distribution. The amplitude vector $\underset{\sim}{b}(r)$ is a slowly varying function of r, and $\text{Re}(\omega)$ the angular velocity of the rotating field pattern. The small number ε or the tightly twisted pattern in both equation (3) and Fig. 1 allows us to make further approximations as,

$$\left|\frac{\partial b_r}{\partial r}\right| \ll \left|\frac{b_r}{\varepsilon r}\right|, \quad \left|\frac{\partial^2 b_r}{\partial r^2}\right| \ll \left|\frac{1}{\varepsilon r} \frac{\partial b_r}{\partial r}\right| \ll \left|\frac{b_r}{\varepsilon^2 r^2}\right|, \tag{4}$$

and similarly for b_ϕ and b_z. Since the magnetic field lines are nearly parallel to the spiral, the following holds approximately,

$$|B_r| \simeq |\varepsilon B_\phi|. \tag{5}$$

Equation (1) is now written by using these conditions and approximations,

$$\frac{\partial B_r}{\kappa \partial t} - \left[\frac{\partial^2 B_r}{\partial r^2} + \frac{\partial^2 B_r}{\partial z^2} - \frac{V}{\kappa r} \frac{\partial B_r}{\partial \phi} - \frac{\alpha}{\kappa} \frac{\partial B_\phi}{\partial z}\right] = 0, \tag{6}$$

$$\frac{\partial B_\phi}{\kappa \partial t} - \left[\frac{\partial^2 B_\phi}{\partial r^2} + \frac{\partial^2 B_\phi}{\partial z^2} - \frac{V}{\kappa r} \frac{\partial B_\phi}{\partial \phi} + \frac{1}{\kappa}\left(\frac{\partial V}{\partial r} - \frac{V}{r}\right) B_r\right] = 0, \tag{7}$$

$$\frac{\partial B_z}{\kappa \partial t} - \left[\frac{\partial^2 B_z}{\partial r^2} + \frac{\partial^2 B_z}{\partial z^2} - \frac{V}{\kappa r} \frac{\partial B_z}{\partial \phi} + \frac{\alpha}{\kappa} \frac{\partial B_\phi}{\partial r}\right] = 0, \tag{8}$$

25

and
$$\frac{\partial B_r}{\partial r} + \frac{1}{r}\frac{\partial B_\phi}{\partial \phi} + \frac{\partial B_z}{\partial z} = 0, \tag{9}$$

where the generation of B_ϕ is considered as due dominantly to the differential rotation $\partial V/\partial r - V/r$ in equation (7) [6]. The substitution of equation (3) into equations (6), (7) and (9) yields

$$\left[m^2 + \frac{1}{\varepsilon^2 r^2} + \frac{i}{\kappa}\left(\frac{V}{r} - \omega\right)\right] b_r + \frac{i\,m\,\alpha}{\kappa} b_\phi = 0, \tag{10}$$

$$-\frac{1}{\kappa}\left(\frac{\partial V}{\partial r} - \frac{V}{r}\right) b_r + \left[m^2 + \frac{1}{\varepsilon^2 r^2} + \frac{i}{\kappa}\left(\frac{V}{r} - \omega\right)\right] b_\phi = 0, \tag{11}$$

and
$$m\, b_z + \frac{b_r}{\varepsilon r} + \frac{b_\phi}{r} = 0. \tag{12}$$

Here it should be noted that r and the functions of r are treated as parameters.

2.2 Gaseous Disk of Finite Thickness and Boundary Conditions

A gaseous disk is bounded by the upper and lower surfaces at $z = \pm h$, outside of which $\kappa = \infty$ or vacuum. We take usual boundary conditions for magnetic fields at $z = 0$ and $\pm h$ except for the jump condition for the dynamo action at $z = 0$ [6],

$$\left(\frac{\partial B_r}{\partial z}\right)_{z=+0} - \left(\frac{\partial B_r}{\partial z}\right)_{z=-0} = \frac{2}{\kappa}\left(\alpha\, B_\phi\right)_{z=+0} \tag{13}$$

which is derived by integrating equation (6) over a vertically infinitesimal spread around $z = 0$. In the space $z > h$ and $z < -h$ the following spiral magnetic fields $\underset{\sim}{C}$ are assumed [6,7,8],

$$\underset{\sim}{C} = \underset{\sim}{c}\, \exp\left\{ i\left[-\omega t + \phi + \frac{1}{\varepsilon}\ln r\right] - \left[\frac{1}{\varepsilon r}\,(1+\varepsilon^2)^{1/2}\,(z \mp h)\right]\right\}, \tag{14}$$

where the amplitudes $\underset{\sim}{c} = (c_r, c_\phi, c_z)$ are so chosen that div $\underset{\sim}{C} = 0$, rot $\underset{\sim}{C} = 0$ and $\underset{\sim}{C} \to 0$ as $|z| \to \infty$: $c_r : c_\phi : c_z = 1 : \varepsilon : i(1+\varepsilon^2)^{1/2}$.

2.3 Bisymmetric Spiral Field Solution

A dispersion relation for ω and m is obtained from equations (10) and (11),

$$\left[-\frac{i\omega}{\kappa} + m^2 + \frac{1}{\varepsilon^2 r^2} + \frac{iV}{\kappa r}\right]^2 + \frac{i\alpha}{\kappa}\left(\frac{\partial V}{\partial r} - \frac{V}{r}\right) m = 0, \tag{15}$$

giving four roots m_j with j = 1 to 4 for ω. A general solution of $\underset{\sim}{B}$ is represented as a superposition of four plane waves characterized by m_j with j = 1 to 4,

26

$$\underline{B}(r,\phi,z;t) = \sum_{j=1}^{4} \underline{b}_{(j)} \exp i\left(-\omega t + \phi + m_j z + \frac{1}{\varepsilon} \ln r\right), \tag{16}$$

where ω and the amplitudes $b_{(j)}$ with $j = 1$ to 4 are determined by the boundary conditions in 2.2 and by joining $\underline{B}$ smoothly to $\underline{C}$ at $z = \pm h$.

Figures 4a and 4b show the behaviour of the BSS fields in the disk whose half-thickness is h and dynamo strength α. The loci of constant grow and decay times (Im(ω): solid lines in units of years) and those of constant angular velocity of the spiral field pattern (Re(ω) $\equiv \Omega_p$: dashed lines in units of km s^{-1} kpc^{-1}). When α and h are on the thick solid lines, the BSS fields are in a steady state. When they are above or below them, the BSS fields grow or decay. It is also shown that the BSS fields can be sustained by the thicker disk when the dynamo action is weaker or α is smaller. The angular velocity of the galactic rotation Ω is taken as 26 km s^{-1} kpc^{-1} in Fig. 4a, corresponding to that of the sun at r = 8.5 kpc. The diffusion constant κ is assumed as 0.3 km s^{-1} kpc, which is estimated from the local turbulence whose root-mean-square velocity v and mean-free-path ℓ are respectively 10 km s^{-1} and 0.1 kpc. According to a relation $\alpha = \gamma\Omega\ell$ with $\gamma = 0.25$–1.0 [6], we have $\alpha = 0.65$–2.6 km s^{-1} in the solar neighbourhood: therefore, if

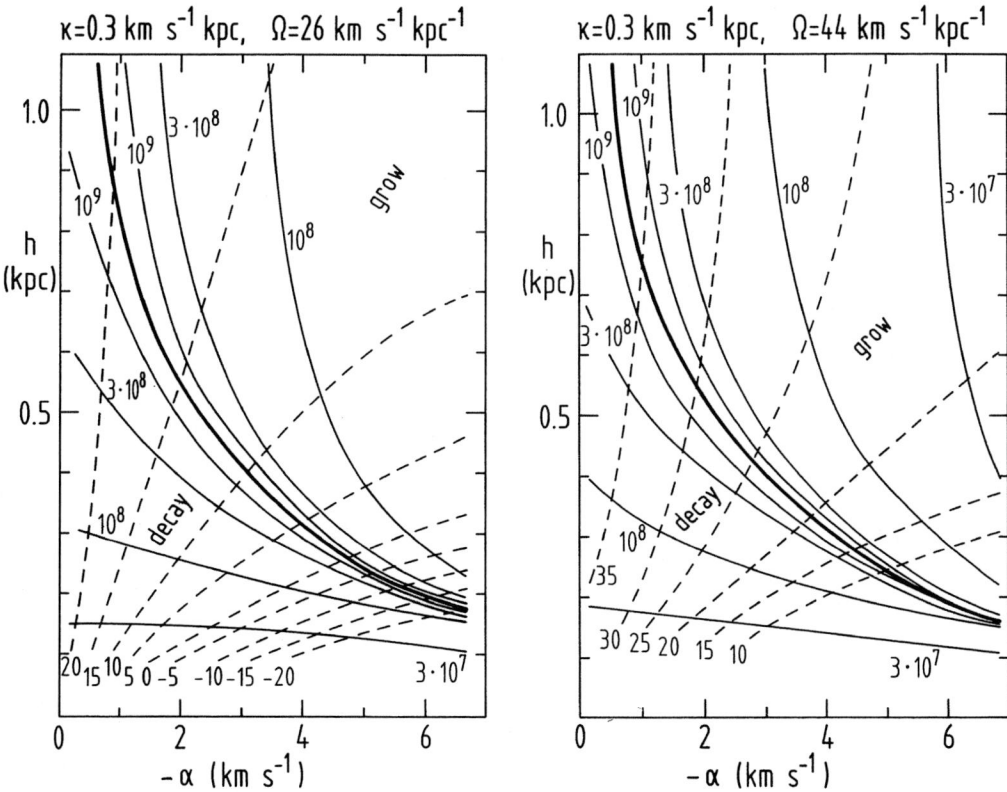

Figures 4a and 4b

27

the half-thickness of the disk h is greater than 0.4–0.7 kpc [9], Fig. 4a guarantees that a BSS field is in a steady state or in a growing state rotating rigidly with $\Omega_p = 13$–20 km s^{-1} kpc^{-1}. Here we mention briefly that γ depends on rotation velocity of each turbulent eddy, produced when it moves up vertically by ℓ and expands. The maximum value, $\gamma = 1$, corresponds to the case in which the expanding eddy increases its horizontal scale by $\sim\ell$ and rotates by the Coriolis force. The minimum value, $\gamma = 0.25$, corresponds to that in which the expanding eddy rotates with the same velocity as the differential rotation of ambient gas [6].

Figure 4b is the same as 4a but for $\Omega = 44$ km s^{-1} kpc^{-1}, corresponding to the galactic rotation at 5 kpc. One finds again that the BSS field can be realized in the inner part of the Galaxy if $h \gtrsim 0.2$–0.4 kpc and $\alpha = 1.4$–4 km s^{-1}, here α is about twice as large as that in Fig. 4a because of the relation $\alpha = \gamma \ell \Omega$ with $\gamma = 0.25$–1. Note that we can choose a single numerical value of Ω_p in a range of 15–20 km s^{-1} kpc^{-1} common to Figs. 4a and 4b. This fact assures that the BSS field rotates rigidly over a wide range of the galactic radius, if the gaseous disk is thicker as mentioned above.

Figure 5 shows the magnetic field distribution on the galactic plane. Figure 6 shows a bird's eye view of the BSS fields in the space above z = 0. In the upper part of the disk and partly in the outer space, the poloidal component becomes dominant, making a helical structure when projected onto the meridional plane.

The field distribution in the lower half of the disk can be reproduced by making the mirror-symmetric transformation of Fig. 6 with respect to z = 0.

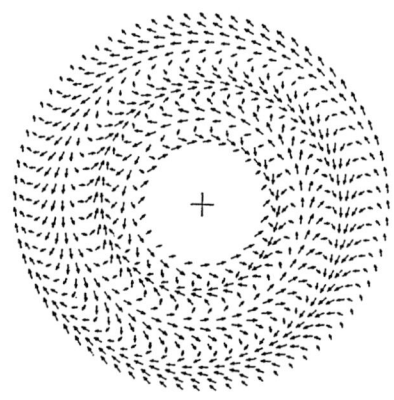

Figure 5

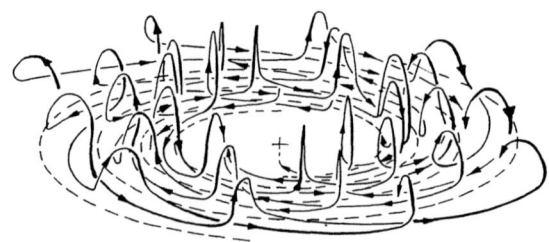

Figure 6

3. Conclusions

A quasi-global analysis is introduced for the kinematic dynamo process in the galactic disk in differential rotation and local cyclonic turbulence. It is shown that the BSS field can rotate rigidly and be maintained in the wide area of the disk, if the gaseous disk is several times thicker than 100 pc which has been obtained for the hydrogen layer. The radio continuum emitting region with a larger scale height [10] could be related to this thick disk for maintaining the BSS fields in our Galaxy and other galaxies [4,5,11-14].

The BSS field is topologically equivalent to a large-scale uniform field twisted by differential rotation; thus it fits very well the primordial origin of the galactic magnetic field. Even if it is so, only the field configuration is of primordial origin and the magnetic fluxes are generated after the formation of the galaxy. See [9] for another idea of the origin of the galactic magnetic field.

Acknowledgement

The author expresses his thanks to the Ministry of Education of Japan for the scientific research fund under Grant No. 60540156 (1985,1986), and to the Asahishinbunsha for the research fund in 1985.

References

1. M. Tosa, M. Fujimoto: Publ. Astron. Soc. Japan 30, 315 (1978)
2. A. Segalovitz, W.W. Shane, A.G. de Bruyn: Nature 264, 222 (1976)
3. A. Sandage: The Hubble Atlas of Galaxies (Carnegie Institution of Washington, Washington 1961), p. 26
4. Y. Sofue, M. Fujimoto, R. Wielebinski: Ann. Rev. Astron. Astrophys. 24, 459 (1986)
5. R. Beck: IEEE Transactions on Plasma Science, Special Issue on Space and Cosmic Plasma, in press (December 1986)
6. E.N. Parker: Astrophys. J. 163, 255 (1971)
7. T. Sawa, M. Fujimoto: Publ. Astron. Soc. Japan 38, 133 (1986)
8. M. Fujimoto, T. Sawa: Publ. Astron. Soc. Japan, submitted (1986)
9. A. Ruzmaikin: this volume
10. E.M. Berkhuijsen: Astron. Astrophys. 140, 431 (1984)
11. Y. Sofue: this volume
12. U.R. Buczilowski: this volume
13. M. Krause et al: this volume
14. E. Hummel et al: this volume

Global Structure of Magnetic Fields in Spiral Galaxies

Y. Sofue

Astronomy Dept., University of Tokyo, 113 Tokyo, Japan and
Nobeyama Radio Observatory, 384-19 Nagano, Japan

Three major components of the large—scale magnetic field in spiral galaxies, the disk
field, the halo field and the poloidal field, are described based on the recently accumulated observational data. A primordial origin of the magnetic field in galaxies is
discussed with a particular concern to the BSS (bisymmetric spiral) field configuration.

1. Introduction

The magnetic field of galactic scale has come to be recognized as an essential concept in understanding the structure and dynamics of galaxies. Observational data,
both optical and radio, have been combined with theoretical models to determine
some definite configurations of the fields in nearby galaxies. Spiral galaxies are
shown to possess either an axisymmetric (ring) or a bisymmetric spiral (BSS) configuration of magnetic field. Recent progress has been reviewed by SOFUE et al. [1],
ASSEO and SOL [2] and BECK [3].

The large—scale magnetic field in a galaxy may be categorized into three major
components: *(1)* The disk field, which has been well studied and shown to be deeply
coupled with the spiral structure and dynamics of the gaseous matter. *(2)* The halo
field, which has been predicted from a magnetohydrodynamic consideration of the
steady—state BSS field in the disk. However, observational data are still too poor to
recognize its thorough structure. *(3)* The poloidal field, which runs perpendicular to
the galactic plane and must be inevitably present if the magnetic fields of galaxies
are of primordial origin. This component may be related to the galactic center activity after being accumulated to the nuclear disk through an accretion of the disk
gas.

In this review we describe these three major components on the basis of recent
observational data. We further discuss the implication of magnetic field on the
galaxy dynamics. We finally discuss the origin of the magnetic field. In particular,
the primordial origin hypothesis will be considered in some detail concerning the BSS
structure and the poloidal field.

2. The Disk Field

Both optical and radio polarization observations provide the opportunity to determine
the magnetic field in the disks of galaxies. Optical observations use the Davis-

Greenstein effect on the starlight. Optical polarization of starlight is caused by the scattering by elongated dust grains in interstellar medium aligned by the magnetic field [4,5,6]. Optical polarization measurements, however, give neither field strength nor direction, but give only the orientation of the field projected on the sky. The predominant magnetic field in spiral galaxies as derived by optical observations seems to be parallel to the spiral arms and to the dust lanes (e.g. [7]). More recent optical polarization measurements made it possible to determine the field configuration in some detail. SCARROTT et al. [8] have shown that M51 possesses a clear spiral configuration of magnetic field. They also obtained results indicating a spiral nature of the field in other galaxies [5].

Radio polarization observations provide a mean to determine the three-dimensional structure as well as the strength of the field. The field strength can be obtained from the assumption of energy-density equipartition between cosmic rays and magnetic field [9,10]. The field strength and density of cosmic-ray electrons are related to the observed intensity and size of a source. The basic premise of the equipartition assumption is open to discussion. However, the results for external galaxies seem to be reasonably consistent with values found in our Galaxy. Good correlations of the field strength inferred from the equipartition assumption with the CO line intensity [1] or with the surface mass density of galaxies [3] support the validity of the use of this simple assumption. The field strengths so far derived for nearby galaxies are between 3 and 15 μG [11].

The field direction can be determined from the rotation measure (RM) and the intrinsic polarization angle (IPA). From IPA we find the field orientation projected on the sky, or the transverse component to the line of sight. From RM we obtain the strength and direction of the magnetic field along the line of sight if an assumption about the thermal electron density is made. On the premise that the magnetic field in a disk is represented either in a BSS or in an axisymmetric configuration we are able to determine the three-dimensional configuration from RM analysis. If the field is axisymmetric (ring or spiral), the RM along an azimuthal circle on the galactic disk varies in a single-peaked sinusoidal way against the azimuthal angle θ. Its variation along the major axis (X) is antisymmetric with respect to the rotation axis (Figure 1). On the other hand, if the field is BSS, the RM variation with θ is in a doubly-sinusoidal way, and its variation with X is symmetric with the rotation axis. Higher order variations of RM with θ may exist, which represent higher order field configurations than axisymmetric or the BSS. However, the observational data are still not good enough to clarify the existence of such higher order configurations.

There have been many galaxies for which the field configurations have been determined by radio observations and RM analysis. For details on individual galaxies the readers may refer to the contributions to this symposium by BERKHUIJSEN et al. [12], LOISEAU et al. [13], GRÄVE and BECK [14], BUCZILOWSKI [15], KRAUSE et al. [16], HUMMEL et al. [17], and to the literature cited in SOFUE et al. [1]. Table 1 summarizes the results from the radio observations. The majority of the nearby

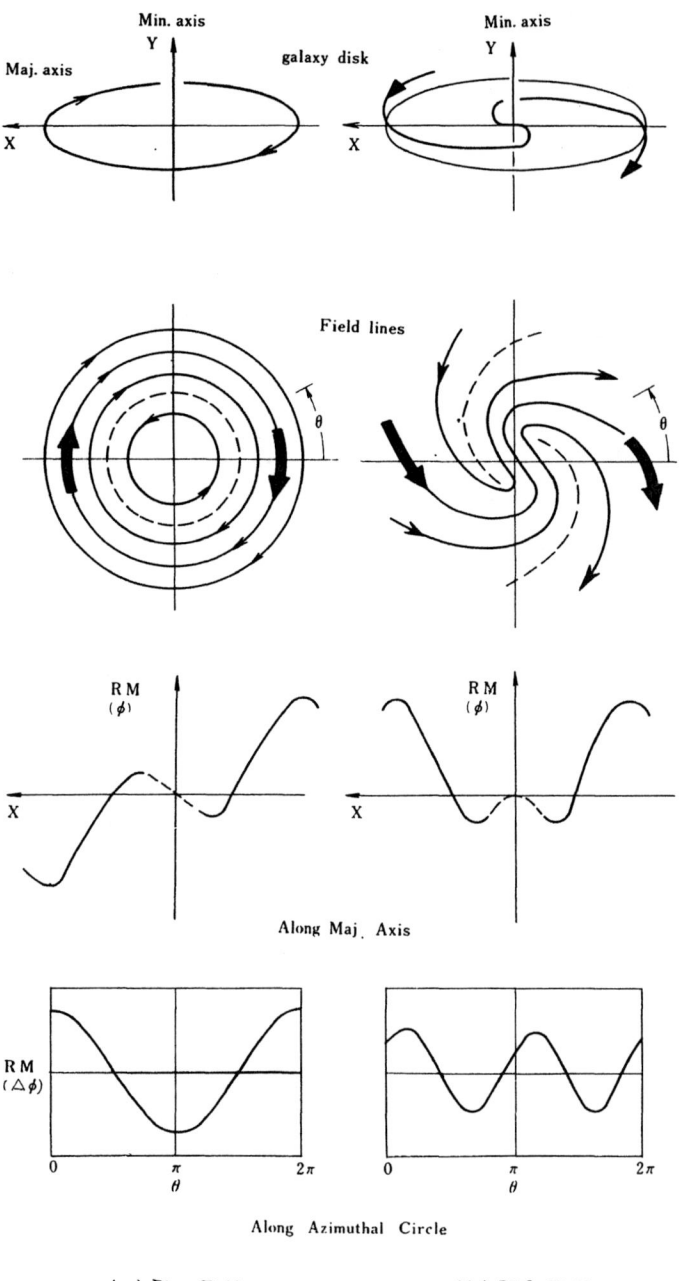

Figure 1: (a) The ring and (b) BSS configurations of magnetic fields in disk galaxies. The characteristic variations of RM are illustrated against the distance X along the major axis, and against the azimuthal angle θ along a concentric circle to the rotation axis

Table 1: Magnetic fields in galaxies derived from radio observations[*]

Galaxy	Type	Field configuration	Field strength (μG)	Remarks
The Galaxy	Sb	BSS	3-4	
M31 (NGC 224)	Sb	axisymmetric	4±1	
M33 (NGC 598)	Scd	BSS	3±1	
M51 (NGC 5194)	Sc	BSS	~10	
M81 (NGC 3031)	Sb	BSS	8±2	
M83 (NGC 5236)	SBc	BSS	—	
NGC 253	Sc(p)	BSS?	13±4	
IC 342	Scd	axisymmetric	7±2	
NGC 6946	Scd	BSS	12±4	
NGC 2903	Sbc	BSS	—	
NGC 5055 (M63)	Sc	BSS?	—	
NGC 4258	SBc	—	—	B along anomalous arms
SMC	Irr	—	—	Polarized E-vectors are perpendicular to the major axis

[*]For references see SOFUE et al. [1].
The field strength was mainly taken from BECK [11]

spiral galaxies are shown to possess a BSS field. Clear examples of the BSS case are found for M81 [16] and for M51 [18]. An example of an axisymmetric "ring" field was shown for M31 (Figure 2a: [10]). However, if we subtract the ring mode variation from the data of M31, there remains a doublysinusoidal component as shown in Figure 2b. This may indicate the existence of a BSS field superposed on the ring field in M31 [19].

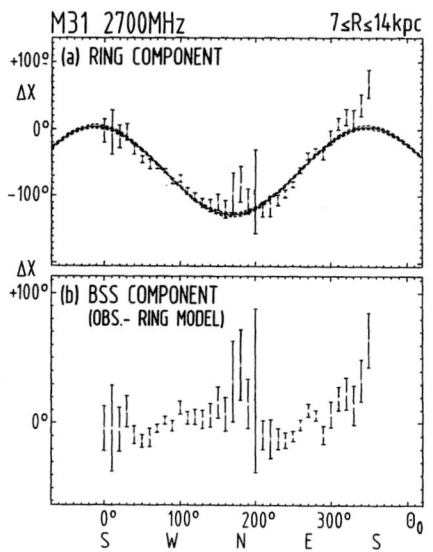

Figure 2: (a) The variation of deviation angle Δχ of polarization vectors from radial vectors for M31 [10], which is an indication of the rotation measure, plotted against azimuthal angle θ. The data are well fitted by a single sinusoidal curve.
(b) The residual after subtracting the best fit sinusoidal curve from (a). A doubly-sinusoidal variation remains, which indicates the existence of a BSS field superimposed on the ring field [19]

3. The Halo Field

The existence of a magnetic halo is a natural consequence of a magnetohydrodynamic treatment of the disk field being maintained in a steady state. Observationally, however, it has been a long-standing question. Evidence for a nonthermal radio halo far beyond the galactic disk has been obtained for several edge-on galaxies. An example is seen for NGC891, for which ALLEN et al. [20] found a radio continuum halo extending over ~ 5 kpc from the galactic plane. The magnetic field strength as estimated from the assumption of equipartition is of the order of a few μG. Similar results have been obtained for several galaxies. They have the same order of field strength in the halos.

A reasonable indication of a halo field in our Galaxy can be obtained by a statistical analysis of RMs of extragalactic radio sources and pulsars [21]. The analysis is performed by plotting their $|RM|$ against cot $|b|$, where b is the galactic latitude. The upper envelope of the distribution of the sources on the $|RM|$–cot $|b|$ plane is well represented by the relation $|RM| = RM_0$ cot $|b|$. This b–dependence of RM is due to a plane parallel distribution of magnetic field and ionized gas in the Galaxy. The coefficient RM_0 for extragalactic objects has been obtained as $RM_0(\text{ext}) \cong 30$ rad m^{-2} [22]. On the other hand, we obtain $RM_0(\text{PSR}) \cong 10$ rad m^{-2} for pulsars from the data of MANCHESTER and TAYLOR [23]. Since pulsars are distributed below 500 pc from the galactic plane, the remaining galactic contribution to the extragalactic sources, ~ 20 cot $|b|$ rad m^{-2}, must be due to a galactic halo beyond 500 pc. If we tentatively take the electron density in the halo as ~ 10^{-3} cm^{-3} and the thickness of the halo to be ~ 3 kpc, we obtain a field strength of roughly a few μG. We note that the total magnetic flux in the halo is comparable to that in the galactic disk. This fact suggests that the magnetic halo plays an important role in the MHD behaviour of a galaxy.

4. The Poloidal Field

The large-scale poloidal field in a galaxy has never been discussed except for the early-day sketch of a primordial field evolution in a galaxy by PIDDINGTON [24]. If we accept the primordial origin hypothesis for the BSS configuration, as discussed in the next section, the poloidal field component plays an important role in the galaxy disk. The component of the primordial magnetic field parallel to the galactic disk, which was fed from the intergalactic space to a protogalaxy, is tightly wound up in the central few kpc and dissipates by the turbulent diffusion or by an exchange process with the halo. In the outer disk it is maintained in a steady-state BSS configuration [25,26,27].

The primordial poloidal field, on the other hand, can never dissipate away from the galaxy disk, as the scale length of the diffusion, or the radius, is large enough to freeze the field lines into the disk. A simple calculation (SOFUE, in preparation) shows that a poloidal field of ~ 10^2 μG can be accumulated to the central few hundred pc by an accretion of the disk gas, if there existed an intergalactic magnetic field of the order of nG when the protogalaxy formed (Figure 3). The accretion of the disk gas is caused by the galactic shock waves and by magnetic braking due to the disk field.

Such a poloidal field was recently discovered in the Galactic center by the detection of strong radio polarization along the radio arc and along its extension to high galactic latitudes [28,29,30]. SOFUE et al. [31] have proposed a cylindrical magnetic field perpendicular to the galactic plane which is twisted by the rotation of the disk gas and forms a semi-helical field structure (see also [32]). The field direction is consistent with the VLA filaments in the radio arc which are perpendicular to the galactic plane [33].

The tightly accumulated poloidal field in the central region of a galaxy may be related to a central activity such as the formation of jets. The galactic center lobe as an ejection phenomenon [34] and some jet-like emission emerging vertically from the centers of edge-on galaxies (e.g. [35]) may be related to the poloidal field in these galaxies. A promising formation mechanism of jets by a twisted poloidal field by the accreting disk has been proposed by UCHIDA et al. [36].

PRIMORDIAL-ORIGIN MAGNETIC FIELD

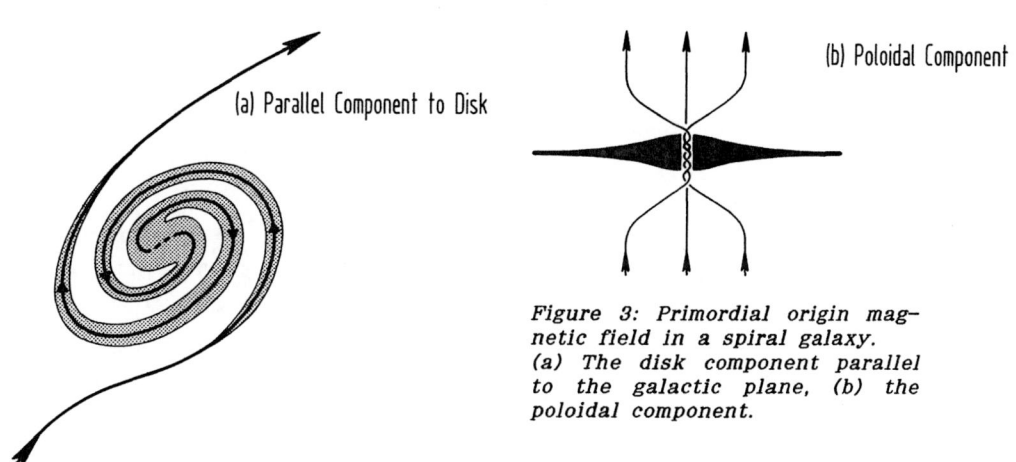

(a) Parallel Component to Disk

(b) Poloidal Component

Figure 3: Primordial origin magnetic field in a spiral galaxy. (a) The disk component parallel to the galactic plane, (b) the poloidal component.

5. Origin of the large-scale magnetic field in galaxies

The origin of the magnetic fields in spiral galaxies is still controversial. However, we suggest that a primordial magnetic field in the intergalactic space fed into a protogalaxy may have easily produced any field configuration so far observed. In particular, the BSS configuration is a natural consequence of the wound-up field from the intergalactic space. The wound-up field can be maintained in a steady state for the galaxy life through the diffusion process in the disk under the existence of the galactic halo as a reservoir [37]. A ring field can also be produced by the primordial field, if there existed any slight asymmetry or inhomogeneity in the primordial field with respect to the rotation axis of the protogalaxy [1]. A poloidal field accumulated in the galactic center is also easily understood as a frozen-in fossil of the poloidal primordial field.

Thus far the primordial origin hypothesis appears to explain the observed magnetic fields in galaxies. A problem may be the origin of the primordial field. The existence of a large-scale intergalactic magnetic field has long been claimed [21], but is still controversial (e.g. [38]). Observational search for such a large-scale structure in the universe is obviously important not only for the problem of the origin of galactic magnetic fields but also from the cosmological point of view, whether the universe is uniform or not.

Finally we stress that the steady-state solution for the BSS field in a galaxy obtained from the primordial origin hypothesis [37], which uses the dynamo action to maintain the field configuration and flux in a steady state, and the BSS solution obtained from the dynamo theory [39] do not conflict each other. They seem to give the same solution in a steady state.

References

1. Y. Sofue, M. Fujimoto, R. Wielebinski: Ann. Rev. Astron. Astrophys. <u>24</u>, 459 (1986)

2. E. Asseo, H. Sol: Physics Reports, in press (1986)

3. R. Beck: IEEE Transactions on Plasma Science, Special Issue on Space and Cosmic Plasma, in press (December 1986)

4. L. Davis, J.L. Greenstein. Astrophys. J. <u>114</u>, 206 (1951)

5. S.M. Scarrott: this volume

6. P. Cugnon: this volume

7. A. Elvius: In The Structure and Properties of Nearby Galaxies, IAU Symp. 77, ed. by E.M. Berkhuijsen and R. Wielebinski (Reidel Publ. Co., Dordrecht 1978), p. 65

8. S.M. Scarrott, D. Ward-Thompson, R.F. Warren-Smith: Monthly Notices Roy. Astron. Soc., in press (1986)

9. A.T. Moffet: In Galaxies and the Universe, ed. by A. Sandage et al., Vol. 9, Chap. 7 (Chicago University Press, Chicago 1973)

10. R. Beck: Astron. Astrophys. <u>106</u>, 121 (1982)

11. R. Beck: In Internal Kinematics and Dynamics of Galaxies, ed. by E. Athanassoula (Reidel Publ. Co., Dordrecht 1983), p. 159

12. E.M. Berkhuijsen, R. Beck, R. Gräve: this volume

13. N. Loiseau, E. Hummel, R. Beck, R. Wielebinski: this volume

14. R. Gräve, R. Beck: this volume

15. U.R. Buczilowski: this volume

16. M. Krause, R. Beck, E. Hummel: this volume

17. E. Hummel, M. Krause, R. Beck: this volume

18. M. Tosa, M. Fujimoto: Publ. Astron. Soc. Japan <u>30</u>, 315 (1978)

19. Y. Sofue, R. Beck: in preparation

20. R.J. Allen, J.E. Baldwin, R. Sancisi: Astron. Astrophys. <u>62</u>, 397 (1978)

21. Y. Sofue, M. Fujimoto, K. Kawabata: Publ. Astron. Soc. Japan <u>31</u>, 125 (1979)

22. M. Inoue, H. Tabara: Publ. Astron. Soc. Japan <u>33</u>, 603 (1981)

23. R.N. Manchester, J.H. Taylor: In <u>Pulsars</u> (Freeman, San Francisco 1977), p. 123

24. J.H. Piddington: Monthly Notices Roy. Astron. Soc. <u>128</u>, 345 (1964)

25. T. Sawa, M. Fujimoto: Publ. Astron. Soc. Japan <u>38</u>, 133 (1986)

26. M. Fujimoto, M. Tosa: Publ. Astron. Soc. Japan <u>32</u>, 567 (1980)

27. M. Fujimoto, T. Sawa: Publ. Astron. Soc. Japan <u>32</u>, 265 (1981)

28. J.H. Seiradakis, A.N. Lasenby, F. Yusef-Zadeh, R. Wielebinski, U. Klein: Nature <u>317</u>, 69 (1985)

29. M. Tsuboi et al.: Astron. J., in press (1986)

30. W. Reich, Y. Sofue, M. Inoue, J.H. Seiradakis: this volume

31. Y. Sofue, W. Reich, M. Inoue, J.H. Seiradakis: Publ. Astron. Soc. Japan, in press (1986)

32. Y. Sofue, W. Reich: this volume

33. F. Yusef-Zadeh, M. Morris, D. Chance: Nature <u>310</u>, 557 (1984)

34. Y. Sofue, T. Handa: Nature <u>310</u>, 568 (1984)

35. E. Hummel et al.: Astrophys. J. Letters <u>267</u>, L5 (1983)

36. Y. Uchida, K. Shibata, Y. Sofue: Nature <u>317</u>, 699 (1985)

37. M. Fujimoto: this volume

38. P.P. Kronberg: this volume

39. A.A. Ruzmaikin: this volume

Rotation Measures and Depolarization near the Minor Axis in M31

E.M. Berkhuijsen, R. Beck, and R. Gräve

Max-Planck-Institut für Radioastronomie, Auf dem Hügel 69,
D-5300 Bonn 1, Fed. Rep. of Germany

1. Introduction

Using the 100-m telescope in Effelsberg a radio continuum survey of an area 170'×
90' centred on M31 was made at $\lambda 6.3$ cm (half-power beamwidth (HPBW) = 2'.4) in
both total power and linear polarization. A preliminary map of the central field is
presented in Fig. 1. Details of the observations and the reduction procedures will be
published elsewhere.

The new survey will enable the study of the structure and the degree of uniformity of the magnetic field in the arms forming the bright ring in M31 on a scale
of about 250 pc × 1200 pc in the plane of the galaxy. A comparison with the earlier
survey at $\lambda 11$ cm [1] will give important information on the distribution of Faraday
rotation measures in the 'ring' and on depolarization effects. Preliminary results of a
comparison of the central fields, smoothed to a HPBW of 5'.0, are described below.

2. Faraday rotation measures and direction of the magnetic field

The rotation measures (RM) between $\lambda 6$ and $\lambda 11$ cm in the central field are typically
-80 to -100 rad m^{-2} with 1σ-errors of ~ 20 rad m^{-2}. These values are in agreement with those inferred from an analysis of the $\lambda 11$ cm survey alone [2]. Using the
foreground RM of -88 rad m^{-2} [2] a map of rotation measures intrinsic to M31
(RM$_i$) was obtained (Fig. 2). RM$_i$ varies between 0 and -50 rad m^{-2} in the W and
NE of the 'ring', but reaches values of $+90$ rad m^{-2} in the SE. Here the direction of
the component of the magnetic field parallel to the line of sight, B$_{||}$, appears to be
opposite to that in the other parts of the 'ring' near the minor axis.

Since RM$_i$ is proportional to the integral of the density of thermal electrons and
B$_{||}$ along the line of sight one would expect large values of |RM$_i$| to coincide with
HII regions. However, no clear correlation between variations in RM$_i$ and positions of
optically detected HII regions [3] was found.

The orientations of the magnetic field component perpendicular to the line of
sight, B$_\perp$, can be obtained from the observed polarization angles after correction for
the rotations derived from the RM. Figure 3 shows that B$_\perp$ runs generally parallel to
the 'ring', thus confirming the model of [2]. In the W the field structure is extremely coherent, whereas in the E several deviations occur. In the region of positive RM in the SE also B$_\perp$ deviates strongly from the general direction. In this area
a large-scale disturbance of the magnetic field is apparently observed. Interestingly,

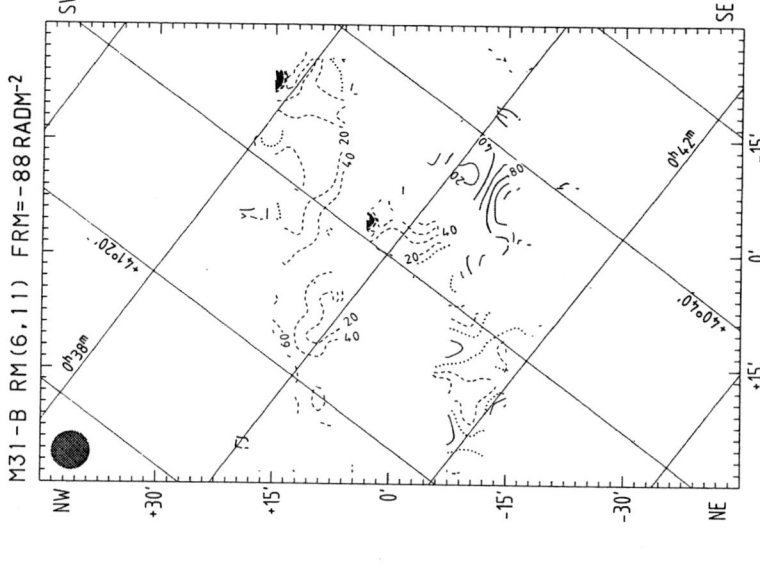

Figure 1: Distribution of the polarized emission ('vectors') at λ6 cm superposed onto that of the total emission (contours) in the central field of M31. The length of a 'vector' is proportional to the polarized intensity (calibrated at the top) and its orientation shows the observed angle of polarization. The contour levels are approximately in mJy/beam. The map was smoothed to a HPBW = 2.5 to improve the signal-to-noise ratio

Figure 2: Distribution of intrinsic rotation measures in the central field of M31 assuming a foreground RM of −88 rad m⁻². Contours of negative RM (dashed), zero RM (dotted) and positive RM (full lines) are shown. RM were only computed if the polarized intensity is larger than twice the noise at both λ6 cm and λ11 cm. The HPBW of 5.'0 is shown in the upper left-hand corner

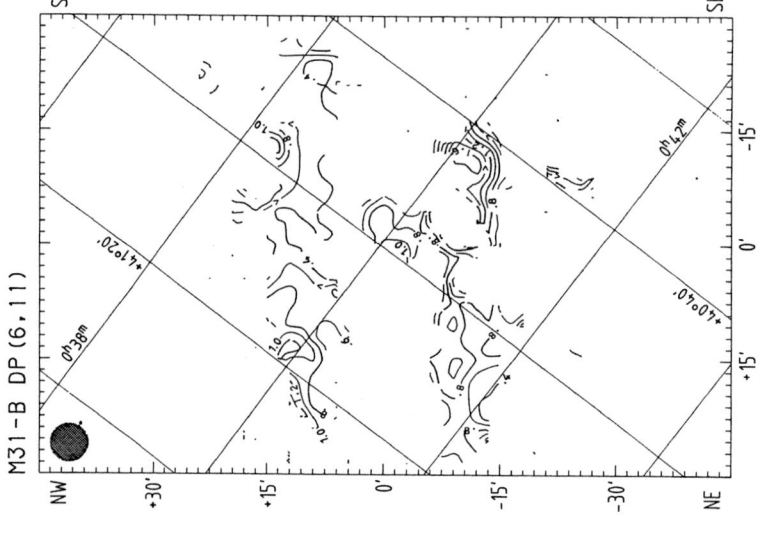

Figure 4: Distribution of the ratio of polarization percentage at $\lambda 11$ cm and $\lambda 6$ cm, $DP(6,11) = p(11)/p(6)$, in the central field of M31. This ratio was only computed if at both wavelengths the polarized intensity exceeds twice the noise

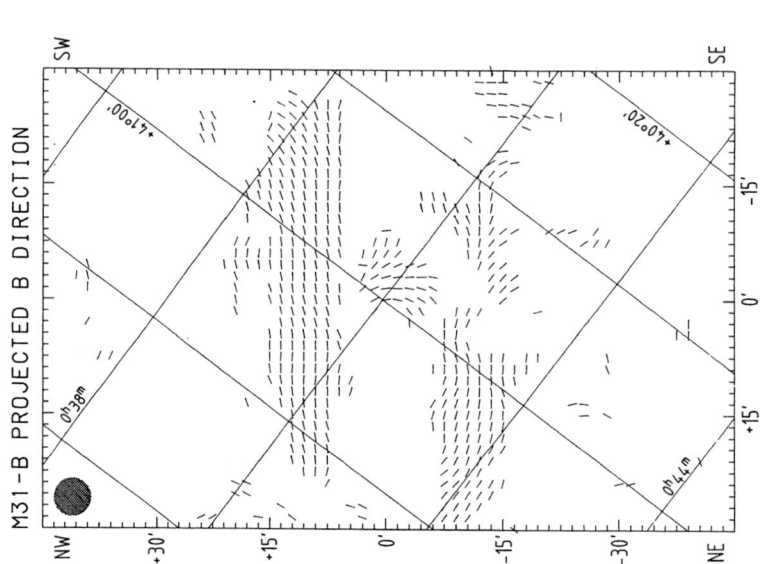

Figure 3: Distribution of the direction of $B_\perp$ projected on the sky in the central field of M31 showing the general alignment of the magnetic field with the 'ring' of emission. The deviation just SE of the nucleus is a large-scale disturbance of the field (see also Figs. 2 and 4). Deviations near the borders of the map need confirmation from neighbouring maps still to be reduced

in this same area three HI arms seem to be splitting up [4] and a spur in the FIR perpendicular to the arms has been detected [5]. A possible connection between the three phenomena needs further study.

3. Depolarization effects

The ratio of the polarization percentages at $\lambda 11$ and $\lambda 6$ cm is plotted in Fig. 4. Compared to $\lambda 11$ cm the decrease in Faraday depolarization at $\lambda 6$ cm is counteracted by a decrease of the nonthermal fraction of the total power emission. The nonthermal fractions derived from the radio spectral index map [6] indicate that these effects approximately cancel in large parts of the ring resulting in percentage ratios of ~ 1. Since depolarization caused by randomness of the magnetic field in the emission regions is independent of wavelength, percentage ratios $\lesssim 0.5$ as observed in the W or $\gtrsim 1.5$ as seen in the SE may indicate that several polarized components with different spectral indices and different intrinsic angles occur along the line of sight. In the central field such components may be associated with different spiral arms seen in the same direction.

References

1. R. Beck, E.M. Berkhuijsen, R. Wielebinski: Nature 283, 272 (1980)
2. R. Beck: Astron. Astrophys. 106, 121 (1982)
3. A. Pellet, N. Astier, A. Viale, G. Courtès, A. Maucherat, G. Monnet, F. Simien: Astron. Astrophys. Suppl. 31, 439 (1978)
4. E. Brinks, W.W. Shane: Astron. Astrophys. Suppl. 55, 179 (1984)
5. R.A.M. Walterbos, P.B.W. Schwering: Astron. Astrophys. (submitted)
6. R. Beck, R. Gräve: Astron. Astrophys. 105, 192 (1982)

High Resolution Polarization Observations of M31

N. Loiseau, E. Hummel, R. Beck, and R. Wielebinski

Max-Planck-Institut für Radioastronomie, Auf dem Hügel 69,
D-5300 Bonn 1, Fed. Rep. of Germany

1. Introduction

The radio continuum emission of M31 is concentrated in a "ring"-like structure which is also seen in many other constituents (e.g. HI, Hα, FIR, UV). The real structure of M31 is difficult to determine because of its high inclination. The first reliable polarization observations of M31 made at λ11 cm with the 100-m telescope of the MPIfR were published by BECK et al. [1], showing that also the polarized intensity is seen in this ring. These data were analysed by SOFUE and TAKANO [2] and BECK [3] who showed that the change of position angle of the E-vector with azimuthal angle was consistent with a ring-like magnetic field configuration. Only very few galaxies show such a distribution (e.g. SOFUE et al. [4], BECK [5]). More recent λ6 cm observations (BERKHUIJSEN et al., this volume), which are nearly free of Faraday effects, indicate again a ring-like field structure.

Already at λ11 cm the E-vectors show some scatter which could be interpreted to be due to local changes of the rotation measure (RM). The causes for such changes can be either changes of the magnetic field strength or direction, clumping of the thermal electrons or variations in the path length. All these possible factors point to the need of high angular resolution observations.

2. Observations

We observed the total intensity and linear polarization with the Very Large Array (VLA) at λ20 cm in the SW region of M31 where the λ11 cm data [1] showed the strongest polarization. The field was centred at RA = $00^h38^m36^s$, DEC = 40°54'00" (1950.0). The VLA was in the D-configuration and two bands of 50 MHz (centred on 1.465 and 1.515 GHz) could be recorded simultaneously. At λ20 cm the synthesized beam had a half-power beamwidth of ~ 40" and the primary beam a half-power beamwidth of ~ 30' so that only a part of M31 could be studied. In order to improve the sensitivity we degraded the resolution to 75". The total power distribution with superposed E-vectors is shown in Figure 1. In Figure 2 we show an overlay of the contours of the linearly polarized intensity onto the optical image of M31.

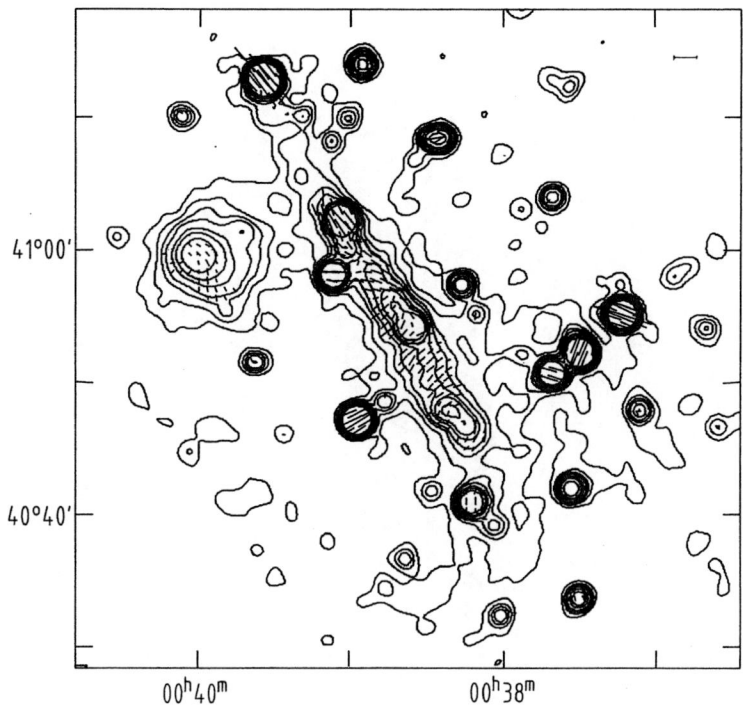

Figure 1: Total emission of the SW region of Andromeda smoothed to 75" resolution, with the observed polarization vectors. Contour levels are: 0.5, 1, 1.5, 2, 3 and 6 mJy/beam. The r.m.s. noise is ~ 0.1 mJy/beam. The bar in the top right corner indicates 1 mJy of polarized intensity

Table 1: Rotation measures and intrinsic polarization angles for five regions along the arm. The positions of the regions (θ_0) are azimuthal angles in the plane of the galaxy. They are at a radius between 7 and 10 kpc. The polarization angles are defined in a Cartesian coordinate system with the x-axis along the major axis of M31. The ring model values for these positions are given for comparison.

θ_0	Polarization angles (°)			RM (rad m^{-2})		Intrinsic angles χ_0 (°)	
(°)	6cm	11cm	20cm	Observed	Model	Observed	Model
53	181	124	163	−95	−59	24	−9
59	175	127	176	−86	−68	14	−7
66	166	121	173	−83	−78	4	−5
77	165	120	143	−94	−95	4	−3
87	159	96	75	−127	−110	5	−1

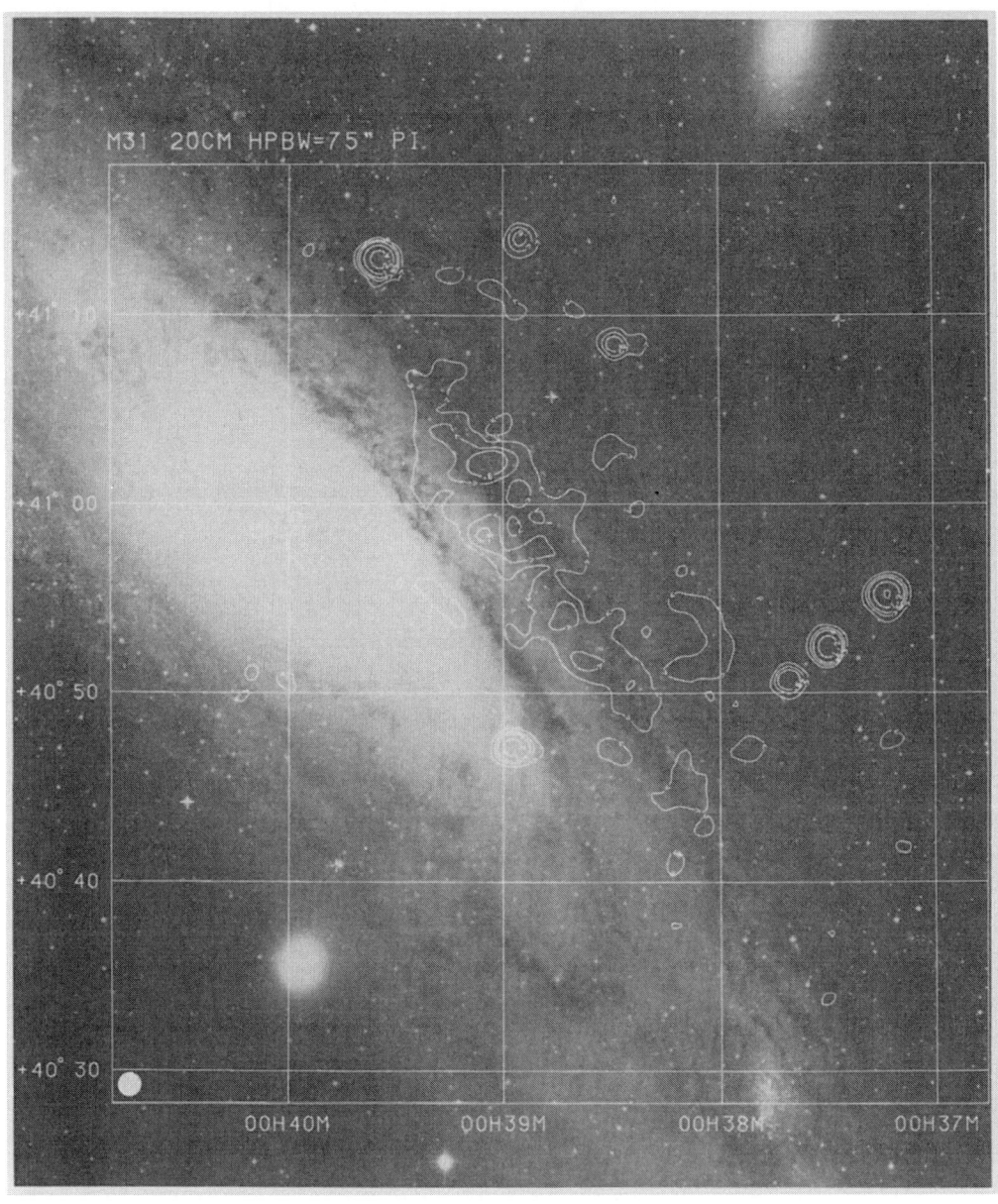

Figure 2: *Polarized intensity at λ20 cm superimposed onto an optical image of M31. Contour levels are: 0.15, 0.3, 0.6, 1.2 and 2.4 mJy/beam. The r.m.s. noise is ~ 0.05 mJy/beam*

3. Results

The total emission as well as the polarized emission follow very closely the dust lane that runs along the most conspicuous arm in this region. The continuum emission peaks also coincide with the maxima of other distributions like HI (BRINKS [6]), HII regions (PELLET et al. [7]), FIR emission (HABING et al. [8]), etc. The inner and outer arms could not be detected but we found weak polarized emission associated with the centre of M31.

The map at $\lambda 20$ cm and the map at $\lambda 6$ cm of BERKHUIJSEN et al. (this volume) were smoothed to the 5' resolution of the map at $\lambda 11$ cm of BECK et al. [1]. The three maps were combined to determine unambiguously the rotation measure distribution along the arm. Table 1 gives average values for the polarization angles at the three frequencies and the derived rotation measures for 5 regions of 5×5 arcmin extent along the arm. The positions of the regions are given in azimuthal angles θ_0 in the plane of the galaxy, as defined by BECK [3], in order to compare our values with the ones of his model. The errors in the polarization angles are less than $7°$ and in the rotation measures less than 6 rad m^{-2}. The distribution of the intrinsic angles in the plane of the sky shows that the magnetic field runs almost parallel to the arm, although there is a deviation from the ring-like field configuration.

From Table 1 it can be seen that the rotation measure (RM) has a maximum at $\theta_0 \sim 65°$, while the model for the magnetic field configuration [3] predicts a constant increase of RM towards $\theta_0 = 53°$. In the analyzed ring section the RM should have an increment of 50 rad m^{-2}, while the observed increment is ~ 30 rad m^{-2}. This difference between the observed value and the model could be due to the effects mentioned in the Introduction. To explain this difference by a variation of the average electron density along the arm, it would be necessary to have a difference of $\sim 30\%$ between both extremes of this region. An inspection of the distribution of HII regions of PELLET et al. [7] shows that this is improbable. The uniform distribution of the continuum emission (Fig. 1) indicates, assuming minimum energy conditions, that there are no large variations in the magnetic field strength. Then at least part of the variation in RM has to be attributed to deviations, most probably in the plane of the galaxy, from a perfect ring configuration of the magnetic field.

A difference of 20 rad m^{-2} can be explained by a pitch angle of about $10°$ of the magnetic field lines. This can be related to a kink in the spiral arm present in that region (Fig. 1). On the other hand, the HI distribution [6] shows that in this region there is a branch splitting from the main arm.

4. Conclusions

The results of the present work can be summarized as follows:

(1) The polarized continuum emission at 1.4 GHz is clearly associated with the dust lane in the main arm in the SW region of M31. Our high resolution data (HPBW =

133 pc) do not show any systematic shift between the radio continuum ridge line and that of other distributions.

(2) The combination of data in three frequencies (6 cm, 11 cm and 20 cm) allowed the unambiguous determination of the intrinsic polarization angles, indicating a magnetic field aligned with the arm.

(3) The rather uniform distribution of the HII regions along the arm indicates that the variation in RM is due to a change in the pitch angle of the magnetic field lines of the order of 10°.

References

1. R. Beck, E.M. Berkhuijsen, R. Wielebinski: Nature 283, 272 (1980)
2. Y. Sofue, T. Takano: Publ. Astron. Soc. Japan 33, 47 (1981)
3. R. Beck: Astron. Astrophys. 106, 121 (1982)
4. Y. Sofue, M. Fujimoto, R. Wielebinski: Ann. Rev. Astron. Astrophys. 24, 459 (1986)
5. R. Beck: IEEE Trans. on Plasma Science, in press (Dec. 1986)
6. E. Brinks: Ph.D. Thesis, Leiden University (1984)
7. A. Pellet, N. Astier, A. Viale, G. Courtès, A. Maucherat, G. Monnet, F. Simien.: Astron. Astrophys. Suppl. 31, 439 (1978)
8. H.J. Habing, G. Miley, E. Young, B. Baud, N. Boggess, P.E. Clegg, T. de Jong, S. Harris, E. Raimond, M. Rowan-Robinson, B.T. Soifer: Astrophys. J. 278, L59 (1984)

The Magnetic Field in IC 342

R. Gräve and R. Beck

Max-Planck-Institut für Radioastronomie, Auf dem Hügel 69,
D-5300 Bonn 1, Fed. Rep. of Germany

1. Introduction

Late-type galaxies feature spiral structures with large pitch angles where phenomena in spiral arms can easily be studied. IC 342 is an outstanding late-type galaxy because of its large angular size (50' in radio continuum, 90' in neutral hydrogen gas) and its favourable inclination of only 25° to the line of sight [1]. This allows a detailed study of the magnetic field in the spiral arms of IC 342 even with the medium resolution achievable with the 100-m radio telescope at centimetre wavelengths. The largescale structure of the magnetic field in IC 342 was earlier suggested to be ring-like [2] similar to M31.

2. Observations

We have carried out observations of IC 342 at wavelengths of 6.3 and 11.1 cm for which the resolving powers were 2.5 and 4.3, respectively. These correspond to 2.2 and 3.9 kpc in the plane of the galaxy, assuming a distance of 3.1 Mpc.

The distributions of the total radio emission will be discussed elsewhere. Here we present maps of the linearly polarized emission (see Figures 1 and 2), which is restricted to the region of optical spiral arms within 13 kpc radius. At λ6.3 cm it is strongest at 6 kpc radius where the degree of polarization p is about 10%. Near the centre p decreases to about 1%. The highest degrees of polarization, around 25%, occur at 12 kpc radius where the rotation curve reaches a maximum [1].

3. Results

The similarity of distributions in Figs. 1 and 2 show that effects of Faraday rotation and depolarization are small in IC 342 as expected if the magnetic fields are mainly parallel to the disk of the galaxy. The different angular resolutions at λ6.3 cm and λ11 cm can account for the smaller average polarization at λ11 cm.

The maximum of polarized emission, several kpc distance from the centre, is consistent with the prediction of the dynamo theory [3] assuming the action of an $\alpha\omega$-dynamo in the disk of IC 342. Near the centre of the galaxy the magnetic field is strong because a high nonthermal intensity is observed. The rotation velocity, however, increases linearly with radius out to 3 kpc [1] so that the lack of differential rotation prevents the generation of magnetic fields by an $\alpha\omega$-dynamo. The low degree of polarization in that region indicates that the action of an α^2-dynamo is

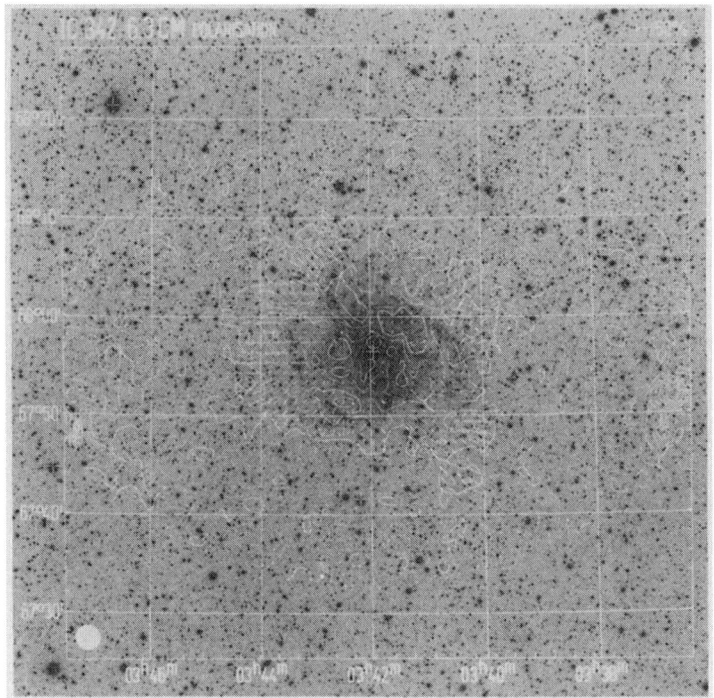

Figure 1: Linearly polarized emission of IC 342 at λ6.3 cm superimposed onto the red POSS photograph. Contour levels are 1, 2, 3, 4 mJy/beam area. The r.m.s. noise is 1.0 mJy/b.a. The lengths of the vectors are proportional to the degree of polarization; the scale is indicated in the upper right-hand corner

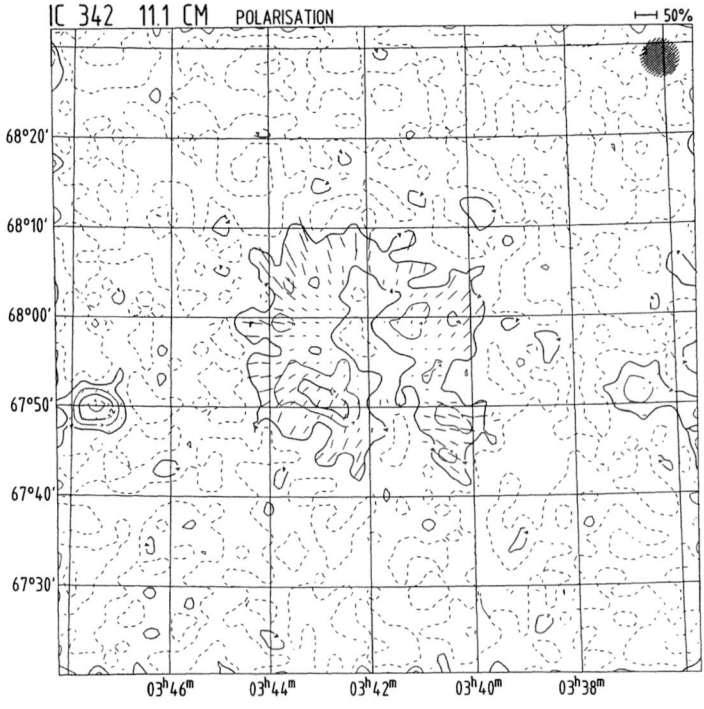

Figure 2: Linearly polarized emission of IC 342 at λ11.1 cm. Contour levels are 4, 8, 12, 16 mJy/b.a. The r.m.s. noise is 2.3 mJy/b.a. The lengths of the vectors are proportional to the degree of polarization, the scale is indicated in the upper right-hand corner

also improbable on scales of more than 2 kpc. However, the origin of the magnetic fields near the centre of IC 342 has to be investigated with higher resolution.

The ground mode (m=0) of the thin-disk dynamo is of axisymmetric type [4] where the azimuthal magnetic field is concentrated in a torus. The polarization vectors should then be orientated in the radial direction. Figure 1 shows that the vectors indeed show such a configuration. A definite statement, however, requires correction for Faraday rotation.

The variation of Faraday rotation allows an independent test of the magnetic field structure. If the field is axisymmetric, the rotation measure RM varies sinusoidally with azimuthal angle in the plane of the galaxy [FUJIMOTO, this volume]. The dynamo mode m=1 produces a double-periodic variation of RM.

The rotation measures RM between λ11.1 cm and λ6.3 cm were determined in two radius intervals and averaged in azimuthal sectors of 20° width (Fig. 3). The azimuthal angle increases counterclockwise from the southern section of the major axis (position angle +39°). RM values vary between +35 and −70 rad m^{-2}, except for one sector in the south. The ambiguity of the vector orientation of $\pm n \cdot \pi$ corresponds to an ambiguity in RM of $\pm n \cdot 376$ rad m^{-2}. The small inclination of IC 342 allows only small magnetic field components parallel to the line of sight so that a rotation measure around zero is the most probable one.

The RM values in Fig. 3 were fitted with sine functions with single or double periodicity along the azimuthal angle in the plane of the galaxy. In the inner ring (5-9 kpc) both curves fit the data equally well, but in the outer ring (9-13 kpc) the curve with one period gives a significantly better fit. Zerolevels and phases of the two curves agree within the errors, whereas the amplitude increases from 12±6 rad m^{-2} in the inner to 26±6 rad m^{-2} in the outer ring. The maxima and minima of

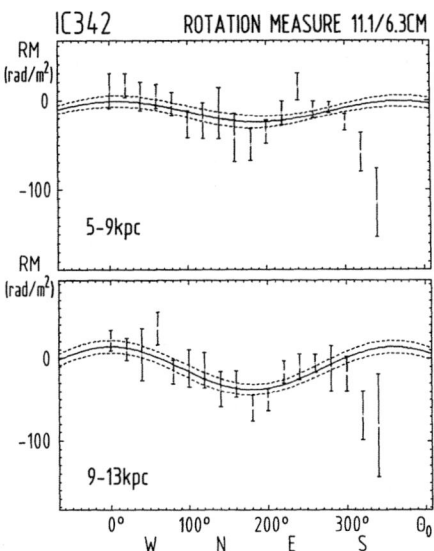

Figure 3: Average rotation measures between 11.1 cm and 6.3 cm wavelength in sectors of azimuthal angle θ_0 of 20° width in the plane of the galaxy in two radial intervals. The best-fit curves are shown together with their range of uncertainty

RM in both rings occur almost on the major axis of IC 342 (azimuthal angles 0° and 180°) which gives further support for the m=0 dynamo mode.

The polarization vectors were corrected for Faraday rotation (as shown in Fig. 3) and averaged in the same sectors. Their pitch angles (Fig. 4) are not zero everywhere, as expected in a solely toroidal magnetic field, but they reveal an average offset of +21°±2°. As the pitch angle of the spiral arms is of the same order [5] the intrinsic magnetic field lines appear to follow the spiral arms.

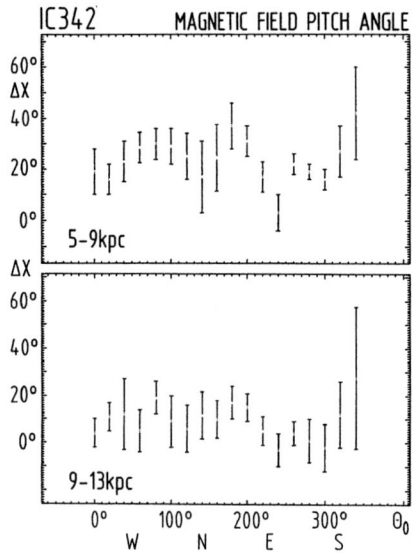

Figure 4: Average pitch angles of the intrinsic B vector in azimuthal sectors of 20° width in two radial intervals

IC 342 is the first clear case with an *axisymmetric spiral magnetic field* structure ("*ASS*"). M31 is most probably another galaxy with an ASS field [BERKHUIJSEN et al. and LOISEAU et al., this volume]. However, many other galaxies appear to show a bisymmetric ("BSS") field.

Observations with higher resolution and lower frequency are required to improve the accuracy of the rotation measure distribution [KRAUSE et al., this volume]. The region of peculiar rotation measure around 340° azimuthal angle (south) should be studied in more detail. The total intensity radio maps give indication neither for an exceptionally high thermal electron density nor for a high magnetic field strength. The field lines in this region could bend strongly out of the plane of the galaxy.

4. Summary

From observations of the linearly polarized emission of IC 342 at wavelengths of 11.1 cm and 6.3 cm the distribution of rotation measures and the orientations of the magnetic field components in the sky plane have been determined. Both distributions give clear indications that the field structure in IC 342 follows the spiral arms, but is axisymmetric, i.e. the magnetic field direction points outwards in all regions of the galaxy. The ground mode of an $\alpha\omega$-dynamo is capable of generating such a field structure.

References

1. K. Newton: Monthly Notices Roy. Astron. Soc. 191, 169 (1980)
2. Y. Sofue, U. Klein, R. Beck, R. Wielebinski: Astron. Astrophys. 144, 257 (1985)
3. A.A. Ruzmaikin, D.D. Sokoloff, A.M. Shukurov: Astron. Astrophys. 148, 335 (1985)
4. E.N. Parker: Astrophys. J. 163, 255 (1971)
5. R.C. Kennicutt: Astron. J. 86, 1847 (1981)

High Resolution Observations
of Linearly Polarized Emissions of IC 342

M. Krause, E. Hummel, and R. Beck

Max-Planck-Institut für Radioastronomie, Auf dem Hügel 69,
D-5300 Bonn 1, Fed. Rep. of Germany

IC 342 recently has been observed in total and polarized intensity at λ20 cm with the VLA in its D-configuration. The angular resolution is 45" HPBW corresponding to a resolution in the plane of the galaxy of 750 × 680 pc (RA × DEC), assuming a distance of 3.1 Mpc [1]. The polarized intensity is shown in Fig. 1. The emission is

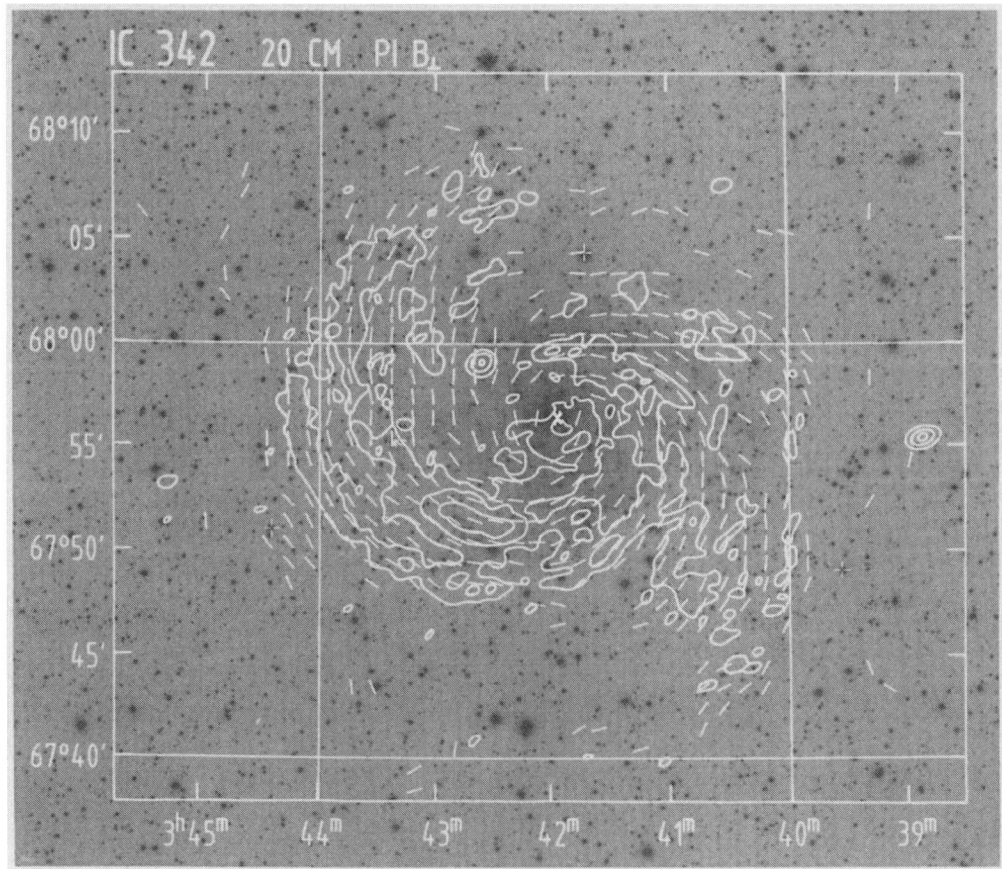

Figure 1: *Contour map of the linearly polarized intensity at 1.4 GHz, superimposed onto the red POSS plate. The half-power beamwidth of this map is 45". The contour levels are: 0.1, 0.3, 0.6, 0.9, 1.2 mJy/b.a. and the r.m.s.-noise is 0.03 mJy/b.a. The position angles of the vectors give the orientation of the magnetic field component perpendicular to the line of sight with an angular resolution of 2.'45 HPBW*

asymmetric with respect to the major axis: In the western part the structure is very fragmentary, but in the eastern part two long ridges of polarized intensity are prominent. The outer one can be traced along about 20 kpc. Both ridges coincide with two optical spiral arms and two HI spiral arms [2]. The region of maximal polarized intensity is near the bifurcation of the optical and HI spiral arms.

The distribution of the intrinsic rotation measure is determined with aid of the $\lambda 20$ cm measurements presented here and $\lambda 6.3$ cm measurements made by Gräve and Beck with the Effelsberg 100-m telescope. It supports the results of GRÄVE and BECK (this volume) that IC 342 has an axisymmetric (dynamo mode m=0) magnetic field structure.

In the region of 5 kpc $<$ R $<$ 9 kpc and $110° <$ θ $< 170°$ (θ = azimuthal angle starting at the southern major axis) in the plane of IC 342 our RM-distribution shows a deviation from the single periodic variation, namely very large $|RM|$. The pitch angle of the magnetic field component perpendicular to the line of sight, $B_\perp$, (Fig. 1) remarkably deviates from the other regions in IC 342, but the field seems to be aligned with two HI arms [2]. Both, the high values of the RM and the deviation in the orientation of $B_\perp$ and HI can be explained by a tilt of the north-western spiral arms of IC 342 of about 30°. NEWTON [3] already suggested warping of the galactic plane in the north-western part to explain the velocity perturbations observed in HI in the outer regions. The warp could originate in tidal interaction with the possible companion galaxy UGCG 2826 about 22' north-west (in the plane of the sky) of the nucleus of IC 342.

References
1. G. de Vaucouleurs: Astrophys. J. 227, 729 (1979)
2. K. Newton: Monthly Notices Roy. Astron. Soc. 191, 615 (1980)
3. K. Newton: Monthly Notices Roy. Astron. Soc. 191, 169 (1980)

Magnetic Fields in M33 and NGC 6946

U.R. Buczilowski

Max-Planck-Institut für Radioastronomie, Auf dem Hügel 69,
D-5300 Bonn 1, Fed. Rep. of Germany

The magnetic fields in two galaxies of type Scd with widely open spiral arms (pitch angle $\approx$ 30°) observed with the Effelsberg 100-m radio telescope are discussed.

Data for NGC 6946 (distance 7 Mpc) at λ2.8 cm have been already published [1,2]. With a resolution of $\approx$ 3 kpc the large-scale structure of the magnetic field can be investigated. It is open within the measurable disk of radius r $\approx$ 10 kpc. Azimuthally in the plane of the galaxy the orientations of the polarization vectors show a doubly periodic deviation of 15° amplitude from the orientation of the radius vector. Possible interpretations are:

(a) the action of density waves where axisymmetric field lines follow the stream lines of the gas,

(b) internal Faraday rotation in a bisymmetric spiral field with both internal rotation measures and total magnetic field strength much higher than in other galaxies [2,3].

To decide between (a) and (b) polarization vectors at another frequency are needed. The high foreground rotation of RM_{fg} = 400±80 rad/m² restricts appropriate observations to the cm wavelength range. New Effelsberg measurements at λ6.3 cm have been recently completed for which the expected foreground rotation of 90°±20° can be corrected unambiguously. Using the polarization angles at the two wavelengths we will obtain the rotation measure distribution and look for azimuthal variations. The determination of the internal Faraday rotations will show which interpretation should be supported.

M33, a nearby spiral at a distance of 720 kpc, is at present the best studied galaxy in polarization. The foreground rotation measure is accurately known [4], allowing the magnetic field to be studied with the three best-resolved Effelsberg measurements [5] at $\lambda\lambda$17.4 cm, 11.1 cm and 6.3 cm.

Using the minimum energy requirement [6] the total non-thermal luminosity in the radio range of $L_{nth}(10^7-10^{11}$ Hz) = 2.2±0.3×10³⁰ W gives a total magnetic field strength of B_t = 4±1 μG. This represents one of the weakest fields in the sample of 9 spirals studied so far, which have values in the range 7±4 μG [4].

The degree of uniformity of the magnetic field can be obtained from the degree of linear polarization of the synchrotron emission [7,8]. P_{nth} = 11±5% at λ6.3 cm corresponds to a ratio between the strengths of the uniform and random field of

$B_u/B_r = 0.4\pm0.1$ on the scale of the resolution element of 500x900 pc. M33 has the lowest field uniformity of 8 galaxies for which comparable estimates are available [4].

At all three frequencies the linearly polarized emission is concentrated to the northern half of the galaxy. There is almost no polarization detectable at the positions of the 35 optically detected supernova remnant candidates [9,10]. They may have disturbed the interstellar magnetic field on scale sizes smaller than our beam, reducing the measured polarized flux. To summarize, rotation measure studies for M33 are hindered by the weak field of low uniformity and the lack of polarization data in the southern half of the galaxy.

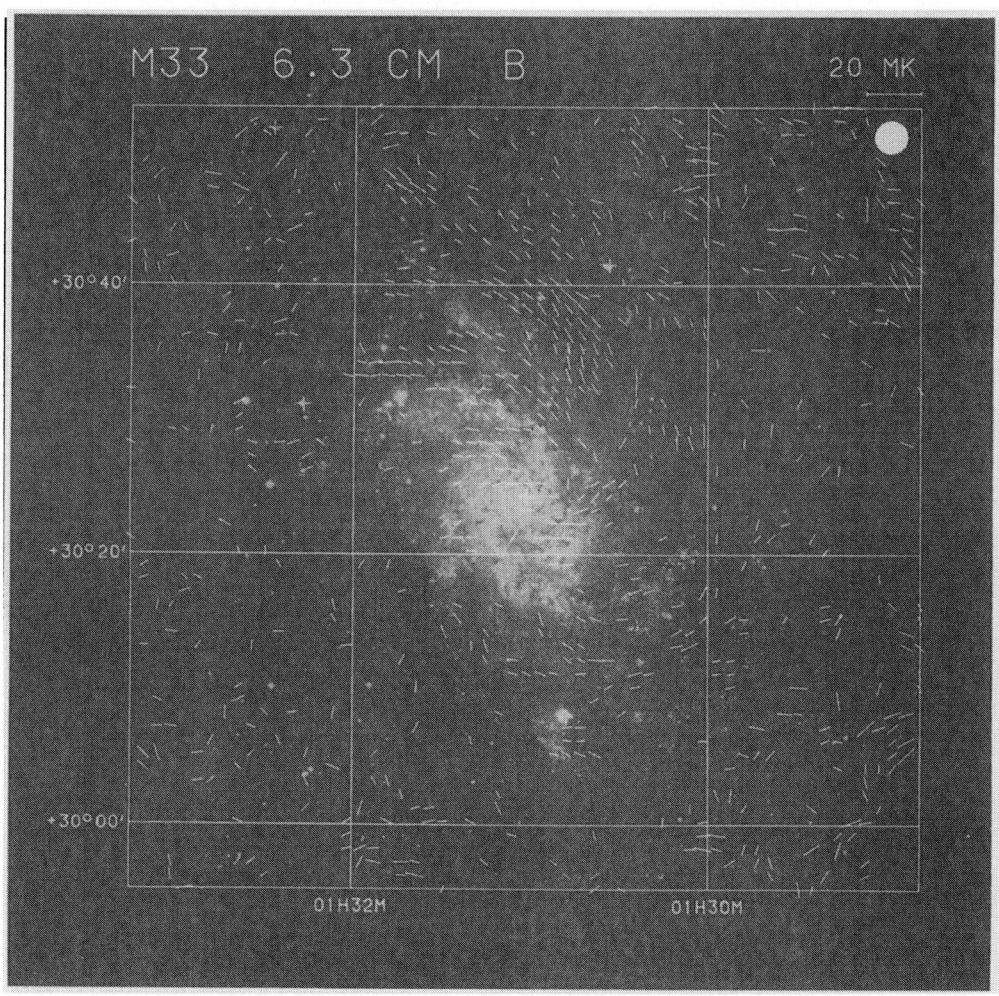

Figure 1: Orientation of the magnetic field in M33 overlayed onto an optical photograph (Lick Observatories). The length of the vectors is proportional to the linearly polarized intensity

However, an unambiguous determination of the rotation measure can be obtained from the 3-frequency data with 1.6 × 2.8 kpc resolution. Although the azimuthal variation of the rotation measures RM(θ) has a low amplitude and a high scatter, it is doubly periodic in all probability and thus compatible with a bisymmetric spiral field. Maxima and minima occur at the same azimuth angles θ when averaged over radial distances in the two ranges 1-4 kpc and 4-7 kpc. There is thus no radial indication of a field reversal on scales larger than 3 kpc. Because spiral arms are typically separated by 1.6 kpc, a change of the field direction from one spiral arm to the other cannot be excluded. The fitted amplitude to the RM(θ) variation corresponds to an internal rotation measure $RM_o = 13\pm3$ rad/m^2, less than half of the typical value for spirals [4].

The polarization angles at $\lambda6.3$ cm have been corrected for foreground rotation. The derived orientation of the magnetic field projected onto the plane of the sky is presented in Figure 1. The spatial resolution of 500×900 pc is comparable to the spiral arm dimensions. Field lines generally follow the spiral arms even in the outer regions of the galaxy. They are not closed within the measurable disk of $r \approx 6$ kpc, a trait already seen in NGC 6946. The magnetic field structure of M33 is consistent with a bisymmetric model first proposed by TOSA and FUJIMOTO [11]. In this model field lines are directed inwards in one half of the disk and outwards in the other half. The predicted reversal of the field direction may occur radially within 2 kpc of the centre.

References

1. U.R. Buczilowski: Diploma Thesis, University of Bonn (1981)
2. U. Klein, R. Beck, U.R. Buczilowski, R. Wielebinski: Astron. Astrophys. 108, 176 (1982)
3. R. Beck, U.R. Buczilowski, U. Klein, R. Wielebinski: Mitt. Astron. Ges. 55, 113 (1982)
4. U.R. Buczilowski: Ph.D. Thesis, University of Bonn (1985)
5. U.R. Buczilowski, R. Beck: Astron. Astrophys. Suppl. Series, in press (1986/87)
6. E. Hummel: Astron. Astrophys. 160, L4 (1986)
7. A. Segalovitz, W.W. Shane: Nature 264, 222 (1976)
8. R. Beck: Astron. Astrophys. 106, 121 (1982)
9. S. D'Odorico, M.A. Dopita, P. Benvenuti: Astron. Astrophys. Suppl. Series 40, 67 (1980)
10. S. D'Odorico, W.M. Goss, M.A. Dopita: Monthly Notices Roy. Astron. Soc. 198, 1059 (1982)
11. M. Tosa, M. Fujimoto: Publ. Astron. Soc. Japan 30, 315 (1978)

The Magnetic Field Structure in M81

M. Krause, R. Beck, and E. Hummel

Max-Planck-Institut für Radioastronomie, Auf dem Hügel 69,
D-5300 Bonn 1, Fed. Rep. of Germany

1. Introduction

M81 (NGC 3031) is a nearby galaxy (d = 3.25 Mpc [1]) of Hubble type Sab with a bright central region and tightly wound spiral arms of low luminosity. The HI spiral arms follow a logarithmic spiral pattern of pitch angle $-15°$ [2] close to the pitch angles of the optical spiral arms [3] and those of the blue light and Hα ($-12°\pm2°$, [4]). The theory of density waves has been applied successfully to explain the clear two-armed spiral structure of M81 [2,5-8].

The galaxy is favourably inclined to the line of sight to observe the perpendicular and parallel component of the magnetic field by means of the linearly polarized intensity and the intrinsic Faraday rotation, respectively. In the following we will assume an inclination of the plane of the disk of 59° (0° = face-on) and a position angle of the major axis of $-31°$.

2. Observations

M81 has been observed in total and linearly polarized intensity at $\lambda 6.3$ cm with the Effelsberg 100-m telescope [9] and at $\lambda 20$ cm with the VLA in its D-configuration. The angular resolution at $\lambda 6.3$ cm is 2.45 HPBW. At $\lambda 20$ cm the HPBW is 45" × 36" (RA × DEC) which corresponds to a linear resolution in the plane of the galaxy of 0.7 × 1.1 kpc. The polarized intensity at $\lambda 20$ cm is mainly confined to the northeast and the south-west near the minor axis of the projected plane of M81 (Fig. 1). The maximum polarized intensity of the disk is in the interarm region on both sides of the galaxy as already indicated by the $\lambda 6.3$ cm observations, whereas the total intensity of the disk is concentrated on the two main spiral arms. The degree of polarization in the western interarm region is about 50% at $\lambda 20$ cm and has even the same value at $\lambda 6.3$ cm. This is the highest degree of polarization detected in a nearby galaxy so far. The fraction of uniform to random fields in this region is unity or more on scales of several kpc.

3. Results

The rotation measure and hence the intrinsic angle of the electric vector of the linearly polarized emission have been determined using the two frequencies. Addition of 90° to these angles yields the orientation of the magnetic field component $B_\perp$ perpendicular to the line of sight. Figure 1 shows that this magnetic field component follows the spiral arms.

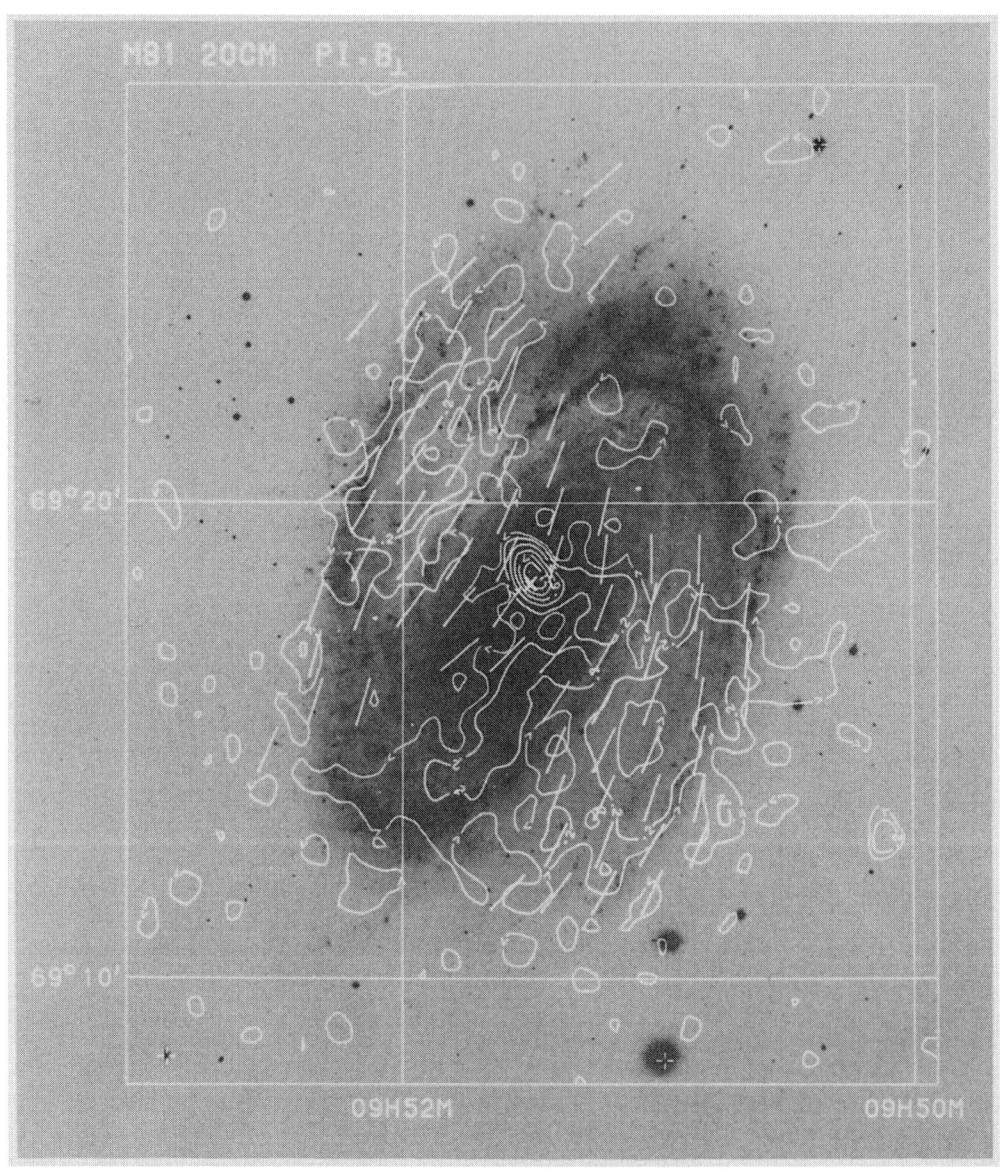

Figure 1: Contour map of the linearly polarized intensity at 1.4 GHz, superimposed onto an optical photograph made by A. Sandage with the Hale Observatories. The half-power beamwidth of this map is 36" × 45" (RA × DEC). The contour levels are: 0.1, 0.2, 0.4, 0.6, 0.8, 1.0 mJy/b.a. and the r.m.s.-noise is 0.05 mJy/b.a. The position angles of the vectors give the orientation of the magnetic field component perpendicular to the line of sight with an angular resolution of 2!45 HPBW

To determine the component of the magnetic field parallel to the line of sight, the plane of M81 has been divided into two rings of 3 kpc width starting at radius R = 6 kpc. Each ring has been split up into 30° sectors. Within each sector the mean rotation measure has been calculated. The distribution of these values with azimuthal angle shows a clear double-periodic variation in both rings around a zero-level of −18±3 rad/m² and with an amplitude of 20±4 rad/m². The observed background sources within 18° around M81 [10] and the low depolarization observed in the western interarm region lead to a Faraday rotation of the foreground of M81 of −9±5 rad/m² [11]. The zerolevel of the distribution of the rotation measure has been shifted by this value. Hence the <u>intrinsic</u> rotation measure varies double periodically around a zerolevel of −9±5 rad/m² reaching positive and negative values. This is strong evidence that in the observed region the magnetic field structure is bisymmetric.

On the assumption that the magnetic field structure in M81 is a bisymmetric <u>logarithmic spiral</u>, we tried to determine its pitch angle ψ and its position angle $\mu(r)$ from the rotation measure distributions. The geometrical projection of a bisymmetric logarithmic spiral field configuration leads to a variation of the observed RM [12]:

$$RM = A \cos(2\theta - \psi - \mu) + A \cos(\mu - \psi) + RM_{fg}.$$

A is the amplitude, θ the azimuthal angle in the plane of the galaxy and RM_{fg} the foreground Faraday rotation. The fit on the observed RM-distribution yields the amplitude A, the zerolevel (including the foreground Faraday rotation) and the phase ϕ of the variation. From these values we tried to determine the free parameters A, ψ, μ, RM_{fg} of the above equation. Due to its nonlinearity only solutions for special values of RM_{fg} and ψ exist for a given ϕ. We considered only those with $\psi < 0$ according to the optical and HI spiral arms.

1. Solutions exist with negative values of the pitch angle in each ring, if RM_{fg} $\gtrsim$ −5 rad/m².
2. From the observed RM-distribution and independently from the distribution of the perpendicular magnetic field component $B_\perp$ the pitch angles of the magnetic field are determined to be −22°±5° in the outer ring and −13°±7° in the inner ring. These independent results are in agreement with each other and with the values of the pitch angles of the optical and HI spiral arms.
3. We also find a good agreement in <u>position</u> of the magnetic field spiral in both rings with its neutral lines being located approximately in the interarm region.

4. Conclusions

We have found strong evidence that the magnetic field structure of M81 is open within the observed region. The observations are compatible with a bisymmetric structure (dynamo mode m=1) of the magnetic field with the neutral lines located approximately in the interarm regions. The fraction of the uniform magnetic field component is higher in the interarm regions than in the spiral arms. These results will be discussed in more detail in a forthcoming paper.

References

1. G.A. Tammann, A. Sandage: Astrophys. J. 151, 825 (1968)
2. A.H. Rots: Astron. Astrophys. 45, 43 (1975)
3. J.H. Oort: In The Formation and Dynamics of Galaxies, IAU Symp. No. 58, ed. by J.R. Shakeshaft (Reidel Publ. Co., Dordrecht 1974) p. 123
4. R.C. Kennicutt: Astron. J. 86, 1847 (1981)
5. H.C.D. Visser: Astron. Astrophys. 88, 149 (1980)
6. H.C.D. Visser: Astron. Astrophys. 88, 159 (1980)
7. F.N. Bash, H.C.D. Visser: Astrophys. J. 247, 488 (1981)
8. M. Kaufman, F.N. Bash: In Star Formation in Galaxies Conference, in press (1986)
9. R. Beck, U. Klein, M. Krause: Astron. Astrophys. 152, 237 (1985)
10. M. Simard-Normandin, P.P. Kronberg, S. Button: Astrophys. J. Suppl. 45, 97 (1981)
11. M. Krause: Ph.D. Thesis, University of Bonn (1987)
12. Y. Sofue, T. Takano, M. Fujimoto: Astron. Astrophys. 91, 335 (1980)

The Magnetic Field in the Anomalous Arms in NGC 4258

E. Hummel, M. Krause, and R. Beck

Max-Planck-Institut für Radioastronomie, Auf dem Hügel 69,
D-5300 Bonn 1, Fed. Rep. of Germany

Introduction

The two anomalous arms in NGC 4258 were first discovered by COURTES and CRUVEL-LIER [1] at H_α. At radio wavelengths they were rediscovered by VAN DER KRUIT et al. [2, KOM] and the real dimensions of these arms became clear. They can be traced out to a radius of ~ 20 kpc. To date more than 100 spiral galaxies have been mapped at radio wavelengths in sufficient detail and it can be stated that galaxy wide structures as seen in NGC 4258 are very rare and the occurrence rate is < 0.01.

Several hypotheses have been put forward to explain the anomalous arms: the original gas expulsion model [2] which has been refined by VAN ALBADA [3] and ICKE [4], jet models [5,6] and the MHD bubble model by SOFUE [7]. Common characteristics of these models are that they seek the ultimate cause in the nucleus, that they more or less describe the radio morphology and that none of them has been worked out in much detail. The latter concerns in particular the more basic properties of the radio continuum emission like magnetic field configuration and particle acceleration and propagation. An important distinction is that the gas expulsion model and the jet model by KUNDT [5] assume that the anomalous arms are in the plane of NGC 4258, while the jet model by SANDERS [6] and the MHD bubble model require that they are not in the plane. The latter two models place the western anomalous arm behind the plane and the eastern in front of the plane. The existing observations favor the "in the disk" models [3] and so do the radio polarization data presented here.

The linearly polarized emission and total emission of NGC 4258 at 4.9 and 1.5 GHz were observed with the Very Large Array in its D and C arrays respectively, resulting in a highest resolution of ~ 14" at both frequencies. To increase the signal to noise in the regions away from the ridges in the anomalous arms we make use of maps with a slighly degraded resolution of 28" x 21" (RA x DEC), similar to the polarization maps presented by VAN ALBADA and VAN DER HULST [8]. Additional polarization measurements at 0.61 GHz with the Westerbork array and at 4.75 and 10.7 GHz with the Effelsberg 100 m telescope are presented by DE BRUYN [9] and KRAUSE et al. [10] respectively.

Results

The new high resolution maps at 4.9 GHz (see Figure) and 1.5 GHz of the total intensity show the same morphology and are also very similar to the maps published by VAN ALBADA and VAN DER HULST [8]. Noticeable is that the eastern and western anomalous arms are not symmetric with respect to the nucleus. The western arm bifurcates about 0.7 north of the nucleus and there are two separate ridges out to the tip of the western arm. Neither of these ridges is symmetric with the eastern arm. The "mirrored" eastern arm would be in between the two western ridges.

Assuming minimum energy condition we find a magnetic field strength of $B_t \simeq 25$ μG at the ridges of the arms and of $B_t \simeq 10$ μG at the "plateaux" (the regions north and south of the eastern and western arm respectively). In the present case these magnetic field strengths are probably lower limits. With the same assumptions the

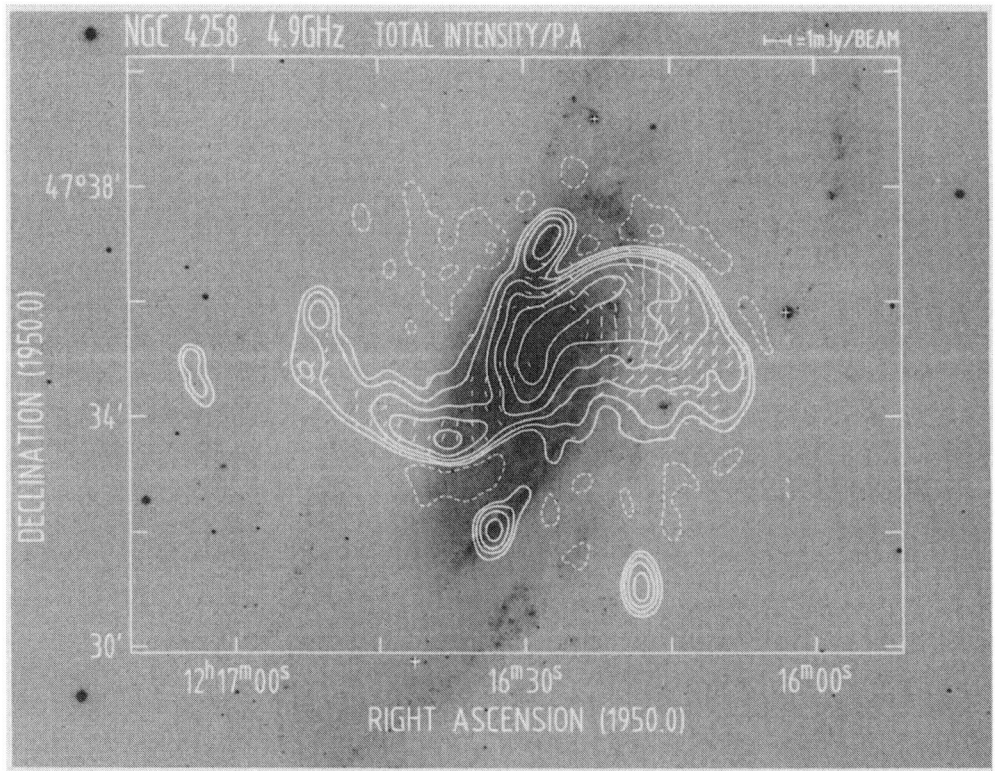

A contour map of the total intensity at 4.9 GHz of the radio continuum emission from NGC 4258 (M106) superimposed on an optical photograph. The half power beam width for this map is 28" x 21" (RA x DEC) and the contours are: −0.4, −0.2, 0.2, 0.4, 0.8, 1.6, 3.2 and 6.4 mJy/beam. The negative (dashed) contours around the emission are due to the missing short spacings for which no correction has been made. The map has not been corrected for the primary beam attenuation. Also superimposed are the <u>observed</u> E−vectors of the linearly polarized radio emission. The length of these vectors is proportional to the polarized intensity. Since the rotation measure is small these observed E−vectors are close to the intrinsic ones.

average magnetic field strength for Sbc galaxies is about $B_t \approx 8 \ \mu G$ [11]. The increase of the magnetic field strength and of the relativistic electron density in the ridges is then a factor 3 and 10 respectively when compared to "normal" disk values. Where the polarization percentages can be estimated accurately, i.e. on the ridges of the anomalous arms, they range from 10 to 30%. This implies that there the ratio of ordered to total magnetic field strength is $B_u/B_t \sim 0.4$ or higher. The change of polarization percentage with distance from the center suggests that the depolarization is symmetric with respect to the center. The same is true for the change of rotation measure (RM) with distance from the center. These two findings are strong additional arguments that the anomalous arms are in the plane of NGC 4258.

Combining the 4.9 and 1.5 GHz maps we obtained a map of the RM. At the tips of the anomalous arms we can solve for the $\pm n\pi$ ambiguity in position angle with aid of published polarization measurements (see [10]). It turns out that the RM is positive everywhere in the anomalous arms where it can be determined. As already found by KRAUSE et al. [10] the foreground RM is very small ($\sim$ 2 rad/m^2), whence the RM internal to NGC 4258 is positive. The measured RMs are small and the corrections at 4.9 GHz to obtain the intrinsic position angles of the E-vectors are at most 13°. The E-vectors displayed in the Figure are the observed ones but they are very close to the intrinsic angle of the E-vector. Since the west side of NGC 4258 is the nearer one we can conclude that in the eastern arm the magnetic field lines are directed toward the center and that in the western arm they are directed away from the center. On both sides the magnetic field lines follow the radio ridges.

Conclusions

The present results strongly suggest that the anomalous arms are in the plane of NGC 4258, hence excluding models that require them to be out of the plane. The observed magnetic field structure is in essence bisymmetric (dynamo mode m=1) and the magnetic field strength is highest in the ridges of the anomalous arms. Usually the bisymmetric field structures found more or less coincide with the optically seen spiral arms (e.g. M33, M81; this volume). This suggests that the present structure is probably the result of compression of a preexisting (not necessarily primordial) bisymmetric field. The KOM expulsion model could give an explanation for this compression.

References

1. G. Courtès, P. Cruvellier: Compt. Rend. Acad. Sci. Paris 253, 218 (1961)
2. P.C. van der Kruit, J.H. Oort, D.S. Mathewson: Astron. Astrophys. 21, 169 (1972)
3. G.D. van Albada: Thesis, University of Leiden (1978)
4. V. Icke: Astron. Astrophys. 74, 42 (1979)
5. W. Kundt: In Extragalactic Radio Sources, IAU Symp. 97, ed. by D.S. Heeschen and C.M. Wade (Reidel Publ. Co., Dordrecht 1982), p. 265

6. R.H. Sanders: In Extragalactic Radio Sources, IAU Symp. 97, ed. by D.S. Heeschen and C.M. Wade (Reidel Publ. Co., Dordrecht 1982), p. 145

7. Y. Sofue: Publ. Astron. Soc. Japan 32, 79 (1980)

8. G.D. van Albada, J.M. van der Hulst: Astron. Astrophys. 115, 263 (1982)

9. A.G. de Bruyn: Astron. Astrophys. 58, 221 (1977)

10. M. Krause, R. Beck, U. Klein: Astron. Astrophys. 138, 385 (1984)

11. E. Hummel: Astron. Astrophys. 160, L4 (1986)

Spiral Arms as Ejection Phenomena

H. Arp

Max-Planck-Institut für Astrophysik, Karl-Schwarzschild-Str. 1,
D-8046 Garching, Fed. Rep. of Germany

Magnetic fields are difficult to measure in galaxies. If we can identify accompanying phenomena perhaps we can trace the fields in a larger number of galaxies and under different conditions, thereby attempting to understand the nature and origin of magnetic fields.

The two general circumstances in which magnetic fields have so far been measured are: a) Radio emission in active nuclei and jets emerging from them. b) In spiral arms and in the clouds and interstellar medium associated with spiral galaxies. Both of these sites represent hospitable environments for magnetic fields because of the presence of an ionized medium. The key point seems to be that we know many galaxies eject ionized material (as shown by radio jets) and that these ejections contain elongated magnetic fields (as measured by polarization). We also know that spiral arms contain magnetic fields running along their length. <u>If the spiral arms are caused by ejection from the interior, this furnishes a natural explanation for the origin of magnetic fields in spiral arms.</u> Only one source of magnetic field in extragalactic astronomy is required, namely in the active, ionized interior of galaxies.

Ejection origins for spiral arms were suggested as early as 1938. In 1956 T. Schmidt–Kaler discussed gaseous outflow from the nucleus as a method for supplying material to spiral arms. Ambartsumian in 1958 proposed that luminous, ejected filaments observed in galaxies were proto spiral arms. Since then a number of astronomers have contributed supporting observations and qualitative models (see review in reference [5]). As for the involvement of magnetic fields, Paris Pismis in 1961 proposed a rotating, dipole magnetic field in the galaxy interior, with its axis in the plane and material in the arms emerging along the axis of the field [1]. In 1963 H. Arp [2] proposed either a dipole field in the nucleus of the galaxy aligned along the axis of rotation or a tangled field [3]. In both latter cases ejections from the interior would stretch out magnetic field lines along the line of ejection. Subsequent differential rotation would shear the magnetic fields into a circle. New ejections could puncture inner rings establishing new magnetic fields running along the arms at a pitch angle or renewing the old field. Therefore both the ring form and the bisymmetric form of the magnetic field that are observed [4] could be explained with a single mechanism.

If these suggestions are correct then filamentary, gaseous filaments (usually star forming) can be generally used to trace the magnetic fields. But, the whole concept must depend on whether, in fact, spiral arms are formed by ejection from interior regions. The prevalent view at present is that spiral arms are formed from material already in the outer regions of galaxy disks by rotating waves of density enhancement. Since the observational difficulties of this density wave theory have been discussed elsewhere [5] I will limit myself to here showing some observational evidence that would support, instead, the identification of spiral arms with ejection phenomena.

1. THE WIDTH OF JETS EJECTED FROM NUCLEI

As the listings below demonstrate, the widths of jets in a sample of well known objects ranges from about 50 to 200 parsecs.

Objects	Width of Jet	Kind of Jet
M87	60 pc	continuum with compact knots
	120	emission, counterjet
NGC 1097	200	narrowest continuum jet
NGC 1068	200	radio jets emerging from Seyfert nucleus
NGC 1808	150	straight line of HII regions
NGC 5128	90	inner X-ray and radio jet
	40-400	narrowest outer filaments (emission) and bundles
M51	200	representative width of spiral arm

At the end of this list we note that a representative width of a spiral arm in the prototype spiral galaxy M51 is just about 200 parsecs. (All these calculations are for H_o=100 km/sec/Mpc.) Therefore the widths of features we know to have been ejected from galactic nuclei are just about equal to, or slightly less than the widths we observe in spiral arms. Therefore jets, if curved by differential rotation in the outer galaxy disk, would be about the right physical size and shape to be identified with spiral arms. The gas present in the ejected filaments would presumably condense to form the bright young stars which mark the spiral arm. The time of duration of these arms, either from winding up by differential rotation or spreading off the arm by internal peculiar velocities is, of course, a problem for both density wave and ejected spiral arm theories but apparently about an order of magnitude more tractable for the ejection theory [5].

2. CONTAINMENT AND EFFECT ON THE INTERSTELLAR MEDIUM

Fig. 1 shows the radio observations in the interior of the Seyfert spiral NGC 1068 (reference [13]). It is clear that the material ejected from the nucleus has a characteristic cross-sectional dimension of about 200 parsecs. But this material does not appear to flattening at its tip or show any signs of impacting on the interstellar medium in the inner regions of NGC 1068. Therefore it is presumably still in the process of flowing out along this jet direction.

What happens if the material cannot get out? Fig. 2 shows a case where two radio lobes have been ejected out on either side of the nucleus. But the explosion has been so violent that it has disintegrated the previously existing spiral arms at their roots. It is presumably also responsible for the supplanting of the normal hydrogen emission in the disk with only shock-excited [NII] (see references [6] and [7]). This example of radio lobes being allowed to escape from a

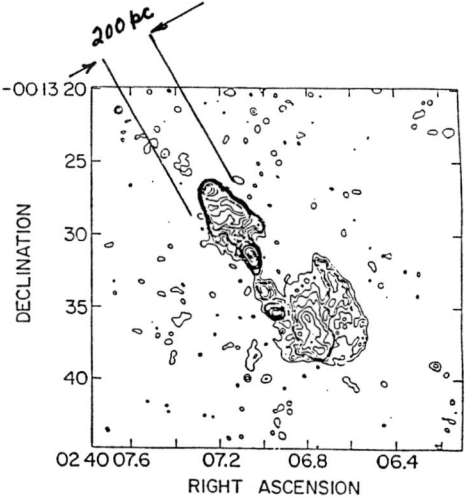

Fig. 1: 4.9 GHz map of NGC 1068 with 0″.4 resolution. See Wilson and Ulvestad (1983) for further details.

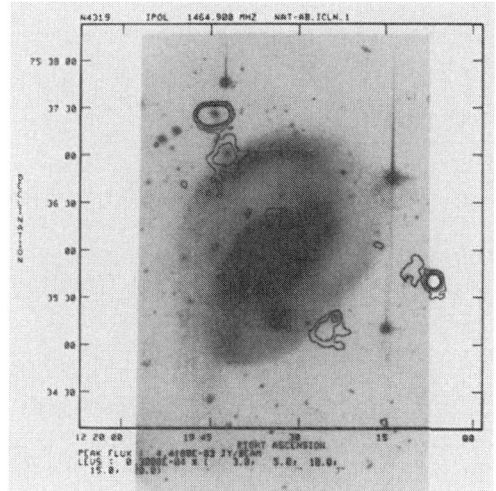

Fig. 2: NGC 4319 with ejected radio lobes.

spiral galaxy is unique, however, and we must conclude that in the usual situation it does not happen. Yet, as in NGC 1068, NGC 4736, NGC 4258, NGC 4151 and a number of other spiral galxies, the radio material does seem to be flowing out of the inner region. I would suggest that it can only do so by flowing along the bisymmetric magnetic field. The outflowing material would then either establish or renew the magnetic tubes leading outward. Of course, as ejection is observed to take place principally in two opposite directions, this naturally accounts for the generally two-armed, grand design spirals which we so commonly see.

3. EXAMPLES OF NARROW, LUMINOUS EJECTIONS FROM GALAXIES

Figure 3 shows number 188 from the Atlas of Peculiar Galaxies [8]. This object is remarkable because the ejected filament is so long, narrow and has all the visual appearance that spiral arms usually have at this distance. The inner spiral from which this filament originates shows "recoil" effects implying the ejection event was fairly powerful. No detailed, large telescope studies have ever been made of this system but just as an empirical demonstration of what kinds of ejection can occur in galaxies it is a very valuable example. Other objects in the Atlas of Peculiar Galaxies furnish further examples of narrow ejection phenomenon in Galaxies and their possible relation to spiral arms.

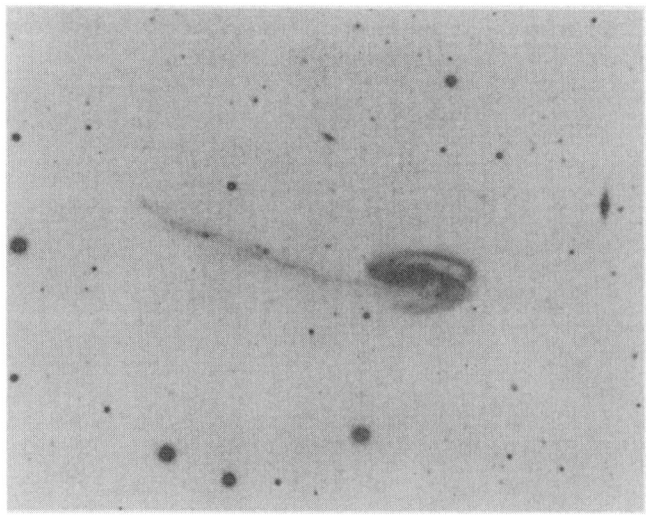

Fig. 3: No. 188 from Atlas of Peculiar Galaxies.

One case of particular interest is NGC 1808, a peculiar spiral in the active class of "hot spot" nuclei galaxies. Studies by G. Schnur [9] show a line of HII regions extending out about 2 kpc from the center and having a width of about 150 parsecs. This feature is straight and shows a linear relation between redshift and distance from the center of NGC 1808. It is most significant that this jet-like feature involves HII regions, that is gaseous regions with young, bright, recently formed stars that are characteristic of spiral arms.

Another example of more or less pure ejection is the radio galaxy 3C277.3 [11]. Along the line of radio ejection are seen HII emission knots. Again these HII emission knots are a tracer of young, spiral arm like stars. Also in [10] are discussed "jets propagating through a rotating gaseous disk" in the radio galaxy 3C293.

4. STAR FORMATION IN AN EJECTED FILAMENT IN NGC 5128

Perhaps the best example of star formation in an ejected filament is shown in Fig. 4. The insert shows the X-ray and radio jet located in the inner regions of NGC 5128. Further out along this line are very narrow filaments which emit H alpha, optical radiation. The spectra imply shock heating of these filaments. Photometric investigations show that young, hot O and B stars have recently formed in the close vicinity of these filaments [12].

Since the inner X-ray and radio jet points out along the line of the Hα filaments and young stars, and since all of this continues on into the extended radio lobes far out on either side of NGC 5128, there can be hardly any doubt that they all are the result of an ejection from the nucleus of NGC 5128. (These features perform a small, continuous change in position angle as if there were some rotation accompanying the ejection. but since the outer radio lobes must be of appreciable age, the ejection track must have retained its integrity for an appreciable time.)

The overwhelming importance of this one case is that it demonstrates unequivocally the possibility of formation of stars associated with a narrow ejection from the nucleus of an active galaxy. If we have this empirical evidence for formation of stars, the presence of hydrogen and gaseous emission in a long linear feature about the width of a spiral arm; then we have empirical evidence that spiral arms can be formed as a result of ejection processes.

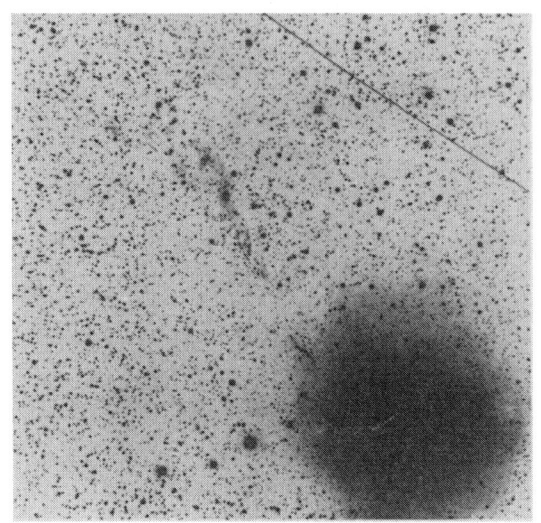

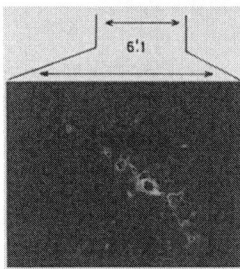

Fig. 4: NGC 5128, inner jet and outer ejected, star forming filament.

REFERENCES

[1] P. Pismis: Boletin de los Observatorios Tonanzintla y Tacubaya No. 23, p. 127 (Jan. 1963)

[2] H. Arp: In "The Evolution of Galaxies", Scientific American, vol. 208, p. 71 (1963)

[3] H. Arp: In "The Galaxy and Magellanic Clouds", IAU Symposium 20, p. 219 (1963)

[4] Y. Sofue, M. Fujimoto and R. Wielebinski: Ann. Rev. Astron. Astrophys. 24, p. 459 (1986)

[5] H. Arp: "The Persistent Problem of Spiral Galaxies". IEE Transactions on Plasma Science, Special Issue on Space and Cosmic Plasma, December 1986; also Max-Planck-Institut für Astrophysik preprint No. 243 (1986)

[6] J.W. Sulentic: "Observational Evidence of Activity of Galaxies", IAU Symposium 121, Yerevan Armenia (1986)

[7] J.W. Sulentic and H. Arp: "The Galaxy-Quasar Connection NGC 4319/ Mark 205", Astrophys. J. submitted (1986)

[8] H. Arp: Atlas of Peculiar Galaxies. Calif. Inst. of Tech. (1966) and Astrophys. J. Supp. 123, p. 1 (1966)

[9] G. Schnur: "Octopus Spectroscopy of Galaxies", to be published in ESO/HP Workshop on CCD Techniques, July (1986)

[10] W. van Breugel, T. Heckman, H. Butcher and G. Miley: Astrophys. J. 277, p. 82 (1984)

[11] W. van Breugel, G. Miley, T. Heckman, H. Butcher and A. Bridle: Astrophys. J. 290, p. 496 (1985)

[12] J.A. Graham and R.M. Price: Astrophys. J. 247, p. 813 (1981)

[13] A.S. Wilson and J.S. Ulvestad, Ap. J. 275, 8 (1983).

Optical Polarisation Studies of Galaxies

S.M. Scarrott, D.W. Ward-Thompson, and R.F. Warren-Smith
Physics Department, The University, South Road, Durham DH1 3LE, UK

Linear optical polarisation maps are presented for the galaxies NGC5194/5 (M51), NGC4594 (M104), NGC1068 (M77) and NGC4565. In each case there is evidence for polarisation induced by magnetically aligned dust grains and the data are used to propose configurations of galactic scale magnetic fields.

1. Introduction

In this paper we attempt to determine the configuration of the magnetic field in a small sample of galaxies by mapping the linear optical polarisation in the light from these objects.

The data presented here are the result of observations with the Durham Imaging CCD Polarimeter [1] on the 1m telescope of the Wise Observatory, Israel during the period 1985 May to 1986 July.

The results will be discussed on an individual basis before more general conclusions are made.

2. NGC5194/5 (M51)

This well known Sbc spiral with nearby interacting companion galaxy has been extensively investigated at most wavelengths and polarisation studies have been made at radio wavelengths [2].

Initially we compared our optical polarisation results with those in the radio by using similar sized integration bins centred on the position of the radio measurements. The optical polarizations are small (typically <1 percent, compared to radio polarizations of ≈20 percent), and in general our vectors are perpendicular to the radio measurements. The perpendicularity between the optical and radio polarization therefore confirms that the optical and radio observations are providing a probe of the magnetic field structure within M51.

In fig 1 we show an optical polarisation map for M51 where we have centred our integration bins along the prominent bright arms and dark lanes respectively. Within the central 4-5kpc the polarisation orientation within the arms and interarm regions follows an open spiral pattern. This pattern persists into the nuclear region to at least 200pc of the centre.

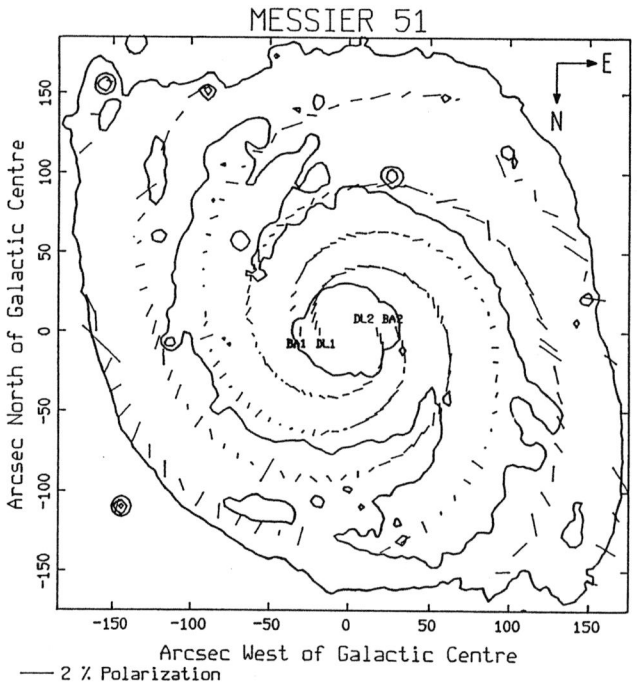

Fig 1. An optical linear polarisation map of NGC 5194/5 (M51). The integration bins are centred on the bright arms (BA1 & 2) and dark lanes (Dl1 & 2) respectively. Measurements were made in the waveband 450-1000nm.

3. NGC1068 (M77)

The large scale optical appearance of NGC1068 (M77) shows a spiral galaxy with well defined arms, copious dust throughout and a bright nucleus. Morphologically it is classified as Sb and it is the archetypal Seyfert galaxy.

In fig 2 we again present a polarisation map of the bright arms and dark lanes and, excluding the very central region for which we have inadequate data at present, we find that throughout the galaxy, in both the arms and interarm regions, that the polarisation orientations follow an open spiral pattern.

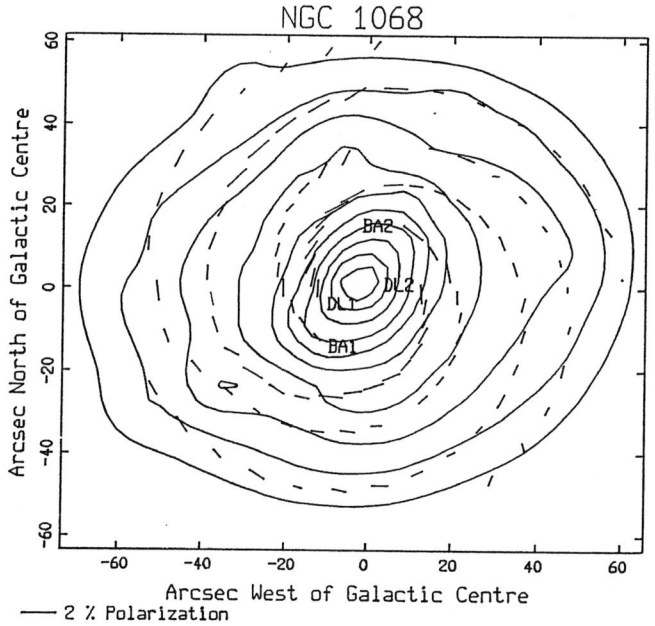

Fig 2. An optical linear polarisation map of NGC 1068 (M77). Otherwise as fig 1.

4. NGC4594 (M104)

This galaxy is generally accepted to be a Sa spiral galaxy seen almost edge-on. The distinctive central dust lane seems to bisect the extensive bulge/halo leading to the characteristic optical appearance which has resulted in the well known name 'The Sombrero Galaxy'.

In fig 3 we present an optical polarisation map of the dust lane regions of NGC4594. These data have been corrected for interstellar polarisation induced by aligned grains in our Galaxy which originally produced a small but significant polarisation (0.4%) over the whole area. The correction reduced the polarisation in the halo/bulge and nuclear regions to an insignificant level while having a minor effect on the dust lane areas due to the higher initial polarisations measured there.

The main features of the corrected map are the very uniform polarisation orientations parallel to the dust lane throughout the central areas of the galaxy. On the periphery of the galaxy where the dust lane appears to turn, to pass behind the galaxy, the orientations become perpendicular to the dust lane.

The origin of this latter feature we attribute to scattering of the light emanating from the inner parts of the galaxy by the dust

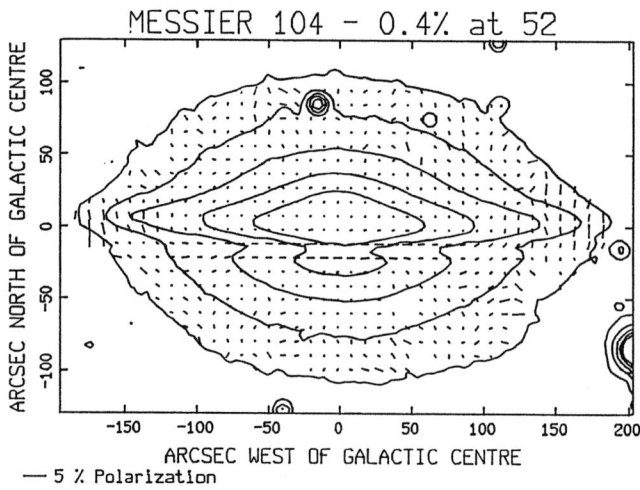

Fig 3. An optical linear polarisation map of NGC 4594 (M104). Observations were made in the waveband 450-1000nm.

concentrated in the galactic plane. In the regions of the dust lane seen approximately parallel to our line of sight we suggest that the polarisation arises from extinction by aligned grains indicating that the magnetic field is uniform on a galactic scale and narrowly confined to the plane of the galaxy. The gradual decrease of level of polarisation with wavelength in this region confirms this interpretation.

Similar features are found for the edge-on Sb galaxy NGC4565 suggesting that the field in this galaxy is of the same genre to that in NGC4594.

5. Conclusion

The polarisation data presented for the four galaxies investigated so far show significant levels of linear polarisations with orientations that form open spiral patterns throughout the plane of the galaxies. The polarisation is also narrowly confined to the galactic plane in edge-on systems.

If one accepts our interpretation that the polarisation in the majority of circumstances arises via extinction by magnetically aligned grains then the inferences are that the fields in galaxies studied so far have open spiral magnetic field configurations.

6. References

1. Scarrott. S.M., Warren-Smith, R.F., Pallister, W.S., Axon, D.J. & Bingham, R.G., 1983. Mon.Not.R.astr.Soc., **204,** 1163.

2. Segalowitz, A., 1976. Ph.D. thesis, University of Leiden.

Nonthermal Radio Emission from Dwarf Galaxies: Do They Host Large-Scale Magnetic Fields?

U. Klein[1] *and R. Gräve*[2]

[1]Radioastronomisches Institut der Universität Bonn, Auf dem Hügel 71,
D-5300 Bonn 1, Fed. Rep. of Germany
[2]Max-Planck-Institut für Radioastronomie, Auf dem Hügel 69,
D-5300 Bonn 1, Fed. Rep. of Germany

Large spiral galaxies are usually characterized by bright nonthermal radio disks, with mostly thermal, but also nonthermal sources related to the most prominent star–forming regions, superimposed. They are massive and exhibit differential rotation, both being prerequisits for spiral structure to persist (either via density waves or via stochastic self–propagating star formation). They also contain a more or less uniform mass fraction of gas, typically 10% or less. Moreover, they were subject to constant birth rates of stars during at least the past 10^9 yr. The situation is fundamentally different in case of dwarf irregular galaxies. They are generally characterized by an irregular morphology, probably as a result of the lack of large–scale processes like density waves which would otherwise govern their overall appearance. They lack nonthermal radio disks, their gas–to–total mass ratios vary over a wide range, from a few percent up to almost unity. And, most characteristically, their star formation rates cannot have been constant over the past ~ 10^9 yr: many of them must have been subject to bursts of star formation.

Nevertheless the bulk of them shows ordered rotation down to rather low masses as revealed by HI line observations. Rigid rotation out to the optical boundaries of dwarf galaxies is rather frequent (see e.g. [1,2,3]). For the lowest–mass galaxies there is a higher probability of finding pure chaotic motions, not ruling out, however, moderate rotation in some cases ([1]). This is clearly the trend continuing the TULLY–FISHER [4] relation down to the low–luminosity end of galaxies.

The lower threshold to rotational velocities is in general around ~ 10 km s^{-1}, this being the typical sound speed in HII regions which are responsible for stirring up the neutral gas. So, at a sufficiently low luminosity level, one could imagine that magnetic field generation and maintenance might assume a character which is rather different from that anticipated for massive spirals. It is not clear, for example, whether large–scale magnetic fields can persist at all in low–luminosity (dwarf) galaxies exhibiting predominantly rigid rotation at low velocities which sometimes exceed the turbulent motions by just a margin. Such motions would still be able to drive an α^2–type dynamo, with a dipole field resulting similar to that of the Earth [5]. (The angular velocity Ω which is an essential ingredient to the dynamo equations is of the same order for dwarf galaxies as for large spirals. Less, however, is known about 'disk' scale heights and length scales of turbulances which also enter the helicity function and hence the additional term in the dynamo equation; see e.g. [6]). Small-scale dynamos would also be able to maintain magnetic fields in such galaxies. Would one then rather expect only small-scale magnetic fields which are continuously regenerated by processes such as suggested by RUZMAIKIN et al. [7]?

What is the observational status? At present the situation is really dismal! As already mentioned, direct observations of magnetic fields via the linearly polarized component of the radio emission require an extraordinary sensitivity because of the weakness of the radio emission from dwarf irregular systems. To our knowledge, the only cases of such direct measurements are those of the Small Magellanic Cloud (SMC) [8] and NGC4449 (see below), where the first is sufficiently nearby and thus strong in terms of flux density, while the latter is among the class of intrinsically bright irregular galaxies of intermediate masses. For a few more dwarf irregular galaxies which are nearby we have instead just total intensity maps, for some of them at several frequencies. For some 40 dwarf galaxies we have integrated flux densities at two or more frequencies, for another 50 or so we have measurements at just one frequency or only upper limits. A review of the nonthermal properties has recently been given by KLEIN [9].

Initial radio continuum measurements of dwarf galaxies indicated that their radio spectra are significantly flatter than those obtained for normal spirals [10]. This trend has been confirmed by the increasing number of measurements. For instance, the histogram of spectral indices for a sample of blue compact galaxies (BCDG) (see [11]) shows a large scatter, probably caused by varying relative amounts of thermal emission. It is difficult to decide, at present, what the underlying nonthermal spectrum is, but there are indications that it is flatter than that established for normal spiral galaxies (e.g. [12,13,14]). If so, then the nonthermal radio emission of such galaxies would probably be governed by supernovae and their remnants, rather than by relativistic electrons which are stored in large-scale magnetic fields through which they are believed to migrate in normal spirals. If the observed nonthermal radio emission in dwarf galaxies is essentially the superposition of sources like SN(R), then one would expect the magnetic field to have a predominantly small-scale structure. Further evidence for this view is delivered by observations of nearby dwarf galaxies which are bright enough and of sufficient angular extent to map them in some detail (see e.g. [10]). Except for the two cases mentioned above the observed degrees of polarization are extremely low, generally of the order of the instrumental response which for the 100-m telescope is ~ 1.5%.

The first case of a dwarf irregular galaxy for which the clear detection of radio polarization has been reported so far is the SMC [8]. The observed distribution of electric field vectors has recently been extensively discussed by LOISEAU et al. [15]. NGC4449, about 4 Mpc distant, is the second case. A deep map with the 100-m telescope at 4.75 GHz was recently obtained in total power and polarization. The aim was exceptional sensitivity to large-scale polarized structure at the expense of resolution (the beam of 2.5 at this frequency corresponds to a 2.8 kpc projected scale). The rms noise in the final map of linearly polarized emission is ~ 0.4 mJy/b.a. after averaging 30 coverages. Previous quick looks at this frequency and at 10.7 GHz as well as a deep map at 24.5 GHz, all obtained with the 100−m telescope, facilitated to accurately establish the integrated radio spectrum of NGC4449, with $\langle\alpha\rangle = -0.33\pm0.03$ $(S_\nu \sim \nu^\alpha)$ [13]. The resulting map (Fig. 1) looks puzzling: the maximum of polarized intensity is − unlike the total power emission which is

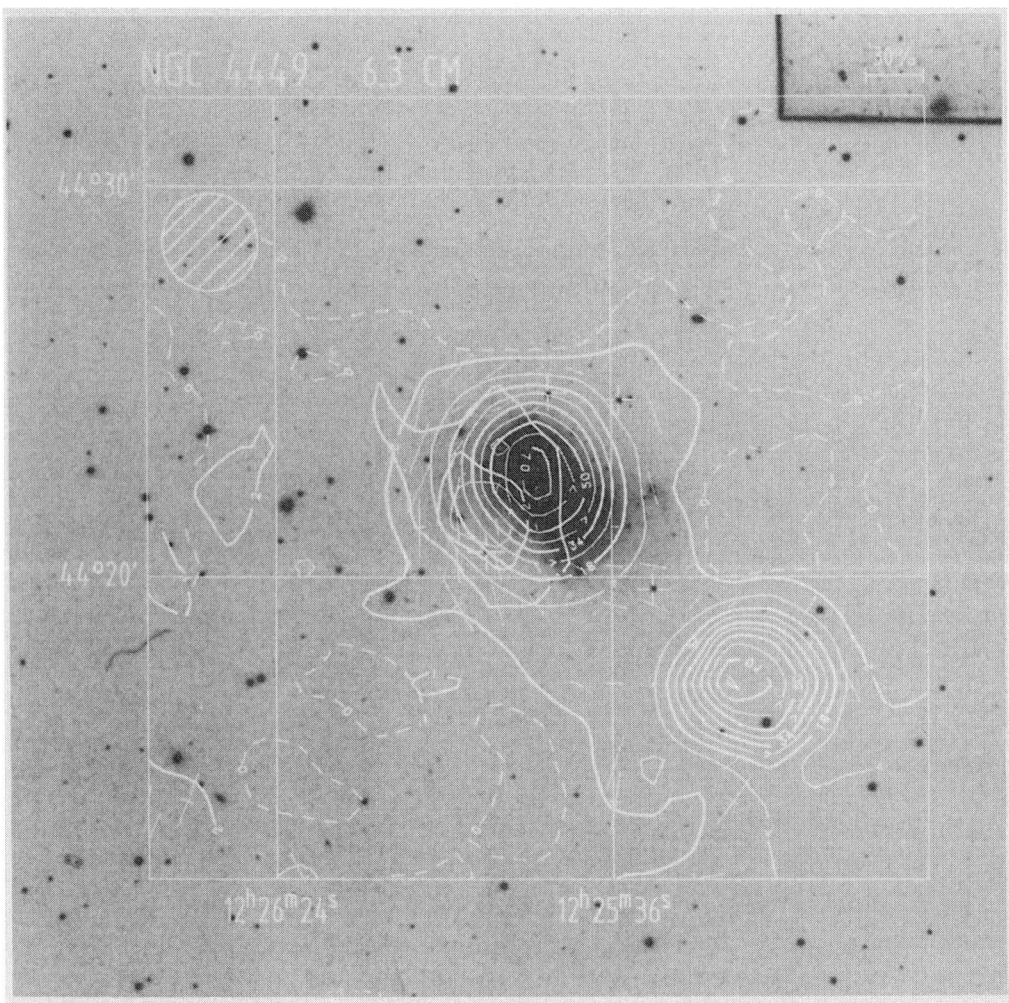

Figure 1: Map of NGC4449 at 4.75 GHz, superimposed onto a reproduction of the optical image from a POSS plate. Total power contours are in units of 2, 8, 16, ... 50, 60, 70 mJy/beam area, contours of the linear polarization in units of 1, 2, 3 mJy/beam area. The bars represent the direction of the magnetic field in case there is no Faraday rotation (i.e. the E−vectors have been rotated by 90°). The beam width at the 3 dB level is indicated in the upper right corner, the bar at the top right edge of the map represents 30% polarization.

78

precisely centered on the optical image of the galaxy – shifted by ~ 1' towards the south–east. The degree of polarization exceeds 30% in the south–eastern region, decreasing to ~ 2% in the central region of the galaxy, and increasing again to ~ 20% in the north–west, but with the bulk of the polarized radiation concentrated towards the south–east. The average degree of polarization is 3.7±0.9%, or ~ 6% if referred to the nonthermal fraction. Shown is the "magnetic field" direction, i.e. the former E–vectors have been rotated by 90°. The structure of this field is entirely unclear, especially in view of the uncertainty of the true spatial orientation of the galaxy. Its kinematic major axis as determined from Hα [16,17] and the HI line [18] is oriented along p.a. = 20°, with the inner (ionized) gas approaching in the north–east and receding in the south–west, and the outer (neutral) gas exhibiting an opposite velocity gradient (see e.g. [17]). The inclination is only poorly determined, i = 30° being a likely value (0° = face–on). The fact that we see the maximum polarization displaced from the total intensity peak and the optical centre of the galaxy must be due to strong depolarization within the galaxy's body which also strongly emits in Hα (see e.g. [16,17]). In the south–eastern region, however, there must be a relatively high degree of field alignment, being responsible for the (relatively) high degree of polarization. An unpublished 21–cm continuum map obtained with the WSRT (Fig. 2) shows a distinguished feature emerging from the north–eastern chain of HII regions which is one of the most conspicuous features of the galaxy, and appears to terminate precisely at the location of the polarization maximum. This extension is indicated even at 24.5 GHz (see the map published by KLEIN and GRÄVE [13]). Furthermore, the magnetic field direction also seems to follow this extension. What is this feature which commences as a string of star formation and ends up as a giant radio blob in a sort of a halo far away (~ 2 kpc) from the galaxy? Maybe one is witnessing here the birth of spiral structure via some process of sequential star formation which could be largely controlled by magnetic fields. This finding warrants higher resolution investigations of the polarized emission at different frequencies, incorporating the VLA and the 100–m.

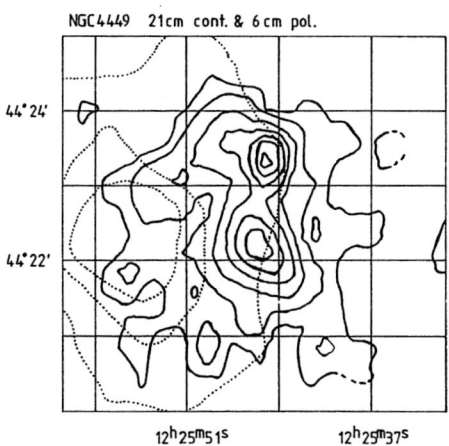

NGC4449 21cm cont. & 6cm pol.

44° 24'

44° 22'

12h25m51s   12h25m37s

Figure 2: Continuum map at 1.4 GHz (unpublished WSRT map), with contours in arbitrary units. Dotted contours mimic the distribution of linearly polarized emission at 4.75 GHz as shown in Fig. 1.

The few cases of dwarf irregular galaxies studied in polarization so far must be considered as a start of a sytematic investigation of such objects in the same way as is being carried out for massive spirals, involving the VLA at low and the 100-m telescope at high radio frequencies. This will possibly enable us to uncover the geometry, spectrum and strength of magnetic fields down to the low-luminosity end of galaxies which may be crucial to comprehend the mechanism of producing and maintaining magnetic fields in galaxies and how they influence the dynamics and evolution of galaxies. At present, no definite answer to the question raised in the title of this paper can be given because we lack statistics. There is some circumstantial evidence for small-scale magnetic fields to prevail in dwarf galaxies, but many more such observations are required to confirm this view.

Acknowledgements

U.K. is grateful to the Max-Planck-Gesellschaft for financial support and for the excellent hospitality at Schloß Ringberg. G. Hutschenreiter is acknowleged for providing tidy reproductions of the figures.

References

1. W.L.W. Sargent, K.-Y. Lo: In Star-forming Dwarf Galaxies and Related Objects, ed. by D. Kunth, T.X. Thuan and J. Tran Thanh Van (Edition Frontieres, Paris, 1986), p. 253

2. E. Skillman, R. Terlevich, H. van Woerden: In Star-forming Dwarf Galaxies and Related Objects, ed. by D. Kunth, T.X. Thuan and J. Tran Thanh Van (Edition Frontieres, Paris, 1986), p. 263

3. G. Comte, J. Lequeux, F. Viallefond: In Star-forming Dwarf Galaxies and Related Objects, ed. by D. Kunth, T.X. Thuan and J. Tran Thanh Van (Edition Frontieres, Paris, 1986), p. 273

4. R.B. Tully, J.R. Fisher: Astron. Astrophys. 51, 661 (1977)

5. Y. Sofue, M. Fujimoto, R. Wielebinski: Ann. Rev. Astron. Astrophys. 24, 259 (1986)

6. A.A. Ruzmaikin, A.M. Shukurov: Sov. Astron. 25, 553 (1981)

7. A.A. Ruzmaikin, D.D. Sokoloff, A.M. Shukurov: In Plasma Astrophysics, Proc. Varenna-Ambastumani Intern. Workshop, Sukhumi USSR, (ESA Scient. Public., 1986)

8. R.F. Haynes, U. Klein, R. Wielebinski, J.D. Murray: Astron. Astrophys. 159, 22 (1986)

9. U. Klein: In Star-forming Dwarf Galaxies and Related Objects, ed. by D. Kunth, T.X. Thuan and J. Tran Thanh Van (Edition Frontieres, Paris, 1986), p. 371

10. U. Klein, R. Gräve, R. Wielebinski: Astron. Astrophys. 117, 332 (1983)

11. U. Klein, E. Wunderlich: In Star Formation in Galaxies, ed. by N. Scoville and G. Neugebauer, Pasadena (1986)

12. U. Klein, R. Wielebinski, T.X. Thuan: Astron. Astrophys. 141, 241 (1984)

13. U. Klein, R. Gräve: Astron. Astrophys. <u>161</u>, 155 (1986)

14. E. Brinks, U. Klein: In <u>Star-forming Dwarf Galaxies and Related Objects</u>, ed. by D. Kunth, T.X. Thuan and J. Tran Thanh Van (Edition Frontieres, Paris, 1986), p. 281

15. N. Loiseau, U. Klein, A. Greybe, R. Wielebinski, R. Haynes: Astron. Astrophys., submitted (1986)

16. R. Crillon, G. Monnet: Astron. Astrophys. <u>119</u>, 301 (1969)

17. F. Sabbadin, S. Ortolani, A. Bianchini: Astron. Astrophys. <u>131</u>, 1 (1984)

18. H. van Woerden, A. Bosma, U. Mebold: In <u>La Dynamique des Galaxies Spirales</u>, ed. by L. Weliachew (Edition du CNRS, 1975), p. 483

Scale Properties of the Turbulent Magnetic Field in the Large Magellanic Cloud

J. Spicker and J.V. Feitzinger

Astronomisches Institut, Ruhr-Universität Bochum, Postfach 10 21 48,
D-4630 Bochum, Fed. Rep. of Germany

1. Introduction

We present some results of the first application of standard statistical methods to the Large Magellanic Cloud (LMC) in order to deduce the turbulent properties of the interstellar medium in that galaxy. Twodimensional autocorrelation (ACF) and structure (STR) functions have been used in the past to investigate turbulence in galactic HII regions, HI and molecular clouds. They have also been applied successfully to the radial velocity field of the LMC by the present authors /1/. This paper is concerned with the application of the statistical treatment to the radio continuum and polarization data available for the LMC and proves the existence of turbulence scales in the magnetic field of that galaxy.

2. Statistical Methods and Physical Assumptions

Throughout our analysis, we assume turbulence to be a stochastic process being describable through two-point correlation tensors /2/. These statistical functions are sensitive to fluctuations of an observable and will reveal to what extent points separated by a certain lag are correlated. The STR is particularly sensitive to small-scale fluctuations in suppressing large-scale trends eventually contained in the data, and is related to the Kolmogorov-Obukhov theory of isotropic subsonic turbulence in incompressible fluids. The ACF of density fluctuations is used to calculate basic correlation scales in terms of typical clump sizes and separations. The statistical vocabulary is summarized in /2/ and references therein.

The study of the turbulent properties of the magnetic field in the LMC is possible via investigating the fluctuations of polarizations of radio continuum emission, the fluctuations in the nonthermal radio emission and the fluctuations of the degree and intensity of optical polarization. Only the two latter data are available for the LMC.

The brightness temperature T_b of synchrotron emission integrated along the line of sight ds is proportional to the magnetic field strength B and the electron density N_e:

$$T_b \sim \int B^{(\gamma + 1)/2} N_e \, ds \, , \tag{1}$$

where $\gamma \simeq 2/5$ is the spectral index of the relativistic electrons. If the relative fluctuations of B and N_e are modest, the ACF and STR can be used to trace the turbulent properties of the magnetic field /3/. The STR should then scale as

$$B(\tau) = \tau^{2/3} , \tag{2}$$

where τ is the lag.

Optical polarization is produced by interstellar dust grains. The degree and orientation of the polarization vectors should roughly reflect the strength and direction of B, so that eventual fluctuations of B are traced by the ACF and STR of these quantities.

3. Data

Two new radio continuum surveys of the LMC became recently available. The 21 cm survey /4/ covers $-6\overset{o}{.}5$ ... $6\overset{o}{.}5$ with $0\overset{o}{.}25$ resolution; the 13 cm survey /5/ covers $-7\overset{o}{.}5$ .. .. $7\overset{o}{.}1$ with $0\overset{o}{.}33$ resolution. A proper analysis in terms of the magnetic field would require the complete separation of the thermal and nonthermal components of the emission, which in turn requires a well-known distribution of the spectral index α of the radio continuum emission over the LMC. A preliminary analysis of α was based on the two surveys and yielded large errors /6/, so that a proper separation is not possible at the moment. The spectral index distribution has, however, an overall mean of $\alpha = -0.6$ /6/, and most of the sources detected are nonthermal /4/, so that the application of the statistical treatment to the T_b maps seems appropriate.

The polarization data were taken from /7/, consisting of the intrinsic polarization of ~ 300 stars. The irregularly spaced data were interpolated to a rectangular grid for the expense of introducing large statistical errors. Almost the complete optical structure of the LMC is covered by the polarization data.

4. Results and Discussion

Figure 1 depicts the twodimensional unbiased ACF of the 21 cm T_b distribution. The map is given in the lag coordinate system $(\Delta y : \Delta y, \Delta x : \Delta x) = (-6\overset{o}{.}25 : 6\overset{o}{.}25, -6\overset{o}{.}25 : 6\overset{o}{.}25)$. Positive correlations are shown as full lines (0.8, 0.6, 0.4, 0.2, 0.1), the zero contour is enhanced, and negative correlations are depicted as dashed contours (-0.1, -0.2). Two striking features emerge:
1) There is a pattern of several recorrelations at different lags. They are identifiable with the most intense sources of continuum emission in the LMC as listed by /4/; most of them are nonthermal in origin and coincide with the positions of the HII supershells in the LMC /8/.
2) A central plateau of the ACF is clearly evident, having two main axes with different orientations. This pattern is obviously not determined by single sources and therefore reflects the clumpy distribution of the overall nonthermal radio emission.

Figure 2 shows radial cuts through the same ACF in the direction of the main pla-
teau (θ_a = 12°, θ_b = 84°) and the main recorrelation feature (θ_c = 35°; θ is counted
from the Δx axis). Direction a is characterized by a more or less exponential decay
of the ACF, the characteristic scale being λ_0 = 1/e ≈ 500 pc. Its form is typical
for emission coming from a wide range of region sizes. Directions b and c show a more
Gaussian distribution of the ACF. Fitting parabolas to the ACF at the origin /1/
allows for the determination of two distinct scale sizes of the emission regions:
λ_{b1} ≈ 450 pc, λ_{b2} ≈ 1200 pc, λ_{c1} ≈ 600 pc, λ_{c2} ≈ 1500 pc. The smaller scales cor-

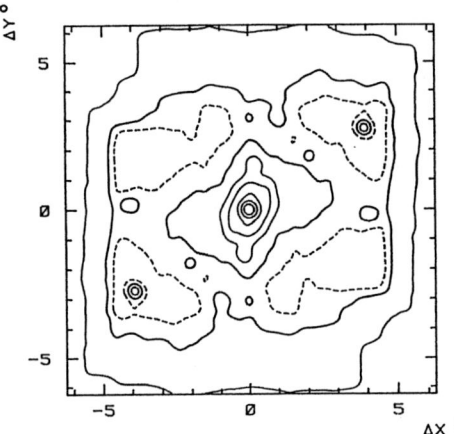

Fig.1

Twodimensional unbiased autocorrelation
function (ACF) of the 21cm radio continuum
brightness temperature distribution. See
text for details.

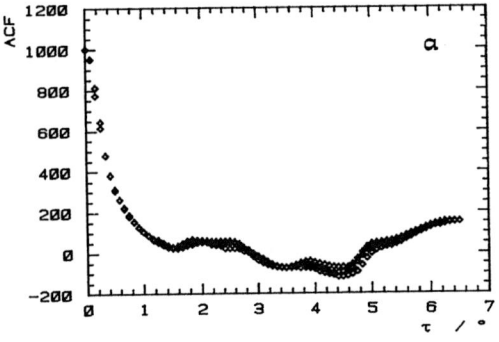

Fig.2

Radial cuts through the autocorrelation
function of Fig. 1 in three different
directions (see text for explanations).
Note the double feature of parabolic
shape near the origin in b and c, which
is not visible in a.

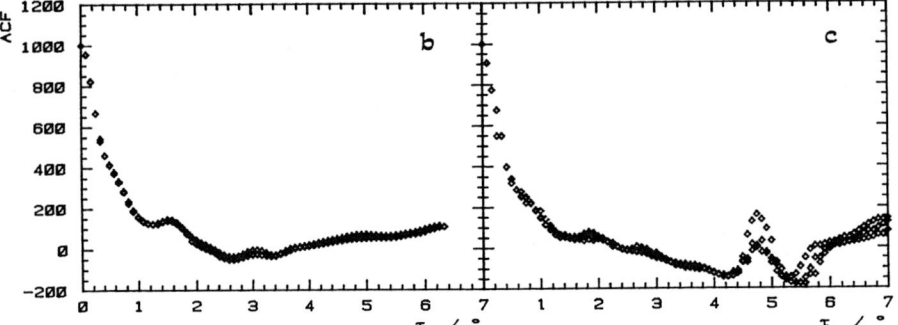

respond to typical sizes of the spheres of influence of star forming regions in the LMC, whereas the larger scales are more representative for the sizes of the supershells in the LMC. These were also found to dominate the structure of the HI radial velocity field of that galaxy /1/, and their influence is again visible as the characteristic scale of the magnetic field fluctuations. This can hardly be a coincidence.

It is also interesting to note that the theory for the fast turbulent dynamo in isotropic mirror symmetric flow predicts a negative tail of the ACF for the magnetic field /9/, which property is also observed for the ACF in the LMC.

The STR (not depicted here) shows a similar distribution of scales, with the characteristic exponent s = 0.6 ± 0.1 being close to the value expected from the idealized Kolmogorov theory (s = 2/3).

The analysis of the 13 cm data gave essentially the same results. The analysis of the degree of polarization was hampered by the large noise of the data; recorrelations indicate typical scales of λ = (1200 ± 400) pc for the strength of the magnetic field. The polarization angles show no correlations at all, so that a general field direction towards the SMC as proposed by /7/ is doubtful. There is, however, too much noise in the data to allow for a firm statement.

5. References

1. Spicker, J., Feitzinger, J. V., Astron. Astrophys. in press (1986 a)
2. Spicker, J., Feitzinger, J. V., Astron. Astrophys. in press (1986 b)
3. Kaplan, S. A., Pikelner, S. B., The Interstellar Medium, Harvard UP, ch. 23, (1970)
4. Haynes, R. F., Klein, U., Wielebinski, R., Murray, R. D., Astron. Astrophys. 159, 22 (1986)
5. Klein, U., Greybe, A., in preparation (1986)
6. Klein, U., in preparation (1986)
7. Schmidt, Th., Astron. Astrophys. Suppl. 24, 357 (1976)
8. Meaburn, J., Mon. Not. Roy. Astron. Soc. 192, 365 (1980)
9. Kleeorin, N., Ruzmaikin, A., in Workshop on Plasma Astrophysics, ESA SP in press (1986)

The support of this research by Deutsche Forschungsgemeinschaft (DFG) under grant Fe 196/3-1 is gratefully acknowledged.

Magnetic Fields and Faraday-Active Clouds out to the Distances of Quasars

P.P. Kronberg

Department of Astronomy, University of Toronto,
Toronto M5S 1A7, Canada

1. Introduction

A reliable determination of the Faraday rotation in our Galaxy at the higher galactic latitudes has enabled us, for the first time, to accurately subtract it from
the measured Faraday rotation measures (RM's) of distant extragalactic radio
sources. The difference, or "residual rotation measure" (RRM) has been used to
investigate two basic phenomena and their interpretation: These are (1) the relationship between the extragalactic Faraday rotation of quasars and the characteristics of their optical spectra, where such exist with sufficient quality, and
(2), the variation with redshift of the RRM of quasars. In this paper, I shall
briefly review the evidence supporting the first detection of large scale magnetic
fields in the universe out to redshifts of 2 and greater. I shall also show some
independent, new evidence which gives us some further, more direct clues concerning the nature of the extragalactic systems containing the magnetic fields which I
and my collaborators have detected.

2. First Attempts to Measure Magnetic Field Strengths in Distant Extragalactic Systems

For an initial sample of 37 quasars, optical spectra of sufficient resolution exist
to make it possible to determine equivalent widths of absorption lines. The
spectra were analysed to estimate the corresponding column density of ionized
hydrogen, N, between us and the QSO. Comparison of this column density with the
RRM of the associated radio emission suggested a correlation between the two in
the sense that strong optical absorption correlates with "excess" RRM among this
sample. This correlation can be illustrated if we simply plot histograms of the
z-corrected RRM's on the assumption that all of the Faraday rotation occurs at the
redshift of the quasar. If the presence of, and/or the redshift of the absorption
line systems have nothing to do with the RRM's then the distribution of RRM for
quasars with and without absorption lines should be similar for otherwise similarly chosen subsamples for each case. i.e. This is the null hypothesis. Fig. 1
(reproduced in modified form from Welter, Perry and Kronberg [1]) shows strikingly
that such is not the case; Quasars with weak or no absorption systems have many
fewer large RRM's. The central histogram in Fig. 1 also shows an intermediate
case in which the RRM's of the same strong absorption line sample are corrected by

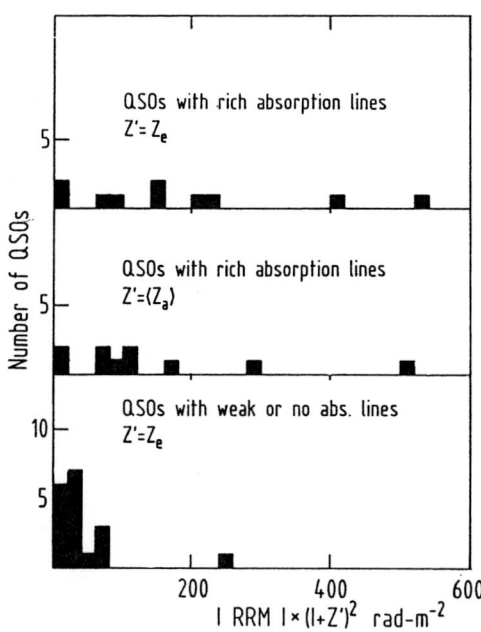

Figure 1. Histograms of the z-corrected RRM's of QSO's: (upper) The observed RRM's of quasars with strong absorption are corrected by $(1+z)^2$. (middle) The same sample are plotted, but this time corrected to the mean <u>absorption line</u> redshift. (lower) The RRM distribution for quasars having weak or no absorption systems, but as above corrected to the reference frame of the emission line redshift. (The data for these samples are published in Welter, Perry and Kronberg [1]).

the mean absorption line redshift instead of the emission line redshift.   Again the contrast with the lower histogram is evident.

On the assumption that the excess Faraday rotation and the absorption occur in the same physical systems, we can also make a first order estimate of the electron density-weighted magnetic field strength in the intervening absorption line clouds. Preliminary results give inferred magnetic field strengths of a few to a few tens of microgauss for some sources (cf. Kronberg and Perry [2]).   For some others we have only upper limits.   It is also important to mention that, among our preliminary magnetic field estimates, there are some very high implied magnetic field strengths, which approach the milligauss level.   Results at this surprisingly high level focus on the question as to whether the column density has not been underestimated.   Such will surely be the case if there is a very hot component ($\geq 10^6$ K) which does not produce transitions in the optical spectra.   Thus one explanation is that very high apparent field strength is due to a hot component of gas in the intervening system which has not yet been detected.   Table 1 shows examples of three categories of magnetic field strengths as deduced by Kronberg and Perry [2].

It may be significant that both sources in category (C) have $z_a \sim z_e$, and, by their proximity to the QSO could reasonably be expected to have a very hot environment which is difficult to detect in the optical spectra.   In each case above

N_x denotes $N \cdot 10^x$ cm^{-2} of free electrons. The value of the subscript is the closest power of 10 of the approximate column density derived from the optical and/or 21 cm absorption spectrum.

TABLE 1

SOURCE	MAGNETIC FIELD ESTIMATE FOR THE INTERVENING SYSTEM
(A) "BEST ESTIMATES"	
3C191	$30\ N_{20}^{-\frac{1}{2}}$ μG
1331+170	$13\ N_{20}^{-\frac{1}{2}}$ μG
(B) UPPER LIMITS AT THE MICROGAUSS LEVEL	
P0119-046	$<6N_{20}^{-\frac{1}{2}}$ μG
OQ172	$<65N_{18}^{-\frac{1}{2}}$ μG
4C 05.34	$<230N_{19}^{-\frac{1}{2}}$ μG
(C) LARGE IMPLIED FIELDS	
P0458-02	$1600N_{19}^{-\frac{1}{2}}$ μG
4C24.61	$400N_{20}^{-\frac{1}{2}}$ μG

In comparing RRM and absorption spectra the angular sizes are also relevant since, for radio-extended sources, the Faraday rotation line of sight does <u>not normally</u> coincide with that for the optical absorption. This means that, strictly speaking, we must allow for such differences in estimating the magnetic field strengths. What is particularly significant in the data is that both radio extended and compact sources are included in Fig. 1, and <u>the RRM-strong absorption correlation holds for radio extended as well as compact sources.</u> This is a most interesting result, in that it strongly suggests that the absorbing systems typically have projected dimensions which are at least as large as the extended radio emission scaled from z_e to z_a. We thus conclude that the intervening systems are typically 40 kpc or greater, and in section 5 I show some recent more direct evidence for the substantial size of the intervenor. Such sizes are consistent with their being halos of large galaxies, or the gaseous medium in distant galaxy groups or clusters.

3. <u>Modelling of Intervening Faraday-Active Systems with a Larger Radio-QSO Sample</u>

The overlapping samples of QSO's with both sensitive, high resolution optical spectra and well-determined, Galaxy-corrected RM's is unfortunately still rather small, so that the extension of this radio-optical study will happen only slowly. However, there is a somewhat larger sample of quasars having well-determined RRM's for which we know just the redshift. Using just such a sample of 116 QSO's, G.L. Welter, J.J. Perry and I [1] established the interesting result that more distant quasars have a statistically larger RRM than those of low z. We have generated models of intervenors in which the intrinsic rms RRM ($\sigma_c(z)$) is z-dependent according to the following relation:

$$\sigma_c(z) = \sigma_{co}(1+z)^{\beta_c - 2} \tag{1}$$

where σ_{co} is the zero-epoch value, and β_c is a parameter which defines the z-dependence of the intrinsic Faraday rotation strength of the intervenors. The probability of striking a cloud at a given redshift is (following the nomenclature in [1], in which the subscript "r" refers to radio, i.e. Faraday rotation, as opposed to "o" for the optical absorption);

$$P_r(z) = \frac{A_{o_r}(1+z)^{1+\eta_r}}{(1+2q_o z)^{\frac{1}{2}}} \tag{2}$$

where $A_{o_r} = c n_{o_r} \Sigma_{o_r}/H_o$ and η_r is the evolutionary parameter for the average cross-section for Faraday rotation of the intervening systems. Here c is the velocity of light, n_{o_r} the zero-epoch density of the intervenors, Σ_{o_r} the cross-section, and H_o the Hubble constant. The variance, $V(z_e)$, can be expressed as

$$V(z_e) = \int_0^{z_e} \sigma_{c_o}^2 A_{o_r}(1+z)^{k-1} dz \tag{3}$$

where $k = 2\beta_c + \eta_r - 2$. (cf [1]). Figure 2(a), reproduced from [1], shows the comparison of model curves of V(z) superimposed on the observed variances. Fig. 2(b) illustrates the "acceptable" portion of the $\sigma_{c_o}^2 A_{o_r}$ - k parameter space. We have constrained the parameters A and η_r from a modeling of the optical absorption spectra by Khare-Joshi and Perry [3], which leads to the following best range for σ_{c_o} and β_c:

$$55 > \sigma_{c_o} > 37 \ \text{rad/m}^2$$

$$0.3 > \beta_c > 1.5$$

The current small data set does not permit me to suggest this as more than a tentative, and possibly not even unique model for the origin of the Faraday rotation measure data from quasars. However, if we take the above model to be correct, then the observed growth of the RRM scatter with increasing z require that the "Faraday strength" of these systems evolves in the past as $(1+z)^{0.9\pm0.6}$.

The quasar RRM data are consistent with intervenor systems whose characteristics are $l \sim 50$ kpc, $B \sim 2$ μG, and $n \sim 0.001$ cm^{-3}. These numbers, and the space density of the systems are entirely consistent with their being halos of large galaxies and/or galaxy groups. It would appear that they are not normally due to

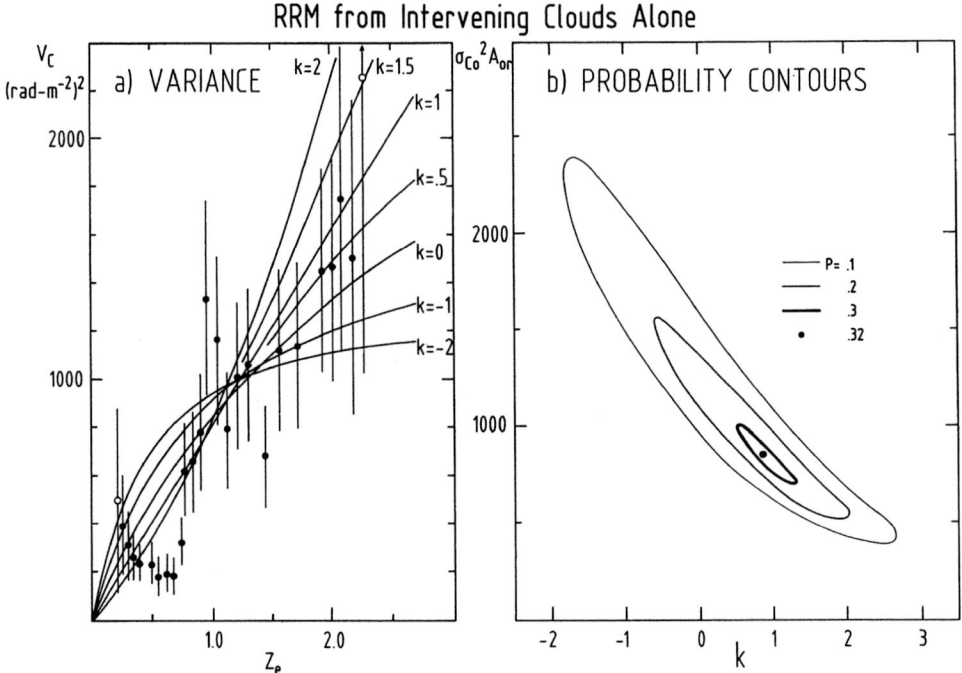

Figure 2. (a) Curves of the variance calculated from (3) compared with the observed RRM variances for different redshift bins for the 116 quasar-sample used [1]. (b) The probability contours for the goodness of fit in (a). Reproduced from Welter, Perry and Kronberg [1], Astrophysical Journal.

large clusters such as Coma, since their space density is too small for P(z) to be significant out to z = 2.5. In the following section we review recent evidence for existence of a magneto-ionic medium from some recent detailed observations.

4. What recent direct evidence is there for systems containing intergalactic gas clouds, and are there any which we can rule out?

I shall begin by discussing a very interesting, recently discovered type of intervening system, namely the "Lyman alpha forest" clouds seen blueward of the Lyman alpha emission profile in some QSO's. (cf. Sargent et al. [4]). At redshifts near 2.5, their space density is $n_c \sim 200$ Mpc^{-3}, and typical inferred cloud parameters are $N_e \sim 10^{19}$ cm^{-2}, $r_c \sim 17$ kpc, and $T \sim 3 \cdot 10^4$ K. [4]. For a quasar at redshift z = 3, $\sigma_c^2 < .05$ rad m^{-2} if we assume the magnetic field to be in equipartition with the thermal energy density of the associated clouds. This is far below the typical values shown in Fig. 1, and well below the current lower detection limit of ~ 20 rad^2/m^4. We can therefore be reasonably certain that the "excess" RM's in Fig. 1 are not associated with the "Lyman-alpha forest" clouds.

An obvious candidate is large galaxies, whose outer disks have a sufficient column density of gas to cause a detectable Faraday rotation. Unfortunately the intervenors are mostly at too large redshift to be seen visually. The first statistical results described above require a substantial evolution of the RRM with epoch, so that if the RRM intervenors are large galaxies we are "seeing" a class of pre-evolved precursors at a significantly earlier epoch!

Another possibility is that we are probing the magneto-ionic medium of clusters or groups of galaxies. Figure 3 shows the most detailed map yet obtained of the synchrotron emission in the gas halo of the Coma cluster. It is very close to the dimensions of the X-ray halo, whose Faraday rotation has been recently detected by K.T. Kim, Dewdney Landecker and myself [5]. Figure 4 shows a preliminary plot of the RM of weak background sources which are at varying angular distances from the centre of the Coma cluster. The halo of the Coma cluster is found to generate about 50 rad. m^{-2} of excess rotation measure. Are Coma-like clusters therefore good candidates for the absorption line/RRM intervenors in front of quasars? Although our new Coma results indicate the existence of an excess RRM, the space density of large clusters of galaxies, $\sim 10^{-9} Mpc^{-3} (10^{12} L_{\odot})^{-1}$, [6] is too low, hence P(z) for large clusters is too low by orders of magnitude at 2 < z < 3.

Coma Cluster of Galaxies at 1380 MHz
(DRAO + VLA Observation)

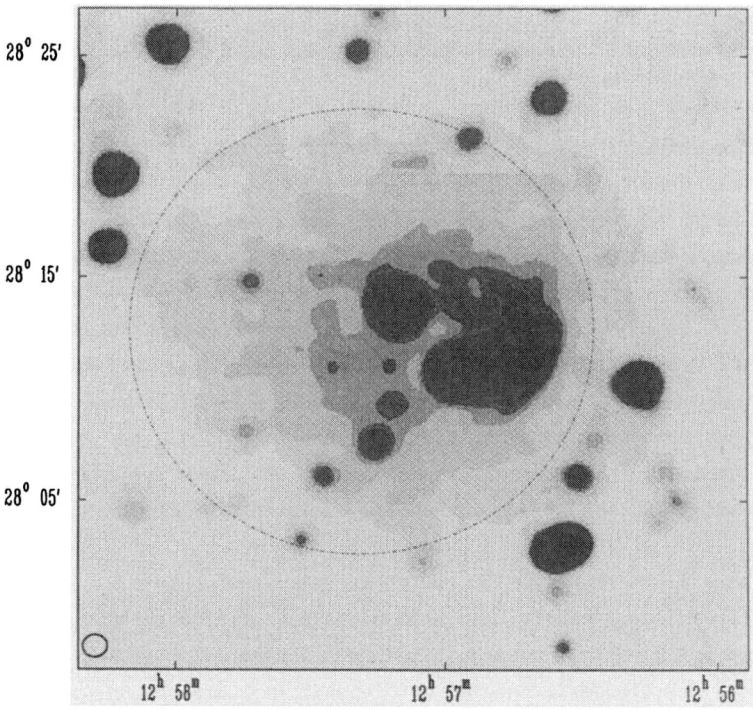

Gray Scale: (1.2,1.8,2.4,2.8,3.2,4.0) mJy/beam, Dot-Dash Circle = 10' Radius Circle

Figure 3. A new map which reveals the detail of the extended radio emission from intracluster gas in the Coma cluster of galaxies (Kim, Kronberg, Dewdney, and Landecker [5]). The resolution is 71" (E-W) x 60" (N-S).

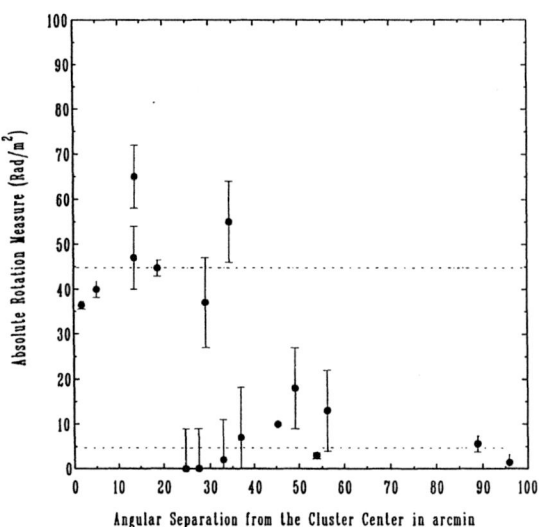

Figure 4. Newly determined rotation measures of faint radio sources at different projected distances from the centre of the Coma cluster, which provides direct evidence for the magneto-ionic medium associated with the visible cluster of galaxies and discrete radio sources. The enhanced RM's are caused by the gas associated with the X-ray source, whose boundary is shown by the dashed line. (Kim, Kronberg, Dewdney, and Landecker [5]).

Smaller groups of galaxies are, however, much more numerous, and some early X-ray maps have recently shown that, at least for some galaxy groups an intergalactic gas halo exists, which is of the approximate dimensions and strength which would explain the observed present-epoch properties of the intervenors. Figure 5 shows a EINSTEIN X-ray map of the group dominated by the SO galaxy NGC 3607 (Biermann, Kronberg and Madore [7]). Because the space density of galaxy groups in this luminosity class is considerably higher than that of large galaxy clusters, the pre-evolved precursors of such objects of galaxy groups would seem to be good candidates for the intervenors whose magnetic field we are detecting in the RRM's of distant quasars. If this is the case, our results imply that such systems must evolve to larger "Faraday strengths" at earlier epochs at which they are being detected as intervenors.

In summary, the Faraday active intervenors, which are also the cause of absorption lines at intervening redshifts, may well be halos of large galaxies, or the intergalactic gas in galaxy groups. Large clusters are also possible, although their lower space density suggests that they do not constitute the majority of intervening objects. The lower column depths of the "Lyman-alpha forest" clouds suggests that they can be ruled out as the cause of the excess RRM's.

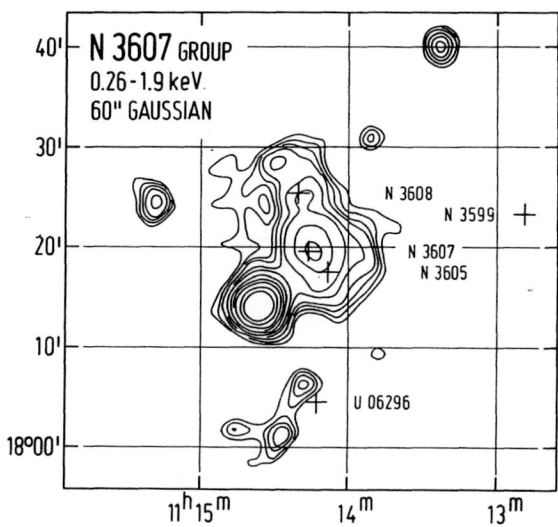

Figure 5. X-ray map of the NGC 3607 group of galaxies as mapped by the IPC of the EINSTEIN X-ray satellite by Biermann, Kronberg and Madore [7]. The data are smoothed with a 60" guassian beam. Contours are shown at 0.2, 0.4, 0.6, 0.8, 1.2, 1.6, 2.4, 3.2, 4.0, 5.6, and 7.2 counts per 32" pixel. (Reproduced from The Astrophysical Journal).

5. Some new preliminary direct evidence for the size of an intervening system

Recently E. Zukowski and I have used the VLA to generate rotation measure maps of three quasars which are known to have (i) strong optical absorption (ii) "excess" Faraday rotation, and (iii) are radio extended. Figure 6 shows a rotation measure map of 3C191 at the lowest common resolution (1.4") of VLA polarization maps at 4 different frequencies. It has a radio jet which is clearly distinguishable at higher resolution, but is only about 3 beamwidths long at the lowest common resolution shown. Nevertheless, the Faraday rotation changes by ~ 100 rad/m^2 (observer's reference frame). At the two closely spaced absorption line redshifts (1.949 and 1.945) [8] the intervenor appears to have dimensions of at least 25 kpc, and an inferred differential RM of ~ 900 rad/m^2 in the reference frame at which the excess RRM occurs. This is a most interesting result, in that it provides one of the first pieces confirming evidence for an extended intervening system. Furthermore, the apparent regularity of the RRM change suggests the presence of a large scale component of magnetic field. Of course we cannot rule out the possibility that the RM is intrinsic to 3C191, although the global correlation evident in Fig. 1 suggests that the strong RM is produced at the intervening redshifts, at which strong absorption occurs.

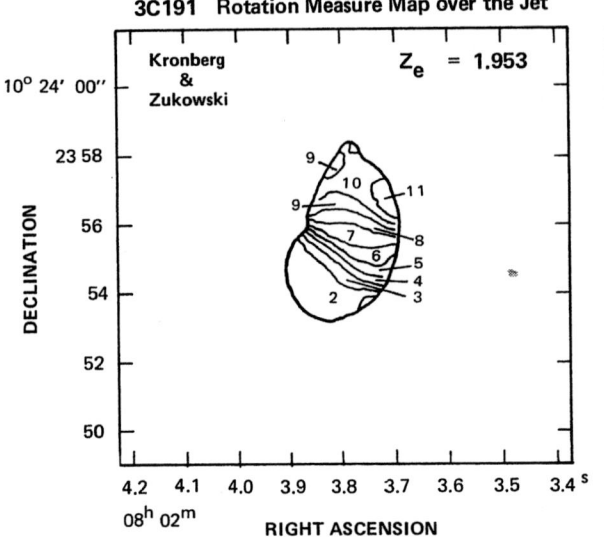

3C191   Rotation Measure Map over the Jet

Level	RM (rad/m^2)
1	0–10
2	10–20
3	20–30
4	30–40
5	40–50
6	50–60
7	60–70
8	70–80
9	80–90
10	90–100
11	100

Figure 6. A recently obtained rotation measure map over the jet of quasar 3C191 at a resolution of 1.4". (Kronberg and Zukowski, in prep. 1987). There is clear evidence for strong spatial variation in the rotation measure. 3C191 has a strong absorption line spectrum [2], which likely to be associated with the RM intervenor.

6. Summary

It is now clear that we can see intervening systems purely in the radio by their Faraday rotation. We have so far been able to make the first detections of magnetic fields in plasma clouds at early epochs in the Universe, up to z ∿2.5. These early results are sufficiently tantalizing to inspire greater effort to tie down the physical properties of pre-evolved galaxy systems out to the redshifts of the most distant quasars.

1. Welter, G.L., Perry, J.J. and Kronberg, P.P., Ap. J., 279, 19, 1984.

2. Kronberg, P.P. and Perry, J.J., Ap. J., 263, 518, 1982.

3. Khare-Joshi, P. and Perry, J.J., M.N.R.A.S., 199, 785, 1982.

4. Sargent, W.L.W., Young, P., Boksenberg, A., and Tytler, D., Ap. J., (Suppl.) 42, 41, 1980.

5. Kim, K.T., Kronberg, P.P., Dewdney, P.D. and Landecker, T. Proceedings of the NRAO Workshop on Galaxy Clusters, 1986.

6. Bahcall, N. Highlights Astr., 5, 699, 1980.

7. Biermann, P., Kronberg, P.P. and Madore, B.F. Ap. J., 256, L37, 1982.

8. Williams, R.E., Strittmatter, P.A., Carswell, R.F. and Craine, E.R., Ap. J., 202, 296, 1975.

Part II

Large-Scale Magnetic Fields in Our Galaxy

Magnetic Fields in the Galaxy

R. Wielebinski

Max-Planck-Institut für Radioastronomie, Auf dem Hügel 69,
D-5300 Bonn 1, Fed. Rep. of Germany

The observations of the optical polarization of starlight [1,2] showed large—scale alignment of the vectors which was interpreted to be due to galactic magnetic fields which align interstellar grains [3]. The data on star polarization accumulated over many years (e.g. [4,5,6]) giving in the end a large sample of objects at known distances. This data was used (e.g. [7]) to derive a model of the local magnetic field. From this observational data a field component along the plane of the Galaxy roughly in the direction of the local spiral arm was deduced with other pertubations that were attributed to the North Polar Spur [8]. Optical data give us information about the field direction and only indirectly about the field strength.

Radio observations give us a number of methods of studying galactic magnetic fields.

1. Measurement of radio polarization from synchrotron emission gives us $B_\perp$.
2. Measurement of rotation measure (RM) of extragalactic sources gives us $n_e \cdot B_\parallel \cdot dl$.
3. Measurement of RM and dispersion measure (DM) of pulsars gives us $n_e \cdot B_\parallel \cdot dl$ and $n_e \cdot dl$ respectively and hence $\overline{B_\parallel}$.
4. The studies of the distribution of synchrotron emission give us after some assumptions the magnetic field intensity.
5. The Zeeman effect can be studied in some HI or molecular clouds.

The first detection of polarized radio emission was by MEYER et al. [9], who observed the Crab Nebula at $\lambda 3.15$ cm wavelength. Earlier predictions of the expected polarization of synchrotron radiation (up to 70%, e.g. [10,11]) have led to many unsuccessful searches of linear polarization of galactic radio waves. Finally detections at 408 MHz were reported by WESTERHOUT et al. [12] and WIELEBINSKI et al. [13]. A northern survey of BERKHUIJSEN and BROUW [14] and WIELEBINSKI and SHAKESHAFT [15] and a southern one by MATHEWSON and MILNE [16] gave us a rather complete picture of the general distribution of radio polarization. The interpretation of these results (e.g. [17,18,8]) indicate a field along the local spiral arm (in good agreement with optical data) and a "Faraday anomaly" towards $\ell = 140°$, $b = 5°$.

The observations of RM of extragalactic sources give us a good and alternative method of probing the galactic magnetic fields. There are over 500 well studied sources with partial data available for some 2000 more. The data are published in numerous papers, the most recent collections being by TABARA and INOUE [19] and SIMARD-NORMANDIN et al. [20]. The RM data has the major disadvantage of giving the product $n_e \cdot B_{||} \cdot dl$ and hence only after some assumptions the magnetic field $B_{||}$. Interpretations of data in agreement with the models discussed previously were given (e.g. [21,22]). On the other hand, a bisymmetrical spiral model is favoured by other authors (e.g. [23,24]).

The use of pulsar RM and DM is one of the methods that gives the magnetic field directly. The structure of the local magnetic fields was discussed by MANCHESTER [25] based on the study of 38 pulsars. The local field is considered to be directed towards $\ell = 94° \pm 11°$ with a longitudinal field strength $B_{||} = 3.0 \pm 0.5$ μG. Also large variations of the field strength for nearby pulsars were observed. The statistics could be improved but of course only by the use of large amounts of observing time. A more recent analysis of HAMILTON and LYNE [26] used data on 163 pulsars.

The distribution of the nonthermal radio continuum emission is of course a signature of the distribution of magnetic fields. Models of the distribution of the 408 MHz radio emission [27] require a model of the magnetic fields. Such work has been published by PHILLIPPS et al. [28,29] and BEUERMANN et al. [30]. Although there are many parameters to be assumed a consistent picture has been given. The arguments of "minimum energy" which are used indicate field strengths of ~ 3 μG. Recent support for the existence of the equipartition (minimum energy) between magnetic fields and the relativistic gas was given by HUMMEL [31] based on the close correlations of radio continuum flux and FIR luminosity of spiral galaxies.

All the above discussion related to the large-scale structure of the galactic magnetic field. Observations with high angular resolution (radio polarization) or the use of the Zeeman effect allow us to probe the fields in smaller volumes. The Zeeman effect (e.g. [32,33,34] can probe fields in individual HI or molecular clouds. Field values of tens of μG have been found. The detailed radio polarization of the region centred on $\ell = 140°$, b = 5° was studied by BAKER and SMITH [35] and SPOELSTRA [36]. The North Polar Spur was studied by SPOELSTRA [37]. Local "bubbles" have been discussed by VALLEE [38], VALLEE et al. [39] and BROTEN et al. [40]. An alternative method to probe the small-scale structure of the galactic magnetic field is to study the rotation measure across radio galaxies [41].

Before completing this introductory talk reference should be made to some reviews of both the magnetic fields in the Galaxy but also of the more recent subject of the large-scale distribution of magnetic fields in nearby galaxies [10,42, 43,44,45,46,47].

References

1. W.A. Hiltner: Science 109, 166 (1949)
2. J.S. Hall: Science 109, 165 (1949)
3. L. Davis, J. Greenstein: Astrophys. J. 144, 206 (1951)
4. A. Behr: Veröffentl. Universitäts-Sternwarte Göttingen No. 126 (1959)
5. D.S. Mathewson, V.L. Ford: Memoirs Roy. Astron. Soc. 74, 139 (1970)
6. D.J. Axon, R.S. Ellis: Monthly Notices Roy. Astron. Soc. 177, 499 (1976)
7. D.S. Mathewson, D.C. Nicholls: Astrophys. J. 154, L11 (1968)
8. D.S. Mathewson: Astrophys. J. 153, L47 (1968)
9. C.H. Meyer, T.P. McCullough, R.M. Sloanaker: Astrophys. J. 126, 468 (1957)
10. F.F. Gardner, J.B. Whiteoak: Ann. Rev. Astron. Astrophys. 3, 245 (1965)
11. V.L. Ginzburg, S.I. Syrovatsky: Ann. Rev. Astron. Astrophys. 7, 375 (1969)
12. G. Westerhout, Ch.L. Seeger, W.N. Brouw, J. Tinbergen: Bull. Astron. Inst. Neth. 16, 187 (1962)
13. R. Wielebinski, J.R. Shakeshaft, I.I.K. Pauliny-Toth: Observatory 82, 158 (1962)
14. E.M. Berkhuijsen, W.N. Brouw: Bull. Astron. Inst. Neth. 17, 185 (1963)
15. R. Wielebinski, J.R. Shakeshaft: Monthly Notices Roy. Astron. Soc. 128, 19 (1964)
16. D.S. Mathewson, D.K. Milne: Australian J. Phys. 18, 635 (1965)
17. R. Wielebinski: Ph.D. Thesis, Cambridge University (1963)
18. J.M. Hornby: Monthly Notices Roy. Astron. Soc. 133, 213 (1966)
19. H. Tabara, M. Inoue: Astron. Astrophys. Suppl. 39, 379 (1980)
20. M. Simard-Normandin, P.P. Kronberg, S. Button: Astrophys. J. 45, 97 (1981)
21. J.B. Whiteoak: In Galactic Radio Astronomy, ed. by F.J. Kerr and S.C. Simonson III (Reidel Publ. Co., Dordrecht 1974), p. 137
22. M. Inoue, H. Tabara: Proc. Astron. Soc. Japan 33, 603 (1981)
23. M. Simard-Normandin, P.P. Kronberg: Astrophys. J. 242, 74 (1980)
24. Y. Sofue, M. Fujimoto: Astrophys. J. 265, 722 (1983)
25. R.N. Manchester: Astrophys. J. 188, 637 (1974)
26. P.A. Hamilton, A.G. Lyne: Monthly Notices Roy. Astron. Soc., in press (October 1986)
27. C.G.T. Haslam, C.J. Salter, H. Stoffel, W.E. Wilson: Astron. Astrophys. Suppl. 47, 1 (1982)
28. S. Phillipps, S. Kearsey, J.L. Osborne, C.G.T. Haslam, H. Stoffel: Astron. Astrophys. 98, 286 (1981a)
29. S. Phillipps, S. Kearsey, J.L. Osborne, C.G.T. Haslam, H. Stoffel: Astron. Astrophys. 103, 405 (1981b)
30. K. Beuermann, G. Kanbach, E.M. Berkhuijsen: Astron. Astrophys. 153, 17 (1985)
31. E. Hummel: Astron. Astrophys. 160, L4 (1986)
32. G.L. Verschuur: Astrophys. J. 156, 861 (1969)
33. G.L. Verschuur: Fund. Cosmic Phys. 5, 113 (1979)
34. T.H. Troland, C. Heiles: Astrophys. J. 301, 339 (1986)

35. J.R. Baker, F.G. Smith: Monthly Notices Roy. Astron. Soc. <u>152</u>, 361 (1971)
36. T.A.Th. Spoelstra: Astron. Astrophys. <u>135</u>, 238 (1984)
37. T.A.Th. Spoelstra: Astron. Astrophys. Suppl. <u>5</u>, 205 (1972)
38. J.P. Vallée: Astron. Astrophys. <u>136</u>, 373 (1984)
39. J.P. Vallée, N.W. Broten, J.M. MacLeod: Astron. Astrophys. <u>134</u>, 199 (1984)
40. W.N. Broten, J.M. MacLeod, J.P. Vallée: Astrophys. Lett. <u>24</u>, 165 (1985)
41. P.P. Kronberg, R. Wielebinski, D.A. Graham: Astron. Astrophys., in press (1986)
42. C. Heiles: Ann. Rev. Astron. Astrophys. <u>14</u>, 1 (1976)
43. J. Vallée: J. Roy. Astron. Soc. Canada <u>77</u>, 117 (1983)
44. Y. Sofue, M. Fujimoto, R. Wielebinski: Ann. Rev. Astron. Astrophys. <u>24</u>, 459 (1986)
45. R. Beck: IEEE Transactions on Plasma Science, Special Issue on Space and Cosmic Plasma, in press (1986)
46. C. Heiles: RAL Berkeley Preprint (August 1986)
47. E. Asseo, H. Sol: Physics Reports, in press (1986)

Interstellar Polarization and Magnetic Fields

P. Cugnon

Observatoire Royal de Belgique, Avenue Circulaire 3, B-1180 Bruxelles, Belgium

Since the work of Davis and Greenstein, polarization of starlight is quite generally considered as one of the most spectacular manifestation of the presence of
magnetic fields in the interstellar space. If polarization measurements are qualitatively able to indicate the general pattern of the magnetic fields, a quantitative analysis seems more difficult to work out. A first purpose of this paper is
therefore to present a synthetic view of the theory, in order to discuss the relationship between the magnetic field strength and the observed quantities. Using
then some recent observations of particular interest, because they correspond to
different interstellar physical conditions, it is shown that any attempt to make a
quantitative determination of the magnetic field from polarization results remains
somewhat hazardous. However, rough estimate and scaling appear possible in some
cases.

1. Introduction

Since its discovery in 1951 by Hall /1/ and Hiltner /2/ there have been so many
observations of interstellar polarization that it would be somewhat tedious to review them completely. This would also bring us quite far from the general frame of
this meeting. Therefore I shall limit my talk to the aspects of the problem which
are in some way connected with interstellar fields, and try to answer the question :
"what kind of informations can we deduce from optical polarization measurements ?"

Most of the "first generation" observations have been collected by Mathewson
and Ford /3, and references therein/, in order to construct the famous fig. 1,
which shows a plot of polarization vectors through the Milky Way. Whether or not
it represents the projection on the sky of the general pattern of the Galactic
magnetic field(s) is certainly no more an open question, since the fundamental theoretical work of Davis and Greenstein /4/ is now generally accepted. Although there have been several subsequent developments and improvements of the theory, the
basic idea has remained the same : a dust grain, elongated or flattened, driven
to high rotational speed, will be submitted to a dissipative torque originating in
the coupling of its induced magnetic momentum with the external field, and will
tend to align its shortest axis of inertia along the field. The grain then scatters
(or absorbs) starlight differently following the direction of polarization of the

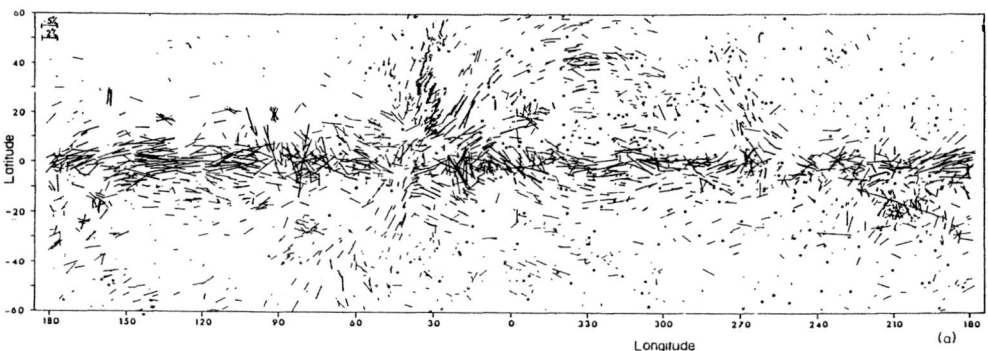

Figure 1 : Optical polarization vectors of 7000 stars (Mathewson & Ford /3/).

incident radiation and because the grains are roughly aligned the same way, a net
large scale effect is observed

Expressing the dependence of optical polarization in terms of the magnetic field
appears then as the crucial point from which we can be able to obtain informations
about the field direction and strength. I shall therefore devote a great part of
my talk to review and summarize the most significant contributions to the theory,
starting with Davis and Greenstein initial paper, in order to discuss the relationship between the degree of alignment and the magnetic field.

The second part will review some recent observations of particular interest,
namely of individual nearby clouds, because of the probable absence in such circumstances of a significant amount of polarizing material in front of the cloud.
Emphasis will also be put on infrared polarization, which is a very promising way
to study molecular clouds. The last part will consist in some conclusive remarks
about the manner of interpreting polarization in terms of the magnetic field.

2. Theory of Polarization

There are two aspects in the theory of polarization :

2.1. The Optical Theory of Dust Grains

This is a very important field of research, so large that it not possible to
review it completely in this paper; furthermore, it would bring us too far from
our purpose (see, for more information, Aannestad and Purcell /5/, Wickramasinghe
and Morgan /6/, Huffman /7/, Greenberg /8/, Rogers and Martin /9/, Hong and
Greenberg /10/, Aannestad and Greenberg /11/, and other references therein).

These contributions to the theory, combined with refined observations of optical extinction, polarization, absorption features, infrared emission from grains,
and laboratory experiments have led to recent important improvements in the knowledge of dust composition, grain size distribution, grain temperature and other
dust properties (Serkowski et al. /12/, Mathis et al. /13/, Aannestad and Kenyon

101

/14/, Mathis /15/, Wilking et al. /16/, Biermann and Harwitt /17/, Mathis and Wallenhorst /18/, Aannestad /19/, Wolstencroft and Greenberg /20/). There remain however several uncertainties in this field, so that for our purpose, which is not the direct knowledge of grain properties, it is not necessary to get into such detailed investigations. I shall therefore limit myself to a simple model, assuming provisionally that :

- in the range of wavelength considered, the particles responsible for polarization are the same as those which cause extinction (this excludes UV observations).

- the determination of the alignment parameter $Q_A = < 1.5 \cos^2 \theta - 0.5 >$, where θ is the angle between the magnetic field and the grain symmetry axis, is sufficient; this quantity is called by Greenberg /8/ "Rayleigh reduction factor" because there is no need to know momenta of higher order of the statistical angular distribution in the case of Rayleigh scattering. Following the same author, the knowledge of this parameter is also sufficient, provided that the appropriate extinction cross-sections are used, even for particles with $2 \pi a/\lambda \gtrsim 1$.

- A mean size of the grain will be used. This is the most restrictive assumption. However, this will be done only in the theoretical discussion which follows, because I am first interested in comparing alignment efficiencies. When interpreting polarization measurements, one should try to derive the local size distribution by means of the extinction and/or polarization curves.

2.2. The Theory of Grain Alignment

In the work of Davis and Greenstein /4/, the dynamical behavior of the magnetic torque was examined in some details, and they emphasized the fundamental role of the imaginary part of the paramagnetic susceptibility, the real part giving rise to a conservative torque without real alignment efficiency. The dissipative torque associated with the imaginary susceptibility, $\chi"$, was shown to induce a slow angular motion in the grain (supposed to be axisymmetric) tending to align its shortest axis of inertia along the magnetic field, this orientation corresponding to a minimum dissipation of energy. The initial rotation of the grain is the result of its collisions with the gas atoms; these collisions also tend to restore random orientation, so that the relative efficiency of the alignment process is roughly measured by the ratio of the collisional damping time to the magnetic characteristic time $\delta = \tau_c / \tau_m$. In the physical conditions of the interstellar medium, the rotational frequency remains low with respect to any magnetic resonance frequency, so that $\chi"$ is in good approximation proportional to the frequency, $\chi" = K \omega$, $K = 2.5 \ 10^{-12} / T_{gra}$ for paramagnetic grains.

The statistical treatment of the problem was however just outlined in Davis' and Greenstein's work; this was also the case in the subsequent extensions of Henry /21/ and Cugnon /22/. A first and quite ignored attempt to solve the statistical problem appeared in an unpublished work due to Miller /23/. At my knowledge, it was the first time that a Fokker-Planck equation was used in this case, implying a modeling of the gas-grain collisions. The treatment was however limited to quasi-spherical grains. A significant step forward was due to Jones and Spitzer /24/, who made a detailed analysis of the possible magnetic behaviors of different types of grains, and suggested superparamagnetism as a mean to increase K by an important factor. They were also the first to consider the magnetic torque from a thermodynamical point of view, observing that a second order derivative term of magnetic origin should be present in the Fokker-Planck equation to account for the role of the internal temperature of the grain. This point is very important, as it will appear in the forthcoming discussion. Jones' and Spitzer's investigation was also limited to quasi-spherical grains.

The simultaneous introduction of realistic collision models and of non-sphericity in the theory was due to Purcell and Spitzer /25/ and Cugnon /26/, using respectively a Monte-Carlo simulation and a Fokker-Planck equation. The agreement between both methods was excellent (Cugnon /27,28/), indicating a very good coherence in the formulation of the theory. These last steps confirmed the dependence of the degree of alignment Q_A upon two important physical parameters, i.e. the ratio of the characteristic times δ defined above and the ratio of the grain internal temperature to its rotational temperature $\xi = T_{gra} / T_{rot}$, implying, as important consequence, that the degree of alignment vanishes at thermal equilibrium ($T_{gra} = T_{rot}$). This fact has been also emphasized by Aannestad and Purcell /5/, Heiles /29/, Greenberg /8/, Rogers and Martin /9/ and Johnson /30/.

Starting from the good agreement between the Monte-Carlo and Fokker-Planck methods, I have shown (Cugnon /28/) that the alignment parameter Q_A can be closely approximated by an ad-hoc modification of Greenberg's /8/ expression of the "Rayleigh reduction factor", at least for prolate grains. Using this expression, I have made some calculations whose results are summarized in Table 1. Here, $C_B = 10^{13} K B^2 / (a\, n_H)$, in the range a/b = 0 (needles) to a/b = 1 (spheres); B is the field in μG, a the transverse radius in μm, n_H the hydrogen density. The maximum interstellar ratio of polarization to extinction has been fixed at 0.03, which seems a more realistic value than 0.06 (Greenberg /8/). The polarizing efficiency of the grains varies from 0.35 (a/b = 0) to 0 (a/b = 1).

Table 1

C_B	a/b	T_{gra}	T_{gaz}	T_{rot}	corresponding type of grain-gas interaction
690	0.4	10	100	55	sticking followed by complete re-evaporation of H atoms.
320	0.45	10	100	300	sticking followed by random emission of H_2 molecules.
4500	0.2	10	20	20	elastic collisions with H_2 molecules.

The two first cases are more or less typical of standard H I clouds, the second situation corresponding to a kind of enhanced thermal rotational excitation process, due to the exothermal character of the formation of H_2 molecules which are here supposed to be randomly emitted from the grain surface (Purcell /31/, Cugnon /32/). Using now n_H = 10 cm^{-3}, a = 0.1 µm, we obtain K B^2 = 0.7 10^{-10} in the first case and about the half in the second case. When ordinary paramagnetism is expected, this gives respectively B = 17 µG and B = 11 µG. Any further increase of the rotational temperature does not change this last value significantly; changing the other temperatures in their allowed ranges (T_{gra} = 7 K, T_{gas} = 80 K, for example) reduces the field by only 10 %. Extinction observations also put drastic constraints upon the grain dimensions. The only way to reduce the field to its observed values (Troland and Heiles /33/) of about 3 µG would be to accept an hydrogen density of about 1 cm^{-3}, but this value has been proposed by Mouschovias /34/ as an average interstellar value between the intercloud medium and the clouds themselves, in which polarization is supposed to be produced. In fact, the difficulty can only be removed by admitting an enhancement of the magnetic properties of the grains by a factor 10 to 100 (Jones and Spitzer /24/, Duley /35/, Mathis /36/). The third case has been chosen to suit in a certain way molecular clouds. It must be noted here that the required degree of alignment cannot be obtained if ξ > 0.6 so that we limited our investigation to ξ = 0.5. B^2/n_H is found to be at least 180. Following Fleck /37/, B^2 scales like n_H but with a proportionality factor equal to 0.09. An enhancement of the magnetic susceptibility by a factor 2000 would then be required here, with also drastic limitations for elongations and temperatures.

All these difficulties have led Purcell /38/ to propose another type of rotational excitation mechanism, which allows the grains to reach a very high rotational speed. The basic idea is that the surface of the grains exhibits a limited number of peculiar sites where energetic phenomena of ejection occur from time to time, giving rise to a mean non vanishing torque accelerating the grain up to a very high stationary angular velocity, provided that the surface remains unaltered during a time of the order of the damping time τ_c. This condition implies that resurfacing due to accretion of new material or to UV photodesorption is not too frequent at this time scale. Alignment is then achieved in a time of the order of the magnetic characteristic time τ_m. A quite attractive possibility of a spin-up mechanism proposed by Purcell is the exothermal formation of molecular hydrogen on privileged sites of the grain surface, because it is very likely that such sites actually exist (Hollenbach and Salpeter /39/).

Following Purcell, the characteristic time related to the spin-up mechanism is the mean interval between two successive crossovers through zero of the angular velocity about the preferential axis of rotation, and is given by the semi-empirical formula t_x = 1.3 (t_L + τ_c), where t_L is approximately the time required to alter completely the grain surface. Spitzer and McGlynn /40/ showed that in the case of frequent resurfacing, the same difficulty for the field requirement as in thermal

spinning alignment is encountered. In fact, very short-lived spin-up can be described as a kind of enhanced thermal spinning alignment (see case 2 in Table 1). This
possibility, however, does not seem to be the rule in standard interstellar conditions; I have shown /32/ that in this case, a suprathermal magnetic torque is always present, provided that H_2 molecules are formed on the grains.

In order to see what happens to the parameter C_B in the case of suprathermal
alignment, I shall use Purcell's simplified linear formula $Q = 0.4 \, t_x/\tau_m$, which is
here sufficient for my purpose. The ratio t_x/τ_m is then 1.3 $(1 + t_L/\tau_c) \, \delta$. Following Aannestad and Greenberg /11/, t_L is approximately the time required for a
grain to accrete a new mononuclear layer. After some calculations, they obtain
$t_L/\tau_c = 5.6/d$, where d is the depletion factor. For a gas not depleted in heavy
elements, d = 1 and we obtain Q = 3.2 δ, this result holding for moderate elongations. The value of C_B needed for a polarizing efficiency of 0.18 and again a ratio
of polarization to extinction of 0.03 is 5.5, about 100 times lower than for thermal spinning alignment, yielding a field of about 1 µG, so that, at my sense, if
this mechanism is effectively working, there is no need to invoke any enhancement
of the magnetic properties of the grains. It is to be noted that the dependence
on the temperature ratio does not exist in this mechanism, so that in regions
completely depleted, alignment would be complete, corresponding to $Q_A = 0.5$ for
prolate grains and $Q_A = 1$ for oblate grains. Such regions are then expected to exhibit a very high ratio of polarization to extinction. The principal difficulty
remaining in the case of suprathermal alignment is the problem of getting enough
hydrogen in atomic form in molecular clouds.

3. Some Observations and their Implications for the Magnetic Fields.

In this section, I shall use abbreviations due to Aannestad and Greenberg /11/
characterizing respectively Davis and Greenstein thermal mechanism and Purcell suprathermal mechanism : TSA (for Thermal Spinning Alignment) and SSA (for Suprathermal Spinning Alignment).·

There is now an enormous amount of polarization data, but very few in fact have
been used to derive the magnetic field strength. My purpose is to review in this
section papers which are exemplary of such attempts, but I would incidently like
to acknowledge the very important work due to many observers in the field of the
wavelength dependence of polarization, of the various correlations found with extinction, distance and other parameters (see for example Serkowski /41/. Appenzeller /42/, Coyne, Gehrels and Serkowski /43/, Serkowski et al. /12/. Their works
allowed significant steps forward in the knowledge of the dust nature and size.
Also, the pioneering work of Martin /44/ in the field of circular polarization, gave a decisive argument in favour of dielectric grains.

A typical HI region with a somewhat larger density than the average is the
α Persei star cluster, studied in great detail by Coyne, Tapia and Vrba /45/.

From the observed ratio of polarization to extinction and a gas density derived
from a correlation law between gas and dust, they obtained a field strength of a-
bout 100 µG, both being however derived from theoretical works on TSA (ref. /5/
and /24/). Using the same method as for Table 1, I estimated the field strength at
50 µG (ordinary TSA) and 30 µG (enhanced TSA), for prolate grains with a/b = 0.4.
With $n_H = 35$ cm^{-3}, we are in the range where the field is independent of the gas
density, following Fleck /37/ and more recently Kazès and Crutcher /46/, so that a
field of about 5 µG is expected in that region. Also, any scaling of B versus n_H
appears inadequate. There are two possibilities for reducing the field : superparamagnetism or suprathermal spinning alignment. This last possibility which puts no
drastic requirement on the grain composition is at my sense the best choice; applying Purcell's formula I obtained values of the order of 3 to 4 µG, again for
grains of moderate elongation.

Another peculiar case is the R-Coronae Australis dark cloud, studied by Vrba et
al. /47/. From the same analysis as above, they proposed a field of about 120 µG
in the outer regions of the cloud, where the density n_H is evaluated at 350 cm^{-3}.
My own calculations (for TSA) give here respectively 170 and 85 µG, but from the
scaling relations a field of only 10 to 20 µG is expected. Once more, SSA gives
the right order of magnitude for $B / n_H^{1/2}$, about 0.7. It must be noted here that
the uncertainties in the above result are quite greater than for TSA, because of
the poor knowledge of the time t_L.

The case of the T Tauri dark cloud, observed by Moneti et al. /48/ is in a certain way similar to R-CrA, at least for the medium surrounding the dense condensations. The principal interest of the paper is that it combines optical (outer medium) and IR (condensations) observations. Except for three sources, there is a
good coherence between both types of observations; this indicates the presence of
an important magnetic field in the embedded collapsing cloudlets, but no derivation of its strength is present in the paper.

A typical and well-studied molecular cloud is OMC 1, for which Beichman and
Chaisson /49/ and Dyck and Beichman /50/ made IR measurements of polarization. They
attributed the large amount of polarization found in the BN and KL objects to TSA,
and obtained a field strength of 7 to 10 µG. This determination does not resist to
deeper analysis, because a ratio of polarization to extinction of 0.19 implies
perfect spinning alignment. Such a possibility must be excluded for TSA, because
the grain temperature is certainly far from zero. Moreover, for perfect spinning
alignment, nothing can be said about the field intensity. Dennison /51/ proposed
reverse TSA, which is actually a theoretical possibility, giving however a poor
alignment, even for $T_{gra}/T_{rot} = 2$. For Johnson et al. /52/, the only way to align
grains in BN and KL is SSA, but this mechanism requires enough hydrogen in atomic
form to make the recombination followed by ejection possible on the grain surface.
The authors attributed this production to shocks sufficiently energetic to break
the molecules of hydrogen. This is an interesting hypothesis, but investigations

are needed about the reality and the efficiency of such shocks. Because it is likely that the medium is rather depleted in heavy elements, relatively small fields can achieve perfect alignment (~ 100 μG). Joyce and Simon /53/, and very recently Mathis /36/ proposed superparamagnetism to reduce the value of the field to an acceptable value in TSA; unfortunately, this does not remove the temperature constraint, which is very drastic in molecular clouds.

Other recent measurements of protostellar IR sources are due to Heckert and Zeilik /54/. They used criteria due to Johnson et al. /52/ to discuss the mechanism of polarization and concluded that five sources at least are polarized by magnetically aligned grains. Quite high degrees of polarization are often observed.

Finally, I would like to mention the study of King et al. /55/ of the Serpens Nebula, because it is an interesting example of mixed polarization of different origins.

4. Concluding Remarks

Many authors have used polarization measurements to derive the magnetic field direction. This is quite legitimate, but it must be kept in mind that we actually observe the direction of the magnetic field projection on the sky, so that we miss the dimension along the line of sight. It is also important to make sure that the polarization observed is actually due to Davis and Greenstein mechanism, and not, for example, to the scattering process of reflection nebulae, both acting however simultaneously in peculiar cases (King et al. /55/).

What concerns coherent derivations of the magnetic field strength, my general impression remains rather pessimistic, but not desperate, for accurate calculations require a good knowledge of many parameters.

There is also a big dilemma : TSA or SSA ? From this paper and from other investigations (Johnson /30/, Aannestad and Greenberg /11/), there are strong arguments favouring SSA in the diffuse medium as well as in moderately dense clouds ($n_H \leq 1000 \, cm^{-3}$). However, I am not convinced that SSA also works in molecular clouds, because of the necessary presence of a mechanism dissociating H_2 (other types of spin-up mechanisms seem unlikely in such dense clouds). On the other hand, TSA combined with enhanced magnetic properties is also difficult to admit, because of the stringent requirement on the temperature ratio ξ. Because of the high ratio of polarization to extinction observed in several molecular clouds, nearly perfect alignment is often required, so that little can be said about the magnetic field strength.

That is not the case for HI clouds, for which the knowledge of the required interstellar parameters can lead to an estimation of the field; moreover, for clouds with $n_H \geq 50 \, cm^{-3}$, a particular knowledge of n_H seems no more necessary, if we admit that B scales like $n_H^{1/2}$.

Acknowledgements

I acknowledge a travel grant of the Belgian "Fonds National de la Recherche Scientifique" (FNRS). I am also very much indebted to the "Max Planck Gesellschaft" for its warm hospitality.

References

1. J.S. Hall, Science 109, 166 (1949)
2. W.A. Hiltner, Science 109, 165 (1949)
3. D.S. Mathewson, V.L. Ford, Mem. Roy. Astr. Soc. 74, 139 (1970)
4. L. Davis Jr., J.L. Greenstein, Astrophys. J. 114, 206 (1951)
5. P.A. Aannestad, E.M. Purcell, A. Rev. Astr. Astrophys. 11, 309 (1973)
6. N.C. Wickramasinghe, D.J. Morgan (eds) Solid State Astrophysics, (Reidel, 1976)
7. D.R. Huffman, Adv. Phys. 26, 129 (1977)
8. J.M. Greenberg, in Cosmic Dust, ed. J.A.M. McDonnel, p. 187 (Wiley, 1978)
9. C. Rogers, P.G. Martin, Astrophys. J. 228, 450 (1979)
10. S.S. Hong, J.M. Greenberg, Astr. Astrophys. 70, 695 (1978)
11. P.A. Aannestad, J.M. Greenberg, Astrophys. J. 272, 551 (1983)
12. K. Serkowski, D.S. Mathewson, V.L. Ford, Astrophys. J. 196, 261 (1975)
13. J.S. Mathis, W. Rumpl, K.H. Nordsieck, Astrophys. J. 244, 483 (1977)
14. P.A. Aannestad, S.J. Kenyon, Astrophys. J. 230, 771 (1979)
15. J.S. Mathis, Astrophys. J. 232, 747 (1979)
16. B.A. Wilking, M. Lebofski, P.G. Martin, G.M. Rieke, J.C. Kemp, Astrophys. J. 235, 905 (1980)
17. P. Biermann, M. Harwit, Astrophys. J. 241, L105 (1980)
18. J.S. Mathis, S.G. Wallenhorst, Astrophys. J. 244, 483 (1981)
19. P.A. Aannestad, Astr. Astrophys. 115, 219 (1982)
20. R.D. Wolstencroft, J.M. Greenberg (eds), Laboratory and Observational Infrared Spectra, Occ. Rep. Roy. Obs. Edimburgh 12 (1984)
21. J. Henry, Astrophys. J. 128, 497 (1958)
22. P. Cugnon, Bull. Soc. Roy. Sci. Liège 32, 228 (1963)
23. R.C. Miller, Ph. D. Thesis, Calif. Inst. Techn. (1962)
24. R.V. Jones, L. Spitzer Jr., Astrophys. J. 147, 943 (1967)
25. E.M. Purcell, L. Spitzer Jr., Astrophys. J. 164, 31 (1971)
26. P. Cugnon, Astr. Astrophys. 12, 398 (1971)
27. P. Cugnon, in Interstellar Grains and Related Topics, Proc. IAU Symp. n° 52, eds J.M. Greenberg & H.C. van de Hulst, p. 187 (Reidel, 1973)
28. P. Cugnon, Astr. Astrophys. 120, 156 (1983)
29. C. Heiles, A. Rev. Astr. Astrophys. 14, 1 (1976)
30. P.E. Johnson, Nature 297, 371 (1982)
31. E.M. Purcell, Astrophys. J. 231, 404 (1979)
32. P. Cugnon, Astr. Astrophys. 152, 1 (1985)
33. T.H. Troland, C. Heiles, Astrophys. J. 252, 179 (1982)
34. T.Ch. Mouschovias, Astrophys. J. 207, 141 (1976)
35. W.W. Duley, Astrophys. J. 219, L129 (1978)
36. J.S. Mathis, Astrophys. J. (in press, 1986)
37. R.C. Fleck Jr., Astrophys. J. 264, 139 (1983)
38. E.M. Purcell, in The Dusty Universe, eds G.B. Field & A.G.W. Cameron, p. 155 (Neale Watson, 1975)
39. D. Hollenbach, E.E. Salpeter, Astrophys. J. 163, 155 (1971)
40. L. Spitzer Jr., T.A. Mc Glynn, Astrophys. J. 231, 417 (1979)
41. K. Serkowski, Astrophys. J. 154, 115 (1968)
42. I. Appenzeller, Astrophys. J. 151, 907 (1968)
43. G.V. Coyne, T. Gehrels, K. Serkowski, Astr. J. 79, 581 (1974)
44. P.G. Martin, Mon. Not. R. Astr. Soc. 159, 179 (1972)
45. G.V. Coyne, S. Tapia, F. Vrba, Astr. J. 84, 356 (1979)
46. I. Kazès, R.M. Crutcher, Astr. Astrophys. 164, 328 (1986)
47. F. Vrba, G.V. Coyne, S. Tapia, Astrophys. J. 243, 489 (1981)
48. A. Moneti, J.L. Pipher, H.L. Helfer, R.S. McMillan, M.L. Perry, Astrophys. J. 282, 508 (1984)

49. C.A. Beichman, E.J. Chaisson, Astrophys. J. <u>190</u>, L21 (1974)
50. H.M. Dyck, C.A. Beichman, Astrophys. J. <u>194</u>, 57 (1974)
51. B. Dennison, Astrophys. J. <u>215</u>, 529 (1977)
52. P.E. Johnson, G.H. Rieke, M.J. Lebofski, J.C. Kemp, Astrophys. J. <u>245</u>, 871 (1981)
53. R.R. Joyce, Th. Simon, Astrophys. J. <u>260</u>, 604 (1982)
54. P.A. Heckert, M. Zeilik, Astr. J. <u>89</u>, 1379 (1984)
55. D.J. King, S.M. Scarrot, K.N.R. Taylor, Mon. Not. R. Astr. Soc. <u>202</u>, 1087 (1983)

The Local Magnetic Field as Determined by Pulsar Rotation Measures

W. Sieber

Max-Planck-Institut für Radioastronomie, Auf dem Hügel 69,
D-5300 Bonn 1, Fed. Rep. of Germany

Pulsars are shown to be well suited to investigate the local magnetic field by means of their rotation and dispersion measures. The existing data sample is presented and different model fits to the data are discussed. The results are promising; major improvements can be expected if the data sample is increased considerably. This has been the case just recently.

1. Pulsar characteristics

Soon after the discovery of pulsars it became evident that pulsars should be ideal objects to study the Galactic magnetic field structure by means of Faraday rotation measures (RMs). Several arguments speak in favour of pulsars:

— Pulsars are known to by highly linearly polarized, the fractional polarization often reaching more than 50% and in rare cases, parts of the pulse profile are even 100% polarized.

— In many measurements time resolution is sufficient to resolve the pulse profile so that independent polarization angle observations are possible for the same pulsar (see Fig. 1). This opens the possibility to determine the rotation measure across the pulse profile. No variation of RM across the pulse profile has been found so far (MANCHESTER [1]) so that there is good evidence that we have no intrinsic contribution to the rotation measure (RMI), in contrast, for instance, to many extragalactic sources where considerable intrinsic contributions have to be taken into account. The assumption RMI = 0 is supported by the fact that high latitude pulsars often show very small rotation measures. Another, weaker argument comes from the fact, that existing rotation measure determinations can be explained well by a smooth distribution of the local magnetic field, as will be shown below, in accordance with RMI = 0.

— Since pulsars are (with few exceptions) Galactic objects no contribution from beyond the Galaxy has influence on the rotation measure. Moreover, pulsars are scattered over the whole Galactic plane so that in principle a detailed analysis not only of the most local features of the magnetic field should be possible. In practice, however, pulsars with known rotation measures are local objects, as demonstrated in Figure 2, mainly due to selection effects which so far favour pulsars with high ap-

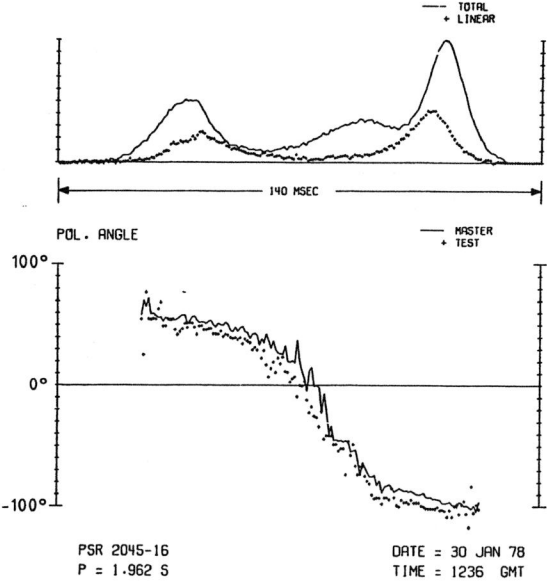

Figure 1:
Mean pulse profile of PSR 2045-16
at λ18 cm

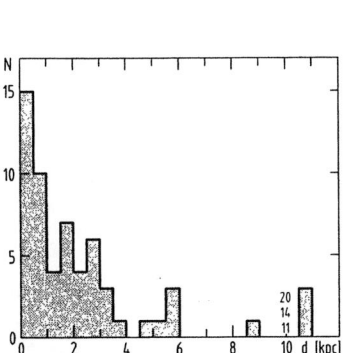

Figure 2: Distance distribution of pulsars with known rotation measures as
given by [9]

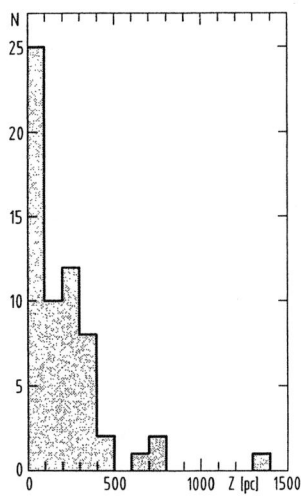

Figure 3: z-distribution of pulsars
with known rotation measures [9]

parent luminosity, i.e. nearby pulsars. Beyond 4 kpc distance only a few pulsars
with known rotation measures are left.

–   Pulsar distances are known with considerable accuracy mainly by means of their
dispersion measures, i.e. the integral over the free electron density on the line of
sight. Distances derived in such a way are good estimates in a statistical sense,
especially when improved by correcting for contributions from HII regions on the line
of sight (MANCHESTER and TAYLOR [2]). The derived distance scale is normally ad-

justed by comparison with known distances from HI absorption measurements and pulsar–SNR associations. Free thermal electron densities around 0.025 cm^{-3} seem to give best results.

— Obviously also z–distances from the Galactic plane can be estimated with sufficient accuracy, from which we know (see Fig. 3) that pulsars are concentrated in a well defined disk of some hundred parsec thickness. LYNE et al. [3] derive a scale hight of 400 pc for a comprehensive sample of pulsars.

2. First investigations

The first, who took advantage of these pulsar properties seems to be MANCHESTER [1], who compiled a list of rotation measures, which contained 19 entries at the time. He computed the line-of-sight components of the magnetic field from the ratio of the rotation and dispersion measures and showed that the observations could be interpreted by the assumption of a uniform, elongated local magnetic field of magnitude 3.5 + 0.5 μG directed towards the Galactic longitude ℓ = 90° (latitude b = 0°), see also Table. He clearly stated that the observations seemed to be inconsistent with a helical field structure, which at that time was very much in discussion.

Author	Data sample	Magnetic field [μG]	Field direction		Characteristics
			ℓ	b	
MANCHESTER [1]	19	3.5 ± 0.5	90°	0°	uniform
MANCHESTER [4]	38	2.2 ± 0.4	94° ± 11°	0°	uniform
RUZMAIKIN and SOKOLOFF [5]	38	2.1 ± 1.1	99° ± 24°	0°	uniform
THOMSON and NELSON [6]	61	3.5 ± 0.3	74° ± 10°	0°	sign reversal Gaussian profile

In 1974 MANCHESTER [4] had doubled the number of known rotation measures so that he could base his statistical analysis on 38 pulsars. He argued that stars farther away than 2 kpc might not be typical for the local field structure and should therefore be excluded from the analysis. This meant that four pulsars were rejected. Furthermore, rotation measures of stars in the North Polar Spur and Spur I might be influenced by local features and were disregarded as well; so that six more pulsars were excluded. The remaining 28 pulsars were weighted by $\cos^2 b$, in order to get the best estimate of the disk component of the field. The fitting procedure gave results similar to those of the first publication, i.e. a magnetic field strength of the uniform component of 2.2 ± 0.4 μG towards ℓ = 94° ± 11° (b = 0°). Superimposed there seemed to be irregularities of comparable strength so that the total field strength was considered to be higher, in the range of 4 – 5 μG. The typical scale size of the irregularities was estimated to be a few hundred parsecs; the irregularities seemed to be rather isolated disturbancies on an otherwise uniform field. And there was certainly no indication of a helical field structure.

3. Modelling

A new approach appeared in 1977, when RUZMAIKIN and SOKOLOFF [5] addressed the problem again. They used as basic data sample the same collection of rotation measure values as published by MANCHESTER [4], and excluded from their analysis eleven pulsars with similar arguments as given by Manchester. This time, however, the rotation measure was analysed directly, not the ratio between RM and DM. They discarded also the $\cos^2 b$ weighting as used by MANCHESTER [4]. Applying a new analysis procedure, which allowed to estimate many hitherto unknown parameters and which had been worked out for the analysis of rotation measures of extragalactic sources, they estimated the uniform component of the magnetic field to be 2.1 ± 1.1 μG in the direction of $\ell = 99° \pm 24°$. Again, the data base was too weak to estimate also the latitude direction of the magnetic field with sufficient reliability, so that $b = 0°$ had to be assumed.

Their analysis allowed them to estimate also the correlation length of the irregular component of the magnetic field, which they computed to be between 100 and 150 pc. By comparison with the results for extragalactic sources they could also derive an estimate for the thickness of the electron layer in the Galactic disk, i.e. $h \approx 400$ pc. From the same data set the mean thermal electron density in the Galactic plane was estimated to be $n_e = 0.035 \pm 0.01$ cm^{-3}.

By 1980 the number of known rotation measures had increased to 61. This was the data set used by THOMSON and NELSON [6] in their 1980 analysis. Again, pulsars further away than 3 kpc were excluded leaving 48 pulsars in the sample. And again the field was assumed to be parallel to the Galactic disk, allowing no latitude component ($b = 0°$). They pointed out that the standard deviation of the polarization angle − produced by the random component of the magnetic field − is proportional to $n_e\sqrt{d}$, i.e. the thermal electron density times the square root of the distance. So that the weighting should be proportional to $1/n_e\sqrt{d}$, provided the whole sample is normalized properly. They tried to fit the data with different models and came to the conclusion that the model with the highest number of independent parameters fitted the data best. This model had the following features:

− The uniform component B_u of the magnetic field was assumed to have a Gaussian profile in the z−direction:

$$B_u = B_O \exp \left\{ - \frac{(z+z_s)^2}{S_h^2} \right\}$$

with S_h scale height of the magnetic field and
      z_s displacement of the sun above the magnetoactive plane.

— The field direction was allowed to reverse. The line at which this reversal takes place was assumed to be at a distance x_0 from the position of the sun.

With these free parameters the data could be modelled very well. THOMSON and NELSON [6] computed the following values:

B_u = 3.5 ± 0.3 μG
ℓ = 74° ± 10°
x_0 = 170 ± 90 pc
S_h = 75 ± 40 pc
z_s = 25 ± 30 pc

There were few discrepancies left between predicted and observed field direction and the residual rotation measures were not dependent on distance. There remains perhaps the slightly suspicious fact that x_0 is very small compared to the distance over which the model is applicable (about 3 kpc), which puts the solar system in a unique position in the Galaxy (which most probably is not the case). THOMSON and NELSON [6] could also derive an estimate for the irregular component of the magnetic field, which they estimate to be $4 \leqslant B_{irr} \leqslant 14$ μG.

4. Conclusions

The work done so far in analyzing pulsar rotation measure values has shown that these observations are potentially most valuable for the determination of the local Galactic magnetic field, i.e. the regular component and the irregular component. Impressive results have been derived by use of a quite limited data sample. Since there are now more than 400 pulsars known, there is the hope that much more information can be gathered once a more comprehensive data sample of rotation measures has been determined. It should then be possible to extent the analysis beyond the nearest surrounding of the solar system to more distant parts of the Galaxy. In this respect it is most promising that the collection of rotation measure determinations has been increased in the past by the work of MORRIS et al. [7] and quite recently by HAMILTON and LYNE [8], which extended the number of known values by 163, so that in the near future a deeper insight into the problem of the structure of the local magnetic field will be possible.

References

1. R.N. Manchester: Astrophys. J. 172, 43 (1972)

2. R.N. Manchester, J.H. Taylor: Astron. J. 86, 1953 (1981)

3. A.G. Lyne, R.N. Manchester, J.H. Taylor: Monthly Notices Roy. Astron. Soc. 213, 613 (1985)

4. R.N. Manchester: Astrophys. J. 188, 637 (1974)

5. A.A. Ruzmaikin, D.D. Sokoloff: Astrophys. Space Sci. 52, 365 (1977)

6. R.C. Thomson, A.H. Nelson: Monthly Notices Roy. Astron. Soc. 191, 863 (1980)

7. D. Morris, D.A. Graham, J.H. Seiradakis, W. Sieber, P. Thomasson, B.B. Jones: Astron. Astrophys. 73, 46 (1973)

8. P.A. Hamilton, A.G. Lyne: Monthly Notices Roy. Astron. Soc., in press (1986)

9. R.N. Manchester, J.H. Taylor: Pulsars (W.H. Freeman & Co., San Francisco 1977)

Extended Polarized Emission at 2695 MHz in the Galactic Plane

N. Junkes, E. Fürst, and W. Reich

Max-Planck-Institut für Radioastronomie, Auf dem Hügel 69,
D-5300 Bonn 1, Fed. Rep. of Germany

1. Introduction

REICH et al. [1] recently gave the results of a radio continuum survey of the Galactic plane ($357°4 < \ell < 76°$, $|b| < 1°5$) at 2695 MHz with the Effelsberg 100-m telescope. These observations were made with a three-channel receiving system. The IF signals were correlated in order to obtain simultaneously total intensity and Stokes parameters U and Q. These data were used to calculate the linearly polarized intensity and the polarization position angle.

2. Observations and Results

The observations and reduction procedures were already described [1,2,3]. At 2695 MHz the system noise of the three-channel receiving system was about 60 K, HPBW = $4°27$ and $T_B/S = 2.51$. 3C286 served as the primary calibration source (assuming a flux density of 10.4 Jy, 9.9% of linear polarization at 33°). Scans were taken in galactic latitude for $|b| \leqslant 1°5$. Each scan was observed at least twice, except for a few areas. Linear fits have been made for the U and Q maps (not considering structures above 80 mK $T_B P$) to define a relative baselevel and to remove large-scale regular foreground polarization.

The instrumental polarization was found to be 0.7%. The maps were not corrected for this effect. Some influence from ground radiation for very low elevations < 15° ($\delta < -25°$, $\ell < 5°$) cannot be ruled out. Therefore the polarization survey is limited to the area $4°9 \leqslant \ell \leqslant 76°$, $|b| \leqslant 1°5$. To increase the signal-to-noise ratio the data were convolved to 6' and regridded to 3'. The rms-noise for polarized intensity was found to be 8 mK $T_B P$. An example of the survey maps is shown in Figure 1, where the contours represent the total intensity distribution (labelled in K T_B). Superposed are bars of linearly polarized intensity in E-field direction. Polarized intensity is shown above 24 mK $T_B P$ ($\approx 3\sigma$).

The data have already been used to identify the nature of some extended sources [2,4,5]. Also a few small diameter sources and polarization data of supernova remnants have been listed [2].

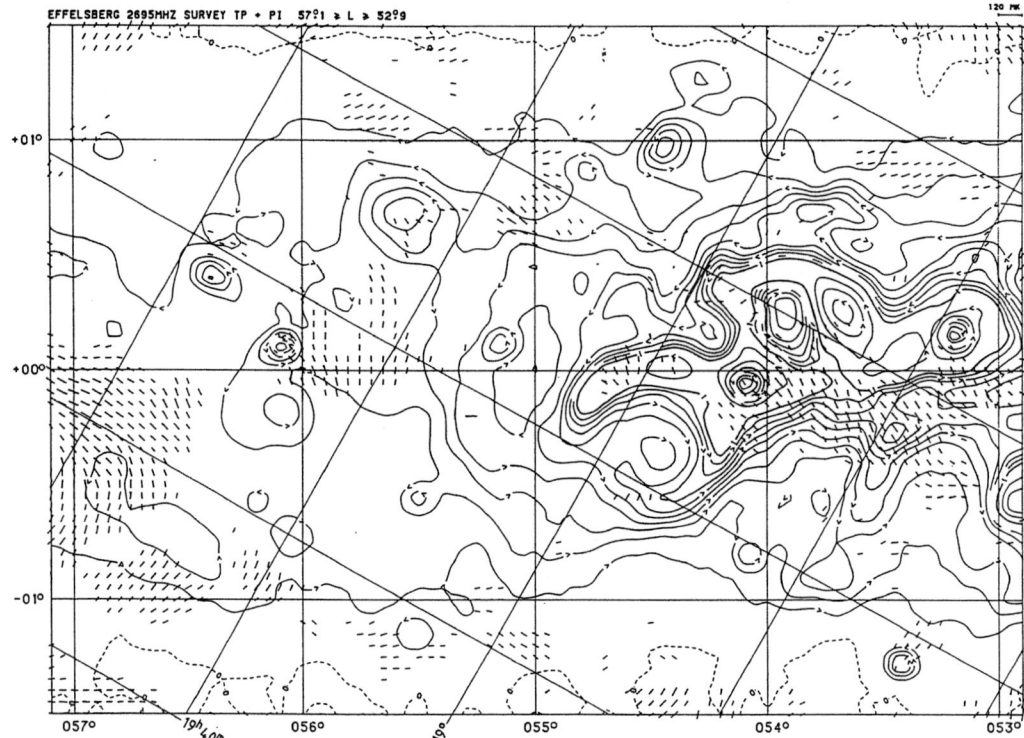

Figure 1:   Example for a survey plate in Galactic coordinates taken from [3]. Equatorial coordinates (Epoch 1950) are shown in addition. Polarized intensity vectors are shown superposed on total intensity contours.

3. Diffuse Polarized Emission

The survey maps show that most of the polarized emission along the Galactic plane is not associated with sources, but is distributed in a rather patchy way. This indicates that polarization from the diffuse background emission is visible, however, we have no distance information to decide on the local or far distant origin. We performed integrations of the polarized intensities as a function of galactic coordinates to find out whether there is a dependence reflecting the characteristics of the local field or the far distant spiral structure of the Galaxy.

Figure 2 shows the average polarized surface brightness as a function of galactic longitude, corrected for instrumental polarization and omitting polarized sources above 80 mK $T_B{}^P$. We attribute the minimum of the integrated emission for $\ell > 70°$ to local depolarization of thermal material near Cygnus X. The variation of intensity for $\ell < 70°$ does not show a dependence as it is expected for a local origin. The direction of the local field is believed to be close to that of the local arm and therefore a systematic increase of polarized emission towards smaller longitudes should result because of the increase of the transverse magnetic field component. In fact the maximum of polarized emission is observed for $50° \leq \ell \leq 60°$ and it decreases for smaller longitudes.

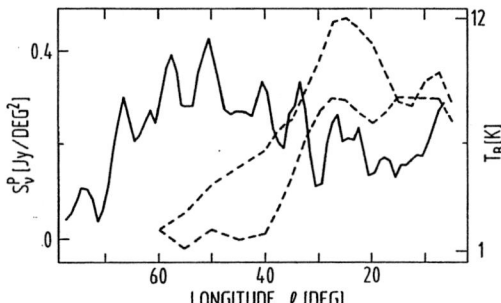

Figure 2: Average polarized surface brightness at 2695 MHz for |b| ≤ 1°.4 as a function of Galactic longitude. The two dashed lines show in addition the thermal background component at 21 cm taken from [6] based on observations of Westerhout [10] (top) and Mathewson et al. [11] (bottom).

Moreover we note an anticorrelation with the thermal component of the radio continuum emission, which is also shown in Figure 2. Because this anticorrelation holds also for the latitude dependence for $\ell < 40°$ (not shown here), we conclude that depolarization of the thermal material causes the observed variation of the polarized emission with Galactic coordinates. GÜSTEN and MEZGER [6] modelled the thermal emission and gave a peak emissivity at 5 kpc galactocentric distance with a sharp decrease for larger distances. This implies that at least the region of maximum emission at $50° \leq \ell \leq 60°$ contains components originating at more than 6 kpc distance. The detection of polarized emission from such large distances gives limits for the depolarization properties of the interstellar medium. For an average electron density of ~ 0.03 cm^{-3} as inferred by pulsar dispersion measures [7], we calculate a uniform magnetic field component in the line of sight direction smaller than 1.6 μG. This is based on the application of the slab model by BURN [8].

If, in future, our 2695 MHz observations are combined with observations at higher frequencies, it seems possible to measure the magnetic field structure of distant spiral arms of the Galaxy directly. However, this means making observations of very small intensities ($\approx$,1 mK $T_B{}^P$ at 10 GHz), which is at present near the technical limits but should be possible in the near future with multibeam polarimeters.

A full discussion of the survey results will be given elsewhere [9].

References

1. W. Reich, E. Fürst, P. Steffen, K. Reif, C.G.T. Haslam: Astron. Astrophys. Suppl. 58, 197 (1984)

2. N. Junkes: Diploma Thesis, Bonn University (1986)

3. N. Junkes, E. Fürst, W. Reich: Astron. Astrophys. Suppl., submitted (1987)

4. W. Reich, E. Fürst, P. Reich, Y. Sofue, T. Handa: Astron. Astrophys. 155, 185 (1986)

5. E. Fürst, T. Handa, W. Reich, P. Reich, Y. Sofue: Astron. Astrophys., in press (1987)

6.   R. Güsten, P.G. Mezger: Vistas in Astronomy <u>26</u>, 159 (1983)

7.   D.S. Harding, A.K. Harding: Astrophys. J. <u>257</u>, 603 (1982)

8.   B.J. Burn: Monthly Notices Roy. Astron. Soc. <u>133</u>, 67 (1966)

9.   N. Junkes, E. Fürst, W. Reich: Astron. Astrophys., in preparation (1987)

10. G. Westerhout: Bull. Astron. Inst. Netherlands <u>14</u>, 215 (1958)

11. D.S. Mathewson, J.R. Healey, J.M. Rome: Australian J. Phys. <u>15</u>, 369 (1962)

The Radio Continuum Brightness Minimum near Polaris: A Hole in the Interstellar Magnetic Field?

K. Reif

Radioastronomisches Institut der Universität Bonn, Auf dem Hügel 71, D-5300 Bonn 1, Fed. Rep. of Germany

1. Summary

Large-scale radio continuum observations at decimetre wavelengths reveal an extended remarkable nonthermal brightness minimum of 10° in diameter close to the equatorial north pole. Its positional coincidence with an enormous arch of neutral hydrogen gas and a similar structure seen in the FIR (100μ) is indicative of it being due to a shell-type phenomenon.

It is shown that a thick shell characterized by an inner and outer radius r and R (where r/R $\approx$ 0.5) with only a small enhancement of the magnetic field strength ($\approx$ 20%) in the shell will account for a distinct brightness minimum with respect to the surrounding. The expected shell emission is much less pronounced and may therefore become undetectable due to fluctuations of the total emission along the whole line of sight.

2. The HI polar arch and the accompanying radio continuum minimum

Radio continuum surveys of intermediate angular resolution (HPBW $\leqslant$ 1°) (BERK-HUIJSEN [1], HASLAM et al. [2], REICH [3], REICH and REICH [4]) have been widely used to investigate large-scale structures of the Galactic plane. Particularly extended loop structures up to several ten degrees in size were detected and are among the favoured high latitude radio continuum features studied in the literature (BERKHUIJSEN [1], BERKHUIJSEN et al. [5], SALTER [6], SOFUE and NAKAI [7]). A large number of loop structures several degrees in diameter has also been found in the distribution of the column density of neutral hydrogen (HEILES [8], HUE [9]). As these HI shells are often suggested to be associated with very old SNRs they were used as guides to search for accompanying faint nonthermal radio continuum loops in the survey data (SOFUE and NAKAI [7], HEITHAUSEN [10]).

One of the most impressive HI shells is found near the equatorial north pole. It forms an enormous arch extending in Galactic longitude from $\approx$120° to $\approx$150° and in latitude from $\approx$25° to $\approx$45° (Fig. 1). This "polar arch" is among the most convincing examples of positional coincidence with a radio continuum feature listed by HEIT-HAUSEN [10]. At 408 MHz (Fig. 2) it appears as an extended brightness minimum. It is also visible in the 820 MHz (BERKHUIJSEN [1]) and 1420 MHz (REICH and REICH [4]) survey data. Further positional coincidence is found with filaments of the infrared "Galactic cirrus" seen in the 100μ IRAS maps.

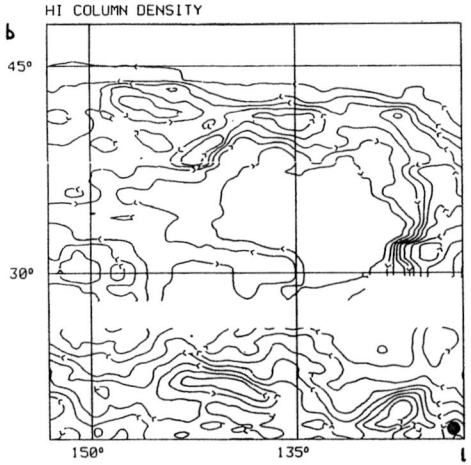

HI COLUMN DENSITY

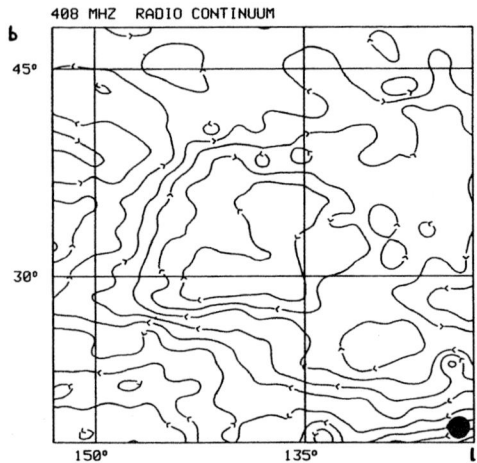

408 MHZ RADIO CONTINUUM

Fig. 1: Neutral hydrogen column density integrated between ±40 km/s. Data are taken from HEILES and HABING [11]. Contours start at 12 10¹⁹ cm⁻² and are in steps of 5 10¹⁹ cm⁻². Contours with the arrow pointing clockwise encircle minima and vice versa. The filled circle in the lower right corner indicates the angular resolution.

Fig. 2: Radio continuum map taken from the 408 MHz survey by HASLAM et al. [2]. The original data were smoothed to 1°.5 resolution. To reduce the confusion due to the intensity increase towards the Galactic plane a gradient of 2 K/10° was subtracted. The minimum in the map centre is at 18 K in absolute brightness temperature. The contour interval is 1.5 K. Symbols as in Fig. 1.

While the HI column density distribution in Fig. 1 reveals the typical shell structure, the radio continuum map shows no indication of a ring-like emission enhancement along the HI arch outside the minimum. On the other hand, the intensities in the minimum are remarkably low even if compared on a larger scale: At Galactic longitude $\ell \simeq 140°$ these intensities are among the lowest in the whole latitude range between the Galactic poles. At Galactic latitude $b \simeq 30°$ lower temperatures are only found around $\ell \simeq 240°$ where the plane emission itself slopes down to its minimum. Instead of a radio continuum loop we find a remarkable minimum in the brightness distribution.

The mean radio spectral index in the field $120° \leqslant \ell \leqslant 150°$ and $25° \leqslant b \leqslant 40°$ has been derived between 408 MHz and 1420 MHz using T-T diagrams. The average index is $ß = 3.0 \pm 0.2$ ($T_b \propto \nu^{-ß}$) with $ß = 2.75$ as the lowest value. The brightness minimum is obviously due to a lack of nonthermal emission.

While this nonthermal emission minimum does not appear as a loop, its remarkable coincidence with loop structures of neutral hydrogen and dust indicates that it might be due to a shell phenomenon.

3. The thick shell "model"

A shell configuration of a slightly amplified magnetic field is used to account for this lack of synchrotron emission. The shell geometry is described by an inner and outer radius r and R respectively. It is assumed that the field is frozen in the in-

120

terstellar matter, and to simplify the calculations it is further assumed that the field is dominated by the random component. The relativistic electrons are coupled to the magnetic field. The magnetic field strength H and the volume density of the relativistic electrons N_e may then be expressed in terms of the volume density of the interstellar matter n

$$H \propto n^{2/3} \quad \text{and} \quad N_e \propto n. \tag{1}$$

For a power law electron energy spectrum with $N(E) \simeq N_e E^{-\gamma}$ the total intensity I varies as

$$I \propto l \, H^{\frac{(\gamma+1)}{2}} \, N_e \, \nu^{\frac{(\gamma-1)}{2}} \tag{2}$$

l – length of the line of sight through the emitting volume

ν is the observing frequency. γ is assumed to take the value $\gamma = 2.4$. Substituting H and N in (2) according to (1) gives

$$I \propto l \, n^{2.13} \tag{3}$$

The emission across the shell is considered with respect to the emission outside from a section l_0 of the line of sight where $l_0 = 2R$. First the ratio I_c/I_0 of the intensity at the shell centre and the intensity outside and its dependence on the shell thickness (R−r) is examined. We assume that the total number of particles inside R is unchanged and – in the simplest case – that the density n_i inside r is zero while the density n_s in the shell increases accordingly. Then the ratio r/R is the only variable and it follows

$$\frac{I_c}{I_0} = \frac{(1 - r/R)}{(1 - (r/R)^3)^{2.13}}$$

This function is displayed in Fig. 2. It shows in fact that *the central intensity decreases below the intensity* outside to a minimum value of about $0.66 \, I_0$ and rises again beyond $r/R \simeq 0.60$. The calculations with $n_i \geq 0$ reveal that the minimum appears at larger r/R and that the minimum intensities increase with increasing n_i/n_0.

To demonstrate the overall distribution of the intensities in the shell area the intensity profile I_r/I_0 starting at the shell centre was calculated. The intensity at any point is the sum of three components along the line of sight i) the emission from inside r ii) the emission from the shell itself and iii) the emission from the remaining portion of the original layer outside the shell.

In Fig. 4 intensity profiles are displayed for two sets of parameters i) r/R·= 0.5, $n_i = 0$ and ii) r/R = 0.5, $n_i/n_0 = 0.5$. The corresponding densities in the shell are $n_s/n_0 = 1.14$ and $n_s/n_0 = 1.07$ respectively. Both intensity profiles show the same overall characteristics. Close to the centre the distribution is flat and significantly below I_0 (0.66/0.70) as expected from Fig. 3. It then rises to intensities above I_0 (1.28/1.14) at r = 0.5 R and decreases slowly to reach I_0 at the shell boundary.

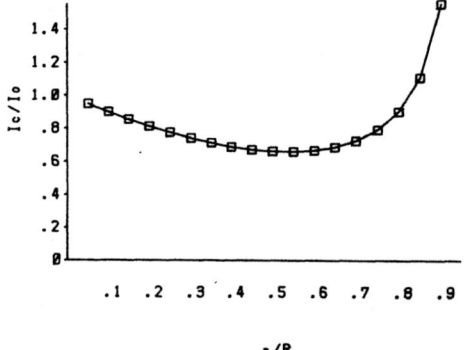

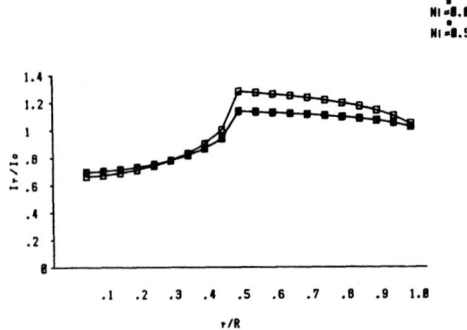

Fig. 3: Intensity observed at the shell centre as function of the shell thickness. See text for details.

Fig. 4: Intensity profiles across the shell for densities $n_i=0$ and $n_i/n_o=0.5$ inside the inner radius. The inner radius is $r=0.5 R$ in both cases.

In the observed intensity distribution the minimum will therefore be more pronounced than the limb brightening and in case of additional intensity fluctuations along the line of sight the minimum might be the only detectable feature. In this case it would look more like a flattened meteor crater instead of exhibiting a loop structure.

4. Literature

1. E.M. Berkhuijsen: Astron. Astrophys. Suppl. Ser. 5, 263 (1972)
2. C.G.T. Haslam, C.J. Salter, H. Stoffel, W.E. Wilson: Astron. Astrophys. Suppl. Ser. 47, 1 (1982)
3. W. Reich: Astron. Astrophys. Suppl. Ser. 48, 219 (1982)
4. W. Reich, P. Reich: Astron. Astrophys. Suppl. Ser. 63, 205 (1986)
5. E.M. Berkhuijsen, C.G.T. Haslam, C.J. Salter: Astron. Astrophys. 14, 252 (1971)
6. C.J. Salter: Bull. Astron. Soc. India, 11, 1 (1983)
7. Y. Sofue, N. Nakai: Astron. Astrophys. Suppl. Ser. 53, 57 (1983)
8. C. Heiles: Astron. J. 229, 533 (1979)
9. E.M. Hu: Astron. J. 248, 119 (1981)
10. A. Heithausen: "Shell structures in HI and in the radio continuum", Diploma thesis, Bonn University (1984)
11. C. Heiles, H.G. Habing: Astron. Astrophys. Suppl. Ser. 14, 1 (1974)

Vertical Expansion of the Galactic Gas Disk Behind the Spiral Shock Waves

M. Tosa

Astronomical Institute, Tohoku University, Sendai, Japan

1. Increase of the Scale Height due to Compression of the Gas Disc

The thickness of the gas disc changes as the gas disc is disturbed by the density waves. Tosa [1] showed that a gas disc with magnetic field expands vertically (in z-direction) by factor 1.3-1.4 behind the galactic shocks.

In this paper it is suggested that further expansion of the diffuse gas is expected as a result of formation of molecular gas clouds in the spiral arm.

The scale height h of an isothermal gas disc in hydrostatic equilibrium with a horizontal magnetic field of intensity B is given by

$$h=[c^2+B^2/8\pi\rho]^{1/2}/\omega_z=c\ [1+\beta]^{1/2}/\omega_z, \tag{1}$$

where c and ρ are the sound velocity and the density of the gas; ω_z is the frequency of the vertical oscillation in the galactic disc; $\beta=B^2/(8\pi\rho c^2)$. Thus, the change of the scale height due to the shock compression is given by

$$h_2/h_1=[(1+\beta_2)/(1+\beta_1)]^{1/2}, \tag{2}$$

where suffixes 1 and 2 refer to the quantities before and after the shock compression.

As the gas is compressed at the spiral shocks, β increases in proportion to the gas density. For a density contrast of $\rho_2/\rho_1=3$, typical for the galactic shock wave, and $\beta_1=0.5$ prior to the shock, the change of the scale height is $h_2/h_1=1.3$ in accordance with the previous calculation by Tosa [1].

2. Formation of Gas Clouds and Expansion of the Diffuse Gas Disc

The formation of molecular gas clouds from the diffuse gas in the spiral arm causes significant reduction of the density of the diffuse gas, and could have significant effect on the vertical equilibrium of the gas disc.

In view of the fact that the molecular clouds are in highly turbulent state with large random velocities, convection and diffusion of magnetic fields could be greatly enhanced in the molecular clouds. If the turbulent diffusion of the magnetic field lines is effective in a gas cloud, the magnetic fields of the cloud are rapidly expelled from the cloud and the cloud is, sooner or later, disconnected with surrounding magnetic fields in the diffuse gas.

If this is the case, formation of gas clouds from the diffuse gas results in a significant reduction of the density of the diffuse gas without significantly changing the flux of the magnetic field.

If we assume that the magnetic flux density and the pressure integrated over the thickness of the disc are constant, then the scale height h_3 of the diffuse gas disc after the formation of the molecular gas clouds is given as

$$h_3/h_2 = (\sigma_2/\sigma_3)^{1/2}, \tag{3}$$

where σ_2 and σ_3 are the surface density of the diffuse gas prior to and after the cloud formation.

If, for example, 3/4 of the diffuse gas is transformed into molecular clouds behind the shock, the scale height further increases by a factor 2. Combined with the increase of the scale height by the shock compression given by (2), the scale height of the diffuse gas disc significantly increases by factor more than 2 after molecular clouds are formed in the spiral arm.

3. Response of the Diffuse Gas Disc to the Density Wave

In order to demonstrate the variation of the thickness of the disk of diffuse gas in the context of the density wave theory, we calculated a response of the gas disc to the density wave.

As a model of the gas disc, we consider a gas composed of two components, each corresponds to the diffuse gas and the ensemble of the gas clouds, the latter is referred to the cloud gas; each component is treated as an isothermal gas. Between these two components exchange of mass and momentum is taken into account that corresponds to the formation and destruction of gas clouds [6]. We assume that the magnetic field is only contained in the diffuse gas and no interaction is taken place through magnetic field between these two components. To describe the dynamics of the gas disc, the following pair of gasdynamical equations are considered. The suffixes 1 and 2 refer to the diffuse gas and the cloud gas respectively.

$$\partial\rho_1/\partial t + \nabla \cdot (\rho_1 U_1) = -Q_{12} + Q_{21}, \tag{4}$$

$$\partial(\rho_1 U_1)/\partial t + \nabla \cdot (\rho_1 U_1 U_1) = -\nabla(c_1{}^2\rho_1 + B^2/8\pi) + B\cdot\nabla B/4\pi - \rho_1\nabla\phi$$

$$-U_1 Q_{12} + U_2 Q_{21} + D, \tag{5}$$

$$\partial B/\partial t = \nabla \times (U_1 \times B),$$ (6)

$$\partial \rho_2/\partial t + \nabla \cdot (\rho_2 U_2) = Q_{12} - Q_{21},$$ (7)

$$\partial (\rho_2 U_2)/\partial t + \nabla \cdot (\rho_2 U_2 U_2) = -\nabla (c_2^2 \rho_2) - \rho_1 \nabla \phi + U_1 Q_{12} - U_2 Q_{21} - D,$$ (8)

where Q_{12} and Q_{21} represent the mass exchange rate from the diffuse gas to the cloud gas and of the opposite direction. We assume $Q_{12} = \text{Minimum}(K\sigma_1^n, \sigma_1/\tau_1)$ and $Q_{21} = \sigma_2/\tau_2$, where K is constant; this form of Q_{12} implies that when the surface density of the diffuse gas exceeds a certain critical value, the diffuse gas is converted into clouds with a time scale of τ_1.

These equations are averaged in z over the thickness of the disk [4] on the assumptions that the gas disc is in a hydrostatic equilibrium in z-direction and that the magnetic lines of force are parallel to the velocity vector; and transformed into the rotating spiral coordinate system with tightly wound approximation [5]; then integrated numerically until a steady state is realized.

The result is shown in Fig. 1 for the parameters $c_1 = c_2 = 7$ km/sec, $\tau_1 = 10^7$ years and $\tau_2 = 10^8$ years. As the initial state, we take an equilibrium state without density wave perturbation; for above values of τ_1 and τ_2 a mass balance is realized with $\rho_1 = 0.6$ H/cm^3 and $\rho_2 = 0.4$ H/cm^3; $B = 3.5$ μG; with these values $h_1 = 108$ pc and $h_2 = 77$ pc. The dragforce D is negleted here.

Figure 1 shows the spatial distirbutions of (a) the surface density of the gases, (b) the scale height of the discs, (c) the magnetic flux surface density obtained by integrating the magnetic flux density (dotted line) over the thickness of the disc. The solid and the dashed lines refer to the diffuse gas and the cloud gas, respectively. The horizontal lines indicate the values in the equilibrium satete. The horizonta axis is the azimuthal angle θ along the stream line; the middle of the axis coincides with the minimum of the density wave potential. The amplitude of the density wave potential is 5 % of the background potential. For comparison, the distributions of the same quantities without mass exchange are shown in Fig. 2.

The effect of the mass exchange is well illustrated in the contrast between the surface density distributions of the diffuse gas with and without mass exchange: the diffuse gas is significantly reduced by efficient formation of gas cloud behind the shock front. Corresponding to the formation of gas clouds, the cloud gas shows higher and broader distribution.

The variation of the scale height shows a large jump at the shock front: the scale height behind the shock is almost twice the value prior to the shock. Compared with the case without mass exchange, the increase of the scale height is more than a twice if formation of gas clouds is considered.

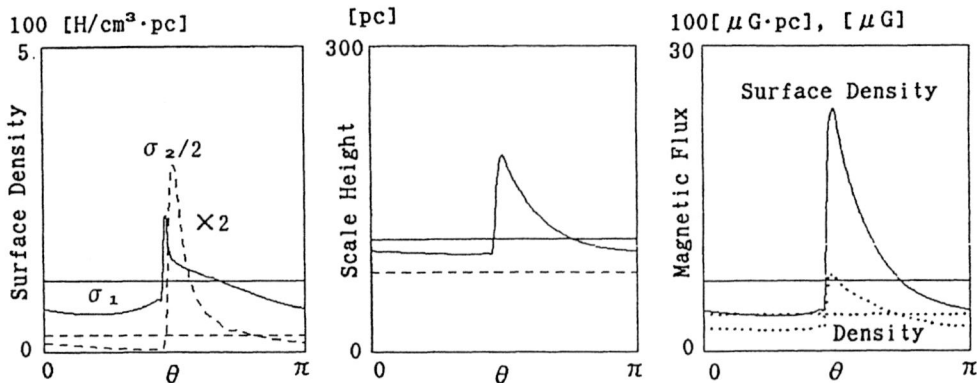

Fig. 1. The response of the gas disc to the density wave with mass exchange

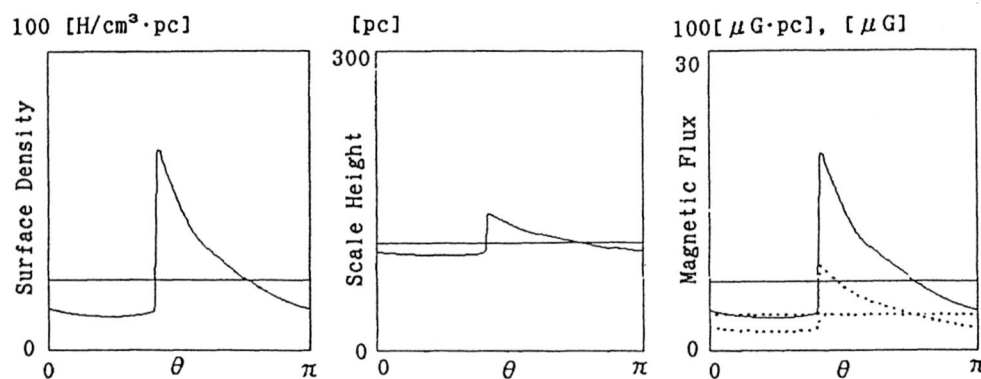

Fig. 2. The same as Fig. 1 but with no mass exchange

4. Observational Implications

A significant expansion of the diffuse gas in z-direction is expected behind the spiral shock waves. The expanded diffuse gases contains compressed magnetic field and relativistic electrons so that these vertical extension of the disc are observed as spurs extending from the galactic plane in directions tangential to the spiral arms by nonthermal emission of radio waves.

A clear correlation has been found between the positions of galactic radio spurs, optically obscured regions, and tangential directions of the spiral arms and it is suggested that the spurs are physically connected with dense interstellar gas along the spiral arms of the Galaxy [2,3].

References

1. M. Tosa: Publ. Astron. Soc. Japan 25, 191 (1973)
2. Y. Sofue: Publ. Astron. Soc. Japan 25, 207 (1973)

3. Y. Sofue, M. Fujimoto, and M. Tosa: Publ Astron. Soc. Japan <u>28</u>, 317 (1976)

4. W.W. Roberts and C. Yuan: Astrophys. J. <u>161</u>, 887 (1970)

5. W.W. Roberts: Astrophys. J. <u>158</u>, 123 (1969)

6. M. Tosa: In <u>Theoretical Aspects on Structure, Activity, and Evolution of Galaxies,II</u>, ed. by S. Aoki and Y. Yoshii (Tokyo Astronomical Observatory, Tokyo), p.58 (1984)

Magnetic Field Effects in Galactic Disks

M.L. Sánchez-Saavedra and E. Battaner

Departamento de Fisica Moderna, Universidad de Granada, Spain

1. Introduction

In a recent paper (BATTANER and SANCHEZ-SAAVEDRA [1]) a steady state theoretical model was presented, in which the MHD equations were solved in order to find the magnetic field responsible of present gaseous rings of some spiral galaxies. Numerical calculations were performed for M31 and the Milky Way. The obtained magnetic field was used to estimate the radial profile of the synchrotron continuum emission, in order to compare computed and observed values. The agreement between our synchrotron profile and that observed by BECK and GRAVE [2] and BECK [3] at 2700 MHz for M31 is noticeable. Therefore a magnetic field distribution possibly explains both the gas distribution and the synchrotron emission.

For the Milky Way our synchrotron profile shows a deep minimum at 6 kpc which is not present in the observed profile of BEUERMANN et al. [4]. Very small changes in some adopted parameters can, however, reconcile both results, as will be shown later.

One of the steady distributions of magnetic field was found to consist in pure azimuthal magnetic field lines. This state corresponds to ring magnetism galaxies (e.g. VALLEE [5]; see the discussion of RUZMAIKIN et al. [6]), examples of which are M31 (SOFUE and TAKANO [7]; BECK [3]) and possibly the Milky Way (VALLEE [5]). It is here suggested that ring magnetism is connected with the presence of gas (HI + 2H$_2$) rings, and that it should be found in other gas-ring galaxies as NGC 2841 and NGC 7331. Gas-ring galaxies have been studied in the near infrared by PRIETO et al. [8] and BATTANER et al. [9].

2. The Free Parameter k

As boundary condition the central pressure was adopted. For a central pressure less than a critical pressure imaginary magnetic fields are obtained at some radii, i.e. magnetic confinement would not be possible. We therefore adopted central pressures higher than or equal to the critical pressure. If we define the parameter k as the relation between adopted and critical pressures, k must be higher than or equal to unity. k was treated as a free parameter. The difference between observed and adopted central pressures would be caused by accretion of gas into a central black hole or ejections from the nucleus. SANCHEZ-SAAVEDRA and BATTANER [10] esti-

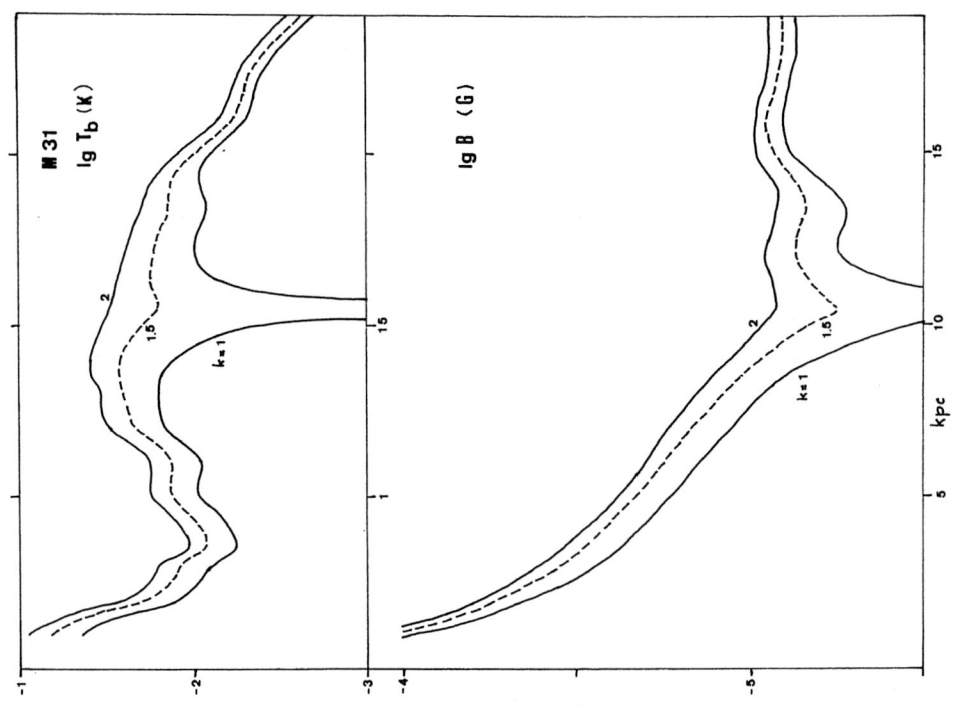

Fig. 2: Prediction for M31 of the radial profiles of the radio synchrotron brightness temperature T_b at 2.7 GHz and azimuthal magnetic field strength B for different values of the parameter *k* (see text)

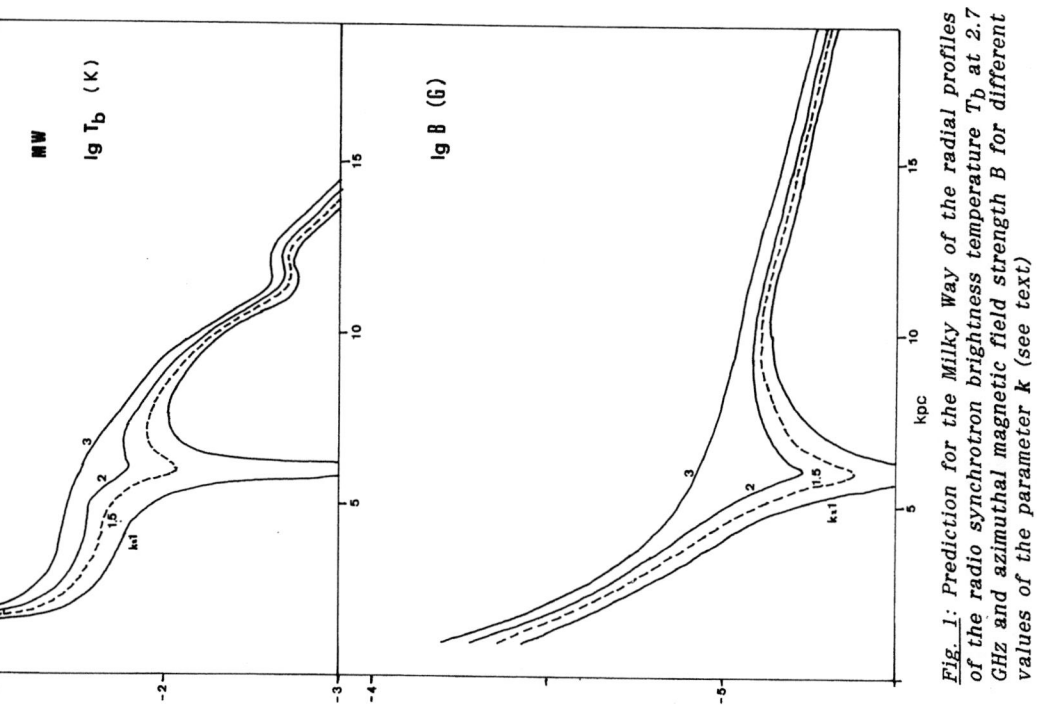

Fig. 1: Prediction for the Milky Way of the radial profiles of the radio synchrotron brightness temperature T_b at 2.7 GHz and azimuthal magnetic field strength B for different values of the parameter *k* (see text)

mated the rate at which the galactic nuclei are fueled by new material, obtaining for the Milky Way a value of $M = 1.2$ $M_\odot$ y^{-1}.

The dependence of our theoretical results upon the free parameter k is shown in the Figures.

For the Milky Way the profiles for $k = 1$ are plotted in BATTANER and SANCHEZ-SAAVEDRA [1]. It seems, however, that the synchrotron profile for $k = 2$ or higher agrees better with the results of BEUERMANN et al. [4]. In these plots B could be interpreted as the total field (uniform + random) as the field able to confine the present gas is actually evaluated in the model. For $k = 2$ at the solar radius a value of B of $8.3\ 10^{-6}$ G would correspond to a uniform field of about 4 μG close to recent estimations of 5 μG (e.g. BECK [11]). For $k = 3$ (not shown in the Figure) a uniform azimuthal field of 5 μG at the solar radius and a synchrotron profile without a minimum at 6 kpc are obtained, in agreement with observations.

For M31 similar values for the uniform magnetic field are obtained of the order of 4 μG. The profile for $k = 2$ agrees better with the results of BECK and GRÄVE [2]. The possibility $k = 1$, however, cannot be rejected, yielding a null magnetic field at 10.5 kpc. Synchrotron emission would be expected at this distance because a low magnetic field and a high gas density would enhance the relativistic electron diffusion.

Bibliography

1. E. Battaner, M.L. Sánchez-Saavedra: Astrophys. J. 304, 450 (1986)
2. R. Beck, R. Gräve: Astron. Astrophys. 105, 192 (1982)
3. R. Beck: Astron. Astrophys. 106, 121 (1982)
4. K. Beuermann, G. Kanbach, E.M. Berkhuijsen: Astron. Astrophys. 153, 17 (1985)
5. J.P. Vallée: Astron. Astrophys. 136, 373 (1984)
6. A.A. Ruzmaikin, D.D. Sokoloff, A.M. Shukurov: Astron. Astrophys. 148, 335 (1985)
7. Y. Sofue, T. Takano: Publ. Astron. Soc. Japan 33, 47 (1981)
8. M. Prieto, E. Battaner, C. Sánchez-Magro, J.E. Beckman: Astron. Astrophys. 146, 297 (1985)
9. E. Battaner, J.E. Beckman, C. Muñoz, M. Prieto, C. Sánchez-Magro, E. Mediavilla, M.L. Sánchez-Saavedra: Astron. Astrophys., in press (1986)
10. M.L. Sánchez-Saavedra, E. Battaner: In Accretion onto Compact Objects (ESA Workshop, Tenerife 1986)
11. R. Beck: IEEE Transactions on Plasma Science, Special Issue on Space and Cosmic Plasma, in press (December 1986)

Magnetic Field Structure of the Galaxy and the Formation of Galactic Winds

D. Breitschwerdt[1], J.F. McKenzie[2], and H.J. Völk[1]

[1]Max-Planck-Institut für Kernphysik, Postfach 103980,
  6900 Heidelberg, Fed. Rep. of Germany
[2]University of Natal, Department of Mathematics and Applied Mathematics,
  King George V Avenue, Durban 4001, South Africa

1. Abstract

Using the basic observational characteristics of the interstellar galactic cosmic ray component, a plausible large—scale magnetic field structure above the gaseous disk of the Galaxy is described. To a considerable extent the field lines should be perpendicular to the galactic disk at heights exceeding several kiloparsec and open towards intergalactic space. This allows cosmic ray loss from the Galaxy, which is largely convective for the lower energy part of the particle spectrum dominating the cosmic ray pressure. The negative outwards cosmic ray pressure gradient gives rise to self—excited MHD—waves due to the streaming instability. As a consequence cosmic ray and wave pressures can drive a galactic mass loss against the gravitational pull of the disk and a massive and extended halo. A typical solution for an adiabatic wind is presented. Obvious areas for observational tests should be the synchrotron structure of galactic halos, as well as spectroscopy of highly ionized species in the UV and soft X—ray regions.

2. Introduction

The question whether the Galaxy accretes or loses mass or whether its matter is practically isolated from the surroundings has not yet been clarified by observations. There are of course mass motions in the halo regions above (and below) the gaseous disk [1]. More recent optical and UV absorption line studies against bright halo stars and QSO's indicate a statistical predominance of negative (= infall) velocities $\leq$ 100km/s at heights below about 5 to 10 kpc above the galactic plane (e.g. [2,3,4]). Radio observations of neutral hydrogen emission in the 21 cm line agree with this trend (e.g. [5]). Thus at least for the high infall velocities we might interpret the observations as infall of extragalactic matter as advocated by OORT [6] and many followers. On the other hand, grossly speaking the observations are consistent with a galactic fountain model [7], where rising hot and compressed material cools radiatively and then falls back onto the disk. Due to its enhanced density it has a high probability to be observed against the faint background gas.

We can now rephrase our question in the following form: is the halo <u>on average</u> static, i.e. with galactic fountains only? Then we need to know also whether the halo magnetic field is corotating or not. Or do we have a halo that is <u>on average</u> convective, with a wind or infalling matter, or both occurring at the same time at different places?

In all cases we must imply that the Galaxy is losing the cosmic ray component of the interstellar medium. The cosmic rays are constantly reproduced in and near the gaseous disk. Their momentum and energy flux are to some degree coupled to that of the thermal gas and will affect the overall dynamics.

In our view the observed preponderance of infall velocities can be explained in a picture where supernova remnants (SNR's) (and to some degree also winds from massive stars) interact with tenuous material at a few kpc height. The argument is roughly as follows: the SNR shock reaches a parcel of gas located at a certain height, heats and compresses it by a factor < 4 and imparts to it a velocity direc- ted obliquely upwards. During this phase the gas is hot and tenuous and difficult to observe in spectral lines corresponding to gas temperatures $\leqslant 10^5$ K, as mostly used. After some time the shocked gas may cool and partly neutralize, thereby losing its own thermal pressure support. Also the hydromagnetic waves, coupling the gas to the cosmic ray high energy component, are damped by ion neutral friction [8] thus weakening cosmic ray pressure support. Therefore the gas welling up is decelerated and begins to fall down towards the disk. Since the gas element is thermally un- stable it can sink down quite far. Thus we expect this later phase both to take a longer time than the upward phase and to be more easily observable.

Clearly, _inside_ this fountain region the magnetic field lines are perturbed by gas motions both from above and below. Thus to lowest order the resulting hydromagnet- ic wave field is expected to be isotropically propagating with equal wave strength in either direction along the field.

Since the z-distribution of SNR's has a scale height exceeding that of the cool gas disk and since supernovae often occur in succession in star clusters there will also be many SNR's that reach heights above the average fountain region and still have considerable strength there. They may even overlap at great heights. Since SNR's are presumably also sources of cosmic rays we must therefore expect that a part of the SNR shells does not cool but rather keeps rising above the disk, effec- tively opening the magnetic field lines locally towards the intergalactic medium. This effect may be enhanced by the independent action of the Parker instability [9] of an assumed stratified equilibrium state of the gaseous disk even though the ultimate opening of the field lines by the cosmic ray pressure in this instability has been questioned [10]. As a result the field topology should be as follows (Fig. 1): Inside the gaseous disk ($|z| \lesssim 100$ pc) and in fact at heights up to several kpc the field is on average parallel to the disk (generated presumably by disk dynamo action). Whether the field lines are actually closed in a ring type topology or open towards the disk edges is of minor importance here. The wave field is isotropic. At heights above several kpc an increasing number of flux tubes should be open in z-direction interspersed with closed loops. Since the gas motions from the fountain are rather below this region the resulting wave field will propagate outwards toward increasing $|z|$. The same is true for the waves excited by the outward streaming cosmic rays. Clearly, in this picture no uniform accretion of cooling, hot extragalactic material is

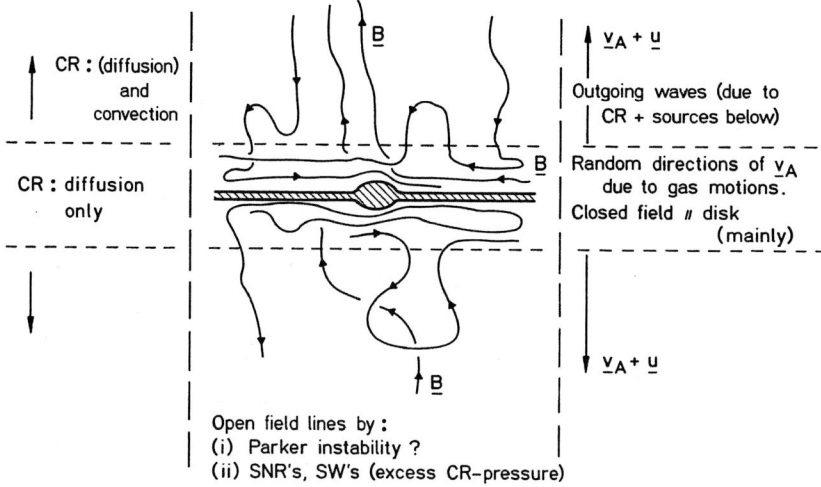

CR : (diffusion)
and
convection

CR : diffusion
only

B

Open field lines by :
(i) Parker instability ?
(ii) SNR's, SW's (excess CR-pressure)

$\underline{v}_A + \underline{u}$

Outgoing waves (due to
CR + sources below)

Random directions of $\underline{v}_A$
due to gas motions.

Closed field // disk
(mainly)

$\underline{v}_A + \underline{u}$

Fig. 1: Topology of the magnetic field in the lower and upper galactic halo and preferential directions of the MHD−wave field

possible but localized infalling clouds from infinity are not excluded. They lead to field distortions in between the open structures.

The reader will readily recognize a number of analogies with the physical situation in the solar corona and this seems quite reasonable indeed. Yet a number of aspects are rather different for galaxies, and in particular our Galaxy: The solar corona has basically a gravitational field ~ r^{-2} and only occasional shock heating and relativistic particle production. Also radiative cooling is not very important. In contrast, at least our Galaxy is presumed to have a very extended gravitational field due to unseen masses distributed rather uniformly out to tens of kpc. This is sometimes called a massive halo. Thus a parcel of fluid needs a long time before it effectively leaves the galactic gravitational field and it may start to cool before doing that. In addition the cosmic ray pressure in the cool gas disk appears to be rather uniform from the γ−ray observations. Thus cosmic ray pressure must be continuously present supporting any outflow motion. Finally the Galaxy is rotating fairly rapidly with centrifugal forces basically balancing gravity. Thus a parcel of matter rising in |z|−direction tends to conserve angular momentum and moves away radially. If the magnetic field has a strong vertical component, a magnetically enhanced wind is possible.

We shall disregard such "magnetic sling-shot" effects although they may be quantitatively quite sizeable. We do this for the time being because these magnetic effects are much more difficult to calculate. In any case the primary agent pushing up material is presumably thermal and cosmic ray pressure together with wave pressure.

We are considering a simple field geometry where magnetic flux tubes have a cross section A(z) increasing with z as the geometry of the disk becomes more and

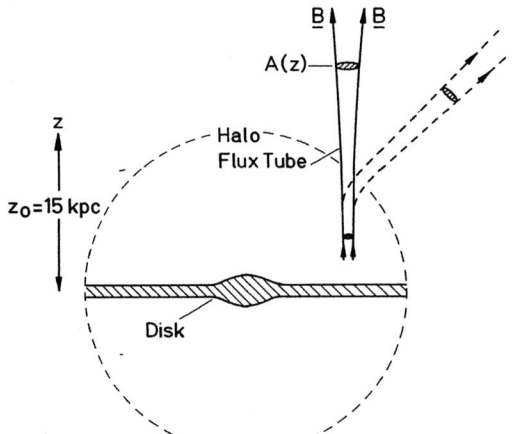

Fig. 2: Geometry of flux tubes with cross section A(z), defined by the magnetic field lines; spherical divergence does not start before z > z_0 = 15 kpc. Field lines may be distorted due to effects of the galactic rotation (dashed lines)

more unimportant. For $z_1 < |z| < z_0$, A(z) ~ constant, where z_1 = few kpc is the base of our halo and $z_0 \simeq$ 15kpc is basically the radius of the cool gas disk, whereas A(z) ~ $1/z^2$ for $|z| \gg z_0$, corresponding to an asymptotically radial field (Fig. 2). The field acts only kinematically.

After a short discussion of galactic cosmic ray properties, we shall describe the overall gas dynamics of the halo and give a typical solution for an adiabatic wind.

3. Cosmic Ray Propagation Characteristics

The galactic cosmic ray component propagates diffusively along the mean magnetic field through the interstellar gas (see e.g. [11]). This is plausible from several lines of evidence. We shall only mention a few key observations here. First of all the observed proportionality between the intensity of the diffuse interstellar high energy γ-radiation and the thermal gas density suggests a rather uniform cosmic ray intensity on a large scale [12]. Secondly, the individual lifetime of (relativistic) GeV-particles reaching the solar system is of the order of 15 million years (e.g. [13] and references therein), more than 1000 times the light crossing time for the galactic disk. Thus cosmic rays must be well scattered directionally. Yet their corresponding scattering mean free path is large compared to their gyroradius in the average magnetic field. This constrains them to move essentially back and forth within magnetic flux tubes.

Furthermore, from the amount of secondary spallation products one can infer that, starting from its source in the disk, a typical primary GeV-particle has on average traversed about 8 g/cm² of interstellar matter during its lifetime. Thus the mean gas density "seen" by e.g. a GeV-proton is given by $\langle \rho \rangle_c \simeq$ 0.3 H-atoms·cm⁻³, corresponding to about 1/3 of the mean mass density $\langle \rho \rangle_{disk} \simeq$ 0.8 H-atoms·cm⁻³ of the gas in the galactic disk [14]. Hence, such a particle ought to have spent considerable time above the gas disk. Finally, the intensity ratio of secondary to primary cosmic ray nuclei decreases with increasing particle energy ~ $E^{-\varepsilon}$, with $\varepsilon \simeq$ 0.6 (e.g.[15]). This implies a cosmic ray confinement volume for GeV-particles (presumably in the form of a thick disk) that is at least three times the volume of the gas disk. In the simplest approximation the inferred mean residence time of the particles decreases with increasing energy beyond some GeV proportional to $E^{-\varepsilon}$. As a consequence their loss from the Galaxy cannot be entirely convective, i.e. energy-independent, but must be partly diffusive, with a diffusion coefficient $\sim E^{\varepsilon}$.

Therefore the removal of the entire cosmic ray population from the disk cannot be due to a convective flow alone, even if it may be driven by cosmic rays in the form of a galactic wind. However the observed galactic particle spectrum, being proportional to $E^{-2.75}$ above about 1 GeV, is steep enough that its energy density and pressure are dominated by particles below about 10 GeV in the sense of a limiting case. This means that at galactic heights $|z| \gg 300$ pc (about 3 times the gas disk thickness) their mode of transport could be entirely convective, whereas for $|z| \ll 300$ pc their transport could be entirely diffusive (cf. Fig. 1). In Section 5 we shall actually make such an approximation, since it simplifies the algebra considerably. Doing this, the cosmic ray pressure gradient which we have available is approximately the one calculated from the observed spectrum. This is less than the theoretically possible pressure gradient which sources in the disk could provide if all particles were scattered like GeV-particles. By the above line of arguments this total pressure gradient would result from a source spectrum $\sim E^{-2.15}$ and would be larger than the observed cosmic ray pressure by a factor of about 4 [16], since now particle energies far above 10 GeV also contribute to the pressure. Since the highest energy particles are the most diffusive, we can disregard their coupling to the gas in a first estimate and assume them to diffuse out so rapidly that their momentum transfer to the gas is negligible. Of course such an approximation must be justified a posteriori. This will be done elsewhere.

The dominant interaction of cosmic rays with the fluctuating wave fields will be resonant scattering due to an anisotropy in pitch-angle distribution in the wave frame. This leads, as a consequence of momentum transfer, to a growth of the waves, which have to travel in the same direction as the particles do on average.

Since the mass loss rate from the Galaxy cannot be excessive the mean upward mass velocity must be negligible compared to the cosmic ray diffusion velocity in the high density gas disk. We assume that the cosmic rays diffuse in the frame of the gas because of a strong isotropically propagating wave field. In contrast, in the low-density upper halo the only source of the waves is due to cosmic rays which diffuse in the wave frame, convected with the gas.

The waves transport energy and momentum. Especially the gradient of the wave pressure tends to accelerate the gas flow away from the disk.

4. Hydrodynamics for Gas Interacting with the Cosmic Rays

The interaction between the thermal plasma (the gas) and the cosmic ray component should in principle be calculated at the kinetic level in order to obtain also the energy spectrum of the particles. A first (hydrodynamic) approximation to this highly nonlinear problem consists in a simplified description of the cosmic ray distribution in terms of moments like pressure and internal energy density (the cosmic ray momentum and kinetic energy density is negligible). In a next iterative step the given dynamics of the overall system can then be used to solve for the kinematics of the energetic particle distribution. IPAVICH [17] has solved the coupled hydrodynamic equations for gas and cosmic rays to calculate galactic winds in spherical symmetry. In contrast OWENS and JOKIPII [18], JONES [19], LERCHE and SCHLICKEISER [20], and others have considered the purely kinematic evolution of the particle distribution in assumed dynamic halo structures.

We shall concentrate here on the basic problem of galactic halo dynamics using a hydrodynamic approximation for the thermal gas, the cosmic rays and the wave field. These equations are derived from the microscopic transport theory (for reviews see [21] and [14]). In an assumed steady state they describe overall balance of mass, momentum, energy and conservation of magnetic flux:

$$\text{div}\{\rho\mathbf{u}\} = 0 \tag{1}$$

$$\text{div}\left\{\rho\mathbf{u}:\mathbf{u} + \left(p_g + p_c + \frac{\langle(\delta\mathbf{B})^2\rangle}{8\pi}\right)\cdot\mathbf{I}\right\} = \rho\frac{\partial V_{grav}}{\partial\mathbf{x}} \tag{2}$$

$$\text{div}\left\{\rho\mathbf{u}\left[\frac{1}{2}u^2 + \frac{\gamma_g}{\gamma_g-1}\frac{p_g}{\rho}\right] + \frac{1}{\gamma_c-1}\left[\gamma_c p_c(\mathbf{u}+\mathbf{v_A}) - \bar{\kappa}\frac{\partial p_c}{\partial\mathbf{x}}\right] + \right.$$
$$\left. + \frac{\langle(\delta\mathbf{B})^2\rangle}{4\pi}\left[\frac{3}{2}u+\mathbf{v_A}\right]\right\} = \Gamma - \Lambda + \rho\mathbf{u}\frac{\partial V_{grav}}{\partial\mathbf{x}} \tag{3}$$

together with an energy balance equation for the cosmic ray component

$$\text{div}\left\{\frac{\gamma_c}{\gamma_c-1}(\mathbf{u}+\mathbf{v_A})p_c - \frac{\bar{\kappa}}{\gamma_c-1}\frac{\partial p_c}{\partial\mathbf{x}}\right\} = (\mathbf{u}+\mathbf{v_A})\frac{\partial p_c}{\partial\mathbf{x}} + Q, \tag{4}$$

and the energy balance for the total wave field, assumed to propagate in a single direction $\underline{v_A}/|\underline{v_A}| = '-(\partial p_c/\partial\underline{x})/|\partial p_c/\partial\underline{x}|$:

$$\text{div}\left\{\frac{\langle(\delta\mathbf{B})^2\rangle}{4\pi}\left[\frac{3}{2}u+\mathbf{v_A}\right]\right\} = u\frac{\partial}{\partial\mathbf{x}}\left(\frac{\langle(\delta\mathbf{B})^2\rangle}{8\pi}\right) - v_A\frac{\partial p_c}{\partial\mathbf{x}} - L \tag{5}$$

and

$$\text{div}\,\mathbf{B} = 0. \tag{6}$$

In eq. (2) p_g, p_c, and $\langle(\delta\underline{B})^2\rangle$ denote the gas pressure, the cosmic ray pressure, and the mean square fluctuating magnetic field amplitude, which can be used to define a wave pressure $p_w = \langle\delta(\underline{B})^2\rangle/8\pi$. The unit tensor is given by I, and $\underline{u}:\underline{u}$ is the dyadic product. In eqs. (2) and (4) γ_g-1 and γ_c-1 denote the ratio of gas pressure, and cosmic ray pressure to their respective internal energy densities. The mean diffusion

136

coefficient $\bar{\kappa}$ determines the diffusive cosmic ray energy flow density $\bar{\kappa}(\partial p_c/\partial \underline{x})/$ (γ_c-1). Γ and Λ are the heating and cooling rate of the gas other than those implied by the explicit contribution from the wave field $\delta\underline{B}$ and the collective interaction with the cosmic rays. The terms Q and L in eqs. (4) and (5) describe sources or sinks for cosmic ray internal energy and fluctuation field energy, respectively, apart from the recoil term $\underline{v}_A \, \partial p_c/\partial \underline{x}$ in eq. (4) and the growth term $-\underline{v}_A \cdot \partial p_c/\partial \underline{x}$ in eq. (5) by the cosmic rays themselves when they stream down their own pressure gradient. Effectively γ_c and $\bar{\kappa}$ are parameters of the theory even though they can be evaluated in an approximate form. With appropriate boundary conditions these equations determine the five dynamical variables ρ, $\underline{u}$, p_g, p_c, and $\langle(\delta\underline{B})^2\rangle$.

5. Galactic Wind Solutions

5.1. Description of the Model

Our first approach to solve the set of eqs. presented in Section 4, is to make the following simplifying assumptions:

(i)   no heating and radiative cooling ($\Gamma=\Lambda=0$);

(ii)   high energy particles that dominate the cosmic ray pressure p_c have an energy below about 10 GeV (see Sec. 3) and are strongly scattered by the waves in the upper halo ($\bar{\kappa}\to 0$);

(iii) no sources or sinks of the cosmic ray internal energy and fluctuation field energy (Q=L=0).

Given the above assumptions it is convenient to cast the system of equations (1)–(6) into a form that demonstrates the connection of the driving forces (i.e. the pressure gradients of the thermal gas, cosmic rays and waves) with the spatial increase of the flow velocity (i.e. the so-called "wind equation"). Furthermore we write the eqs. in flux tube geometry (s. Sec. 2), where the area cross section A(z) = $A_0(1 + z^2/z_0^2)$ and div $\equiv$ 1/A (dA/dz).

The thermal pressure gradient is simply given by

$$\frac{dp_g}{dz} = \gamma_g \frac{p_g}{\rho} \frac{d\rho}{dz},\tag{7}$$

the cosmic ray pressure gradient by

$$\frac{dp_c}{dz} = \gamma_c \frac{p_c}{\rho}\left(\frac{M_A + \frac{1}{2}}{M_A + 1}\right)\frac{d\rho}{dz},\tag{8}$$

where $M_A = u/v_A$ is the Alfvénic Mach number and the wave pressure gradient by

$$\frac{dp_w}{dz} = \frac{1}{2(M_A + 1)}\left[(3M_A + 1)\frac{p_w}{\rho}\frac{d\rho}{dz} - \frac{dp_c}{dz}\right]\tag{9}$$

where $d\rho/dz$ can be obtained from continuity as a function of dA/dz and du/dz. These equations have to be solved simultaneously with the "wind equation", which reads

$$\frac{1}{u}\left(u^2 - c^2\right)\frac{du}{dz} = c^2\frac{1}{A}\frac{dA}{dz} - g_{eff}(z) \tag{10}$$

where the gravitational acceleration is $g_{eff}(z) = -\dfrac{\partial V_{grav}}{\partial z}$.

Now

$$\begin{aligned} c^2 &= \gamma_g\frac{p_g}{\rho} + \gamma_c\frac{p_c}{\rho}\frac{(M_A + \frac{1}{2})^2}{(M_A + 1)^2} + \frac{p_w}{\rho}\frac{3M_A + 1}{2(M_A + 1)} \\ &= c_g^2 + c_{CR}^2 + c_w^2. \end{aligned} \tag{11}$$

We call c the "compound sound speed".

The structure of the "wind equation" is strikingly similar to that describing the flow produced in a convergent–divergent device like a Laval nozzle and we are in fact looking for solutions that pass smoothly from the subsonic to the supersonic regime.

It is instructive to have a look at the different behaviour of the components of the "compound sound speed" as a function of $\rho(z)$. Taking γ_g = 5/3 and γ_c = 4/3 (ultra–relativistic particles) we get the following dependencies:

$$c^2_g \sim \rho^{2/3}, \quad c^2_{CR} \sim \rho^{1/3} \quad \text{and} \quad c^2_w \sim \rho^{1/2}.$$

This nicely demonstrates that the energetic particle pressure and the wave pressure contribute substantially to the total pressure and moreover keep it up over considerably larger scale heights. Hence, driving a galactic outflow does not require unrealistic high gas temperatures in the halo like in thermal pressure driven models [22]. This difference becomes even more obvious when cooling is taken into account which primarily affects the thermal gas but not the particles and the waves (we tacitly assume the halo to be fully ionized).

The next step towards a more realistic model of the galactic halo consists in abandoning the purely spherical geometry and in particular the treatment of the Galaxy as a point source with respect to the gravitational potential as was assumed in previous models [17].

We used a model for the gravitational potential $V_{grav}(r,z)$ given in the paper by HABE and IKEUCHI [22] which consists of a bulge and disk component resulting from an assumed axisymmetric mass distribution. In order to produce conservative results a massive spherical halo component was also included, extending over the large scale of 100 kpc. The major effect is a g_{eff} that falls off much more slowly than $1/z^2$ (point source model).

5.2. Outflow Solutions

We look for solutions of eqs. (7)–(11) as a boundary value problem with parameters given at the base of the upper halo (which we call "reference level"). These solutions should undergo a smooth subsonic–supersonic transition at a critical point, where $u = c$. A glance at eq. (11) shows that smooth solutions with $du/dz > 0$ through this singular point require that the right hand side is zero, when $u = c$. It is important to emphasize that the outflow velocity at the reference level u_0, and therefore also the mass loss rate, is uniquely determined by the above requirements.

5.3. Results — a Sample Solution

The flux tube is located at the Sun's position ($r_0 = 10$ kpc) where $V_{grav}(r_0, z)$ is evaluated. The reference level has been fixed at $z = 1$ kpc, where we prescribe the mass density of the gas $\rho_0 = 1.67 \cdot 10^{-27}$ g/cm³, the gas pressure $p_{go} = 2.8 \cdot 10^{-13}$ erg/cm³ the cosmic ray pressure $p_{Co} = 10^{-13}$ erg/cm³, the magnetic field strength $B_0 = 10^{-6}$ G and the wave pressure $p_{Wo} = 10^{-2}$ $B^2_0/8\pi = 4 \cdot 10^{-16}$ erg/cm³.

Again in order to be conservative, we have chosen the initial wave pressure to be very low. This enables us to study the effect of wave growth on the driving forces of a galactic wind. Figure 3 shows the build-up of the wave pressure (at the expense of energetic particle momentum flux), reaching a maximum at $z \approx 8$ kpc. This is in contrast to the continuous decrease of both cosmic ray and thermal pressure. The latter one decreases very rapidly as can be inferred from Fig. 4 with $p_g \sim \rho^\gamma_g$.

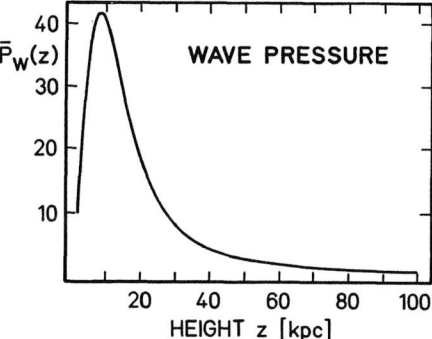

Fig. 3: Wave pressure $\bar{p}_W(z)$ in the halo, normalized to its value at the reference level (z=1kpc) $p_{Wo}=4 \cdot 10^{-16}$ [erg/cm³] as a function of the height z (perpendicular to the disk)

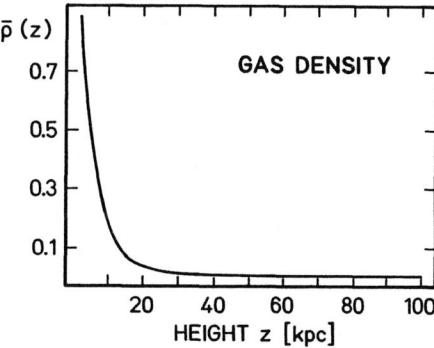

Fig. 4: Gas density $\bar{\rho}(z)$ in the halo, normalized to its value at the reference level (z=1kpc) $\rho_0 = 1.67 \cdot 10^{-27}$ [g/cm³] as a function of the height z

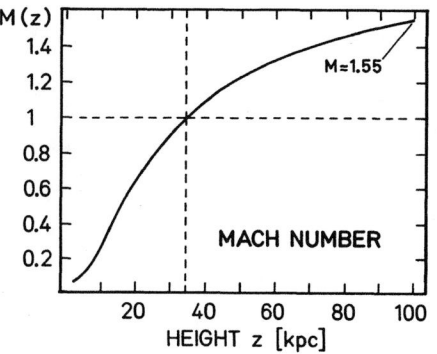

Fig. 5: Typical outflow solution passing through the critical point at z=34kpc, where the Mach number M(z)=1; the asymptotic value at height z=100kpc is M(z)=1.55

Figure 5 shows a galactic wind outflow solution in terms of the Mach number $M(z)$, referring to the "compound sound speed" c. Note that the critical point is located at $z = 34$ kpc. The asymptotic value of the outflow velocity (at $z \gtrsim 100$ kpc) turned out to be roughly 240 km/s. From the initial value $u_0 = 9.89$ km/sec which is determined by the outflow solution conditions, the mass loss rate can be deduced to be $\dot{M}_{gal} = 0.33$ $M_{\odot}$/yr.

6. Discussion

The model presented in the previous sections can be improved in a number of ways, even after a more detailed study of the dependence of the present approximations on the various observational parameters (like the supernova rate, the field strength and density in the halo and the wave energy flux from below) is done. Important modifications are the inclusion of (i) radiative cooling and partial recombination of the thermal plasma, (ii) a finite $\bar{\kappa}$, i.e. a diffusive cosmic ray energy flux, (iii) wave dissipation due to nonlinear Landau damping [23], ion neutral friction [8], and — for $v_A > c_g$ — Alfvén wave decay [24].

However, we believe that the basic characteristics of the dynamics of galactic halos are contained in the present model. Therefore it seems that a number of very general predictions could be made, in particular regarding synchrotron observations of disk type galaxies.

First of all we expect the synchrotron halo to be roughly spherical with a radius of about 100 kpc or more. To calculate the synchrotron spectrum one needs to take into account synchrotron and inverse Compton losses which we have not done here, assuming the cosmic ray nucleon component to dominate the energetic particle pressure everywhere.

Secondly the polarization of the synchrotron emission should be quite small since the strongly turbulent wave fields δB are large, $\delta B/B = O(1)$, making the total field quite irregular.

Regarding irregular (dwarf) galaxies, observed spectral indices α [25] are on average much smaller ($\alpha \lesssim 0.5$) than for disk type galaxies ($\alpha \simeq 0.75$) [26]. Small α-values ($\alpha \simeq 0.5$) are characteristic of SNR's and thus presumably for the spectrum at the cosmic ray sources. Large α-values correspond more to the spectrum observed in our Galaxy. Within a model of the type discussed above, such a distinction in spectral indices could be attributed to a purely convective cosmic ray transport for small $\alpha \simeq 0.5$, and a diffusion convection-type transport in the case of $\alpha \simeq 0.7$–0.8 where the source spectrum $\alpha \simeq 0.5$ has been steepened by energy dependent diffusion in the disk and lower halo. Thus we may speculate that in irregular dwarf galaxies we see directly the sources and that a sizeable diffusive disk around them is absent. In this case SNR's (and possibly other sources) may blow out their envelopes directly to intergalactic space with diffusion being a minor effect.

Acknowledgements

We thank L. Drury and R. Schlickeiser for discussions on some aspects of the present work, and A. Baron and G. Breuer for the careful preparation of the manuscript.

References

1. G. Münch, H. Zirin: Astrophys. J. $\underline{133}$, 11 (1961)
2. F.P. Keenan, P.L. Dufton, C.D. McKeith, J.C. Blades: Monthly Notices Roy. Astron. Soc. $\underline{203}$, 963 (1983)
3. K.S. de Boer, B.D. Savage: Astrophys. J. $\underline{265}$, 210 (1983)
4. K.S. de Boer, B.D. Savage: Astron. Astrophys. $\underline{136}$, L7 (1984)
5. A. Blaauw, C.R. Tolbert: Bull. Astron. Inst. Netherlands $\underline{18}$, 405 (1966)
6. J.H. Oort: Bull. Astron. Inst. Netherlands $\underline{18}$, 421 (1966)
7. P.R. Shapiro, G.B. Field: Astrophys. J. $\underline{205}$, 762 (1976)
8. R.M. Kulsrud, W.P. Pearce: Astrophys. J. $\underline{156}$, 445 (1969)
9. E.N. Parker: Astrophys. J. $\underline{145}$, 811 (1966)
10. T.C. Mouschovias: Astron. Astrophys. $\underline{40}$, 191 (1975)
11. C.J. Cesarsky: Rev. Astron. Astrophys. $\underline{18}$, 289 (1980)
12. F. Lebrun, K. Bennett, G.F. Bignami, J.B.G.M. Bloemen, R. Buccheri, P.A. Caraveo, M. Gottwald, W. Hermsen, G. Kanbach, H.A. Mayer-Hasselwander, T. Montmerle, J.A. Paul, B. Sacco, A.W. Strong, R.D. Wills, T.M. Dame, R.S. Cohen, P. Thaddeus: Astrophys. J. $\underline{274}$, 231 (1983)
13. J.A. Simpson: Ann. Rev. Nucl. Part. Sci. $\underline{33}$, 323 (1983)
14. H.J. Völk: In High Energy Astrophysics, Proc. 19th Rencontre de Moriond, ed. by J. Tran Thanh Van (Editions Frontières, Gif-sur-Yvette, France 1984), p. 281
15. L. Koch-Miramond: Proc. 17th Int. Cosmic Ray Conf. $\underline{12}$, 21 (1981)
16. H.J. Völk, L.O'C. Drury, E. Dorfi: Proc. 19th Int. Cosmic Ray Conf. $\underline{3}$, 148 (1985)
17. F. Ipavich: Astrophys. J. $\underline{196}$, 107 (1975)
18. A.J. Owens, J.R. Jokipii: Astrophys. J. $\underline{215}$, 677 (1977)
19. F.C. Jones: Astrophys. J. $\underline{229}$, 747 (1979)
20. I. Lerche, R. Schlickeiser: Astrophys. Letters $\underline{22}$, 31 (1981)
21. L.O'C. Drury: Rep. Progr. Phys. $\underline{46}$, 973 (1983)
22. A. Habe, S. Ikeuchi: Progress in Theor. Phys. $\underline{64}$, 1995 (1980)
23. M.A. Lee, H.J. Völk: Astrophys. Space Sci. $\underline{24}$, 31 (1973)
24. R.Z. Sagdeev, A.A. Galeev: Nonlinear Plasma Theory (Benjamin Press, New York 1969)
25. U. Klein, R. Gräve: these proceedings
26. I.M. Gioia, L. Gregorini, U. Klein: Astron. Astrophys. $\underline{116}$, 164 (1982)

Galaxy Magnetic Fields and Hidden Matter in Galaxies

A.H. Nelson

Dept. of Applied Mathematics and Astronomy, University College, Cardiff, UK

Since there have been several suggestions at this meeting that galaxy magnetic fields might lie at the origin of spiral structure in galaxies, and since there have been a few doubting voices raised concerning the density wave theory of spiral structure, I feel it is appropriate to start by reminding you that there is in fact a fairly well founded theory for the origin of spiral structure. This is based on the generation of a large scale spiral shock in the gas disc, an idea first put forward by Professor Fujimoto, among others, nearly twenty years ago. Since then several groups have performed computations which show that it is rather easy to produce such large scale spiral shocks in the gas by almost any non-axisymmetry in the galactic potential [1,2,3,4]. This could be caused for instance by a stellar density wave, or by a stellar bar, or by a companion galaxy, or simply by a relative tilt or warp between the stellar and gas discs. The idea is also supported by an increasing body of observational evidence, for instance density features and velocity kinks in HI and CO maps of nearby galaxies, and the relative positions of HII, dust and HI in the arms. What these strong shocks will do to the magnetic field is to bend it into a spiral shape, since the component parallel to the shock is strongly enhanced across the shock while the perpendicular component is unchanged. The exact structure and strength of the shock will be modified by the presence of the magnetic field, since it will be an MHD shock, but the field does not play a primary role in the formation of the spiral arms.

However the aim of this paper is to point out an area where the galactic magnetic field may play a primary role in large scale galaxy dynamics, viz. the flat rotation curves seen in HI in the outer parts of many galaxies [5]. For these galaxies the HI rotation curves remain flat well outside the radius of the optical stellar disc, and if we interpret the rotation as being due to circular orbits in the galactic potential then the flatness of the curve must imply large amounts of hidden matter, since the orbital velocity should be roughly Keplerian outside the optical disc if most of the matter is in the luminous stars. My purpose here is to suggest that such an interpretation may be wrong. The crucial point here is that where the rotation curves are anomolously high, the density of the HI gas drops to very low values and this significantly alters the energy balance between magnetic energy and rotational energy. HI observations show that the surface HI density drops suddenly by a factor fo 10-50 from the edge of the optical disc to about twice this radius [6], and taking into account the increase in the thickness of the gas disc of perhaps a factor of 10 we may expect a decrease in the volume density of gas by a factor of 100-500. We could therefore divide the gas disc of galaxies into two parts, a dense inner disc and a tenuous outer disc which I shall call the disc corona. Now in the dense disc with a field of say 3 $[\mu G]$, and a density of 1 HI atom per cm^3 and a rotational velocity of 200 [Km/sec] the rotational energy density is 400 times the magnetic energy density, therefore the field here has little dynamical influence; but in the disc corona, if the field remains as high as 3 $[\mu G]$, and the density drops by 500 then the field energy would become comparable to the rotational energy density, and magnetic stresses would significantly affect the dynamics of the corona.

To quantify this idea I have developed a simple model of the field in the outer corona which allows the motion of the disc to be calculated. The assumptions of the model are as follows :-

1) The magnetic field **B** is generated in the dense disc and is spiral in form

2) Turbulence in the corona is small, therefore there is no magnetic dynamo action or diffusion here, simply a frozen-in field

3) The field pattern rotates with fixed angular velocity, and in the rest frame of the field the field lines have the equation

$$r_B \ = \ r_0 \exp(\theta \ / \propto \ ) \qquad \text{with} \quad \tan i = 1/ |\propto|$$

with $\propto$ a function of radius   (the pitch angle i increases outwards).

4) $B = (B_r , B_\theta , 0)$, i.e. **B** is wholly in the plane

5) all variables independent of θ and time in the rest frame of **B**.

6) $|B| \propto \exp(-z^2 / h^2)$ where z is the vertical coordinate.

7) Gravitational potential $\propto 1/r$ (i.e. no massive halo)

The equations that have to be solved are the continuity and momentum equations for the gas, and the frozen-in field equations with zero electric field. With the above assumptions these can be reduced to a single ordinary differential equation for the function $\propto$, which can be easily solved numerically. The results shown in figs.1 and 2 start at an initial corona radius of 10 Kpc, and extend out to 50 Kpc. The initial values of the Alfven velocity, the gravitational orbital velocity, and the actual rotational velocity at 10 [Kpc] are 50, 200, 199 [Km/sec] respectively, and the four curves in fig.

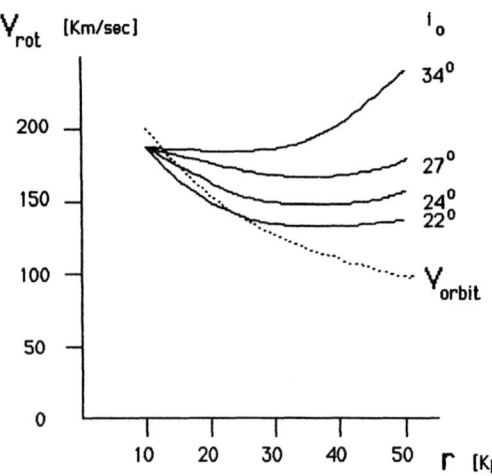

Figure 1. Rotational velocity versus r for various values of i_0. The dotted curve shows the orbital velocity in the gravitational field alone.

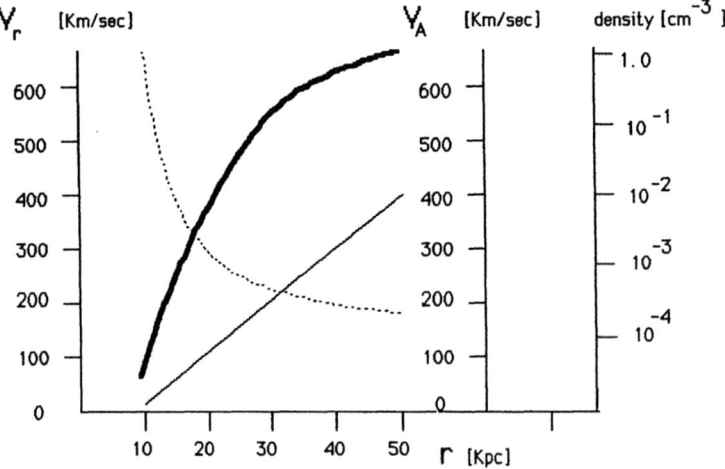

Figure 2. Radial velocity (continuous line), Alfven velocity (bold line), and gas density (broken curve) versus r for the case of $i_0 = 24^0$.

1 correspond to initial pitch angles of 22^0, 24^0, 27^0, and 34^0. As can be seen the rotational velocities are maintained significantly higher than the gravitational orbital velocity, although the shapes of the curves are not flat. It may be that further refinement of the model (eg. finite magnetic diffusion, varying thickness of the disc and field) might produce a flatter curve. At any rate it can be concluded from this calculation that it is very unwise to ignore the effects of **B** when considering the dynamics of gas in the disc corona, particularly when one notes that the values of the Alfven and rotational velocities used at the initial radius correspond to a magnetic energy density 16 times lower than the rotational energy density. Further out this rapidly reverses as the density of the gas drops off, with the magnetic field dropping much less rapidly so that the Alfven velocity increases outwards (see fig. 2). In addition the radial velocity increases outwards from an initial value of approximately 1 [Km/sec] at 10 [Kpc] to about 400 [Km/sec] at 50 [Kpc], so that a by-product of this model is a magnetically driven wind along the galactic plane.

References
1. Sorensen, S.-A., Matsuda, T., Fujimoto, M. : Astrophys. Space Sci., 54, 491 (1976)
2. Sanders, R.H.: Astrophys. J., 217, 916 (1977)
3. van Albada, G.D., Roberts, W.W.: Astrophys. J., 246, 740 (1981)
4. Johns, T.C., Nelson, A.H.: M.N.R.A.S., 220, 165 (1986)
5. Bosma, A.: The Distribution and Kinematics of Neutral Hydrogen in Spiral Galaxies of various Morphological types, Ph. D. Thesis, University of Groningen (1978)
6. Sancisi, R. : In Internal Kinematics and Dynamics of Galaxies, IAU Symposium no. 100 (Reidel, Dordrecht, 1983)

Observations of Linear Polarization in the Galactic Centre Region

W. Reich[1], Y. Sofue[2,3], M. Inoue[2], and J.H. Seiradakis[4]

[1]Max-Planck-Institut für Radioastronomie, Auf dem Hügel 69,
 D-5300 Bonn 1, Fed. Rep. of Germany
[2]Nobeyama Radio Observatory, Tokyo Astronomical Observatory,
 Minamisaku, Nagano 384-13, Japan
[3]Dept. of Astronomy, University of Tokyo, Bunkyo-ku, 113 Toyko, Japan
[4]Dept. of Astronomy, University of Thessaloniki, GR-54006 Thessaloniki, Greece

Introduction

Polarized radio emission has been detected near Sgr A along the radio Arc and its extension above and below the galactic plane [1,2,3,4,5,6]. There is a compact source on the radio Arc (A) and the north-western (B) and south-eastern (C) "plumes" extending up to $|b| = 0°.6$. Rotation measures (RM) at 10 GHz exceed 1000 rad m^{-2} at some places. The magnetic field direction is perpendicular to the galactic plane.

6 cm Observations

We made new multi-channel, narrow-band polarimetric observations with the 100-m telescope. The centre frequencies were 4550, 4670, 4806 and 4950 MHz, the bandwidth was 40 MHz, the HPBW ~ 2'.4, and the calibration source was 3C 286 (7.5 Jy, 11.5% polarized at 33°). We corrected for the spurious instrumental polarization and estimate remaining effects to be < 0.05%.

Because of the narrow bandwidth band depolarization ($|RM| \leq 16000$ rad m^{-2}) is negligible and the small channel separation allows the determination of RM unambiguously. The rms-noise (10 mK T_B) gives a typical error of $\Delta |RM| \sim 200$ rad m^{-2}, this error causes an uncertainty of ~ 50° for the magnetic field direction. Therefore the data give the RM distribution only.

Figure 1 shows the averaged polarized intensity distribution. Significant polarization is detected in many regions including sources B and C. At 10 GHz source A is very little polarized, while a region 3' south (source A') is polarized by 1%. Table 1 lists the polarization properties of A, A', B and C.

Results

Figure 2 shows the RM distribution for the ridge ($\ell \sim 0°.16$). A significantly different RM behaviour for A and A' is found (Table 1). The proximity of both sources suggests a physical relation, but their anomalous RM behaviour is far from allowing modelling yet. The RM distribution for B and C agrees with that at 10 GHz [3].

Because the RM variations in B and C are antisymmetric with respect to the galactic plane, the line of sight component of the magnetic field has a reversal near the galactic plane and most likely near A or A', where a Faraday window appears.

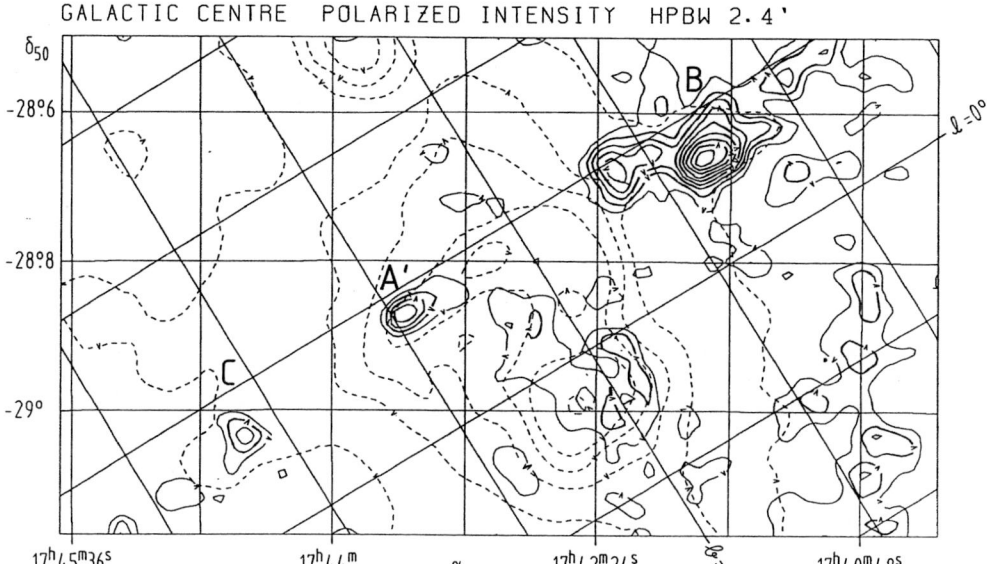

Figure 1: *Averaged polarized intensity of the four channels near 6 cm. Steps are 15 mJy/b.a. starting from 18 mJy/b.a. (2.5 x r.m.s. noise). Dashed lines represent some total intensity contours.*

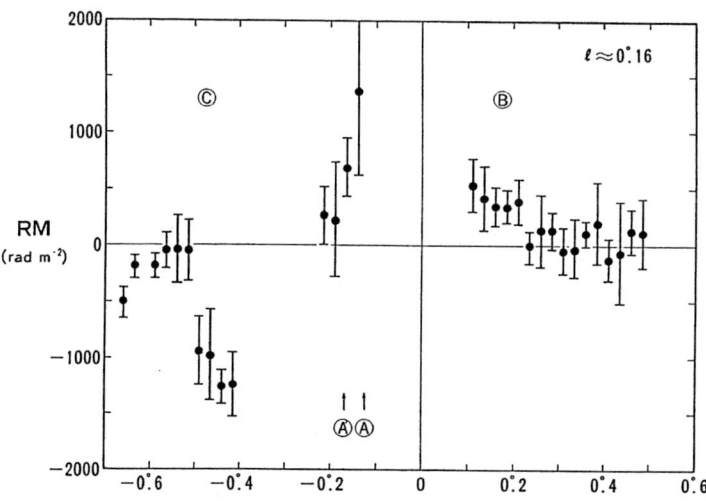

Figure 2: *The distribution of RM at 5 GHz at l = 0.16° across the galactic plane.*

TSUBOI et al. [3] have shown that the transverse component of the magnetic field is nearly perpendicular to the galactic plane.

Polarization is detected near the Bridge area, but no polarization is seen on the total intensity ridge. A distribution of RM is shown in Figure 3. The coherent increase of |RM| towards the Bridge centre and the depolarization at the maximum indicate the existence of a strong magnetic field along the Bridge and a large internal

Table 1: Polarized Properties along the Radio Arc at 5 and 10 GHz[*]

Source	ℓ	b	pol. degree (%)		RM (rad m^{-2})	
			5 GHz	10 GHz	5 GHz	10 GHz
A	0:16	-0:13	<0.5	~10	—— (-2880)[+]	~-1000
A'	0:16	-0:18	~1	0:5	500~1000 (-200)[+]	——
B	~0:15	0:15~0:5	10~20	20~40	0~500	0~1000
C	~0:14	-0:4~-0:6	5~10	20~30	-100~-1000	-500~-1000

[*]5 GHz: 100-m telescope, HPBW 2:4; 10 GHz: 45-m telescope, HPBW 2:7
[+]VLA observations, HPBW 20", Yusef-Zadeh et al. [6].

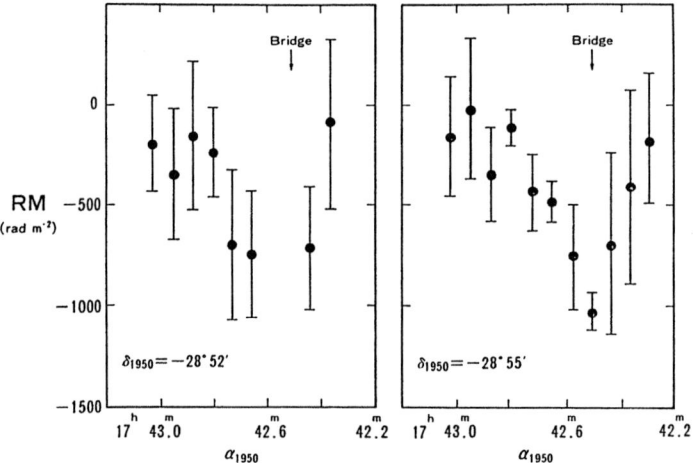

Figure 3: Distributions of RM across the radio Bridge. The coherent variation shows that the magnetic field is highly ordered in the Bridge.

Faraday rotation. From the largest |RM| and n_e ~ 10^2 cm^{-3} [7], we obtain ~ 10 μG at the edge for a path length of 2 pc. In the central part the field strength is estimated to ~10^2 μG by SOFUE and FUJIMOTO [8].

A further discussion of the magnetic field structure is given by SOFUE [9], while a full discussion of the 5 GHz observations and results will be published elsewhere [10].

References

1. M. Inoue, T. Takahashi, H. Tabara, T. Kato, M. Tsuboi: Publ. Astron. Soc. Japan **36**, 633 (1984)

2. M. Tsuboi, M. Inoue, T. Handa, H. Tabara, T. Kato: Publ. Astron. Soc. Japan **37**, 359 (1985)

3. M. Tsuboi, M. Inoue, T. Handa, H. Tabara, T. Kato, Y. Sofue, N. Kaifu: Astron. J., submitted (1986)

4. J.H. Seiradakis, A.N. Lasenby, F. Yusef–Zadeh, R. Wielebinski, U. Klein: Nature 317, 69 (1985)

5. F. Yusef–Zadeh, M. Morris, D. Chance: Nature 310, 557 (1984)

6. F. Yusef–Zadeh, M. Morris, O.B. Slee, G.J. Nelson: Astrophys. J. Letters, in press (1986)

7. T. Pauls, D. Downes, P.G. Mezger, E. Churchwell: Astron. Astrophys. 46, 407 (1976)

8. Y. Sofue, M. Fujimoto: Nature, submitted (1986)

9. Y. Sofue, W. Reich: this volume

10. Y. Sofue, W. Reich, M. Inoue, J.H. Seiradakis: Publ. Astron. Soc. Japan, in press (1986)

The Magnetic Field in the Galactic Center Region

Y. Sofue[1,2] and W. Reich[3]

[1]Dept. of Astronomy, University of Tokyo, Bunkyo-ku, 113 Tokyo, Japan
[2]Nobeyama Radio Observatory, Tokyo Astronomical Observatory,
  Minamisaku, 384-13 Nagano, Japan
[3]Max-Planck-Institut für Radioastronomie, Auf dem Hügel 69,
  D-5300 Bonn 1, Fed. Rep. of Germany

1. Introduction

Recent multi-channel, narrow-band polarimetric observations with the 45-m and 100-m telescopes have shown the existence of highly polarized radio emission with an extremely high rotation measure (RM) along the radio arc and its extensions to higher galactic latitudes [1,2,3,4,5,6]. The available data allow to model a possible magnetic field structure in the central region of a few ten's of pc in extent of our Galaxy.

2. Poloidal Field – Transverse Component to the Line of Sight

The field component transverse to the line of sight can be obtained from intrinsic polarization angle (PA) after correction for the Faraday rotation. TSUBOI et al. [3] found that the field runs perpendicularly to the galactic plane along the strongly polarized features coinciding with the radio arc and its extensions to higher galactic latitudes (Fig. 1). The direction is consistent with the VLA filamentary structure found in the arc [7]. This fact indicates that the field in this region has a poloidal component.

3. Toroidal Field – Parallel Component to the Line of Sight

The RM along the polarized features [5] is negative at $b \lesssim -0°.4$ attaining the minimum value of RM $\lesssim -2000$ rad m^{-2} at $b \approx -0°.3$. Then RM increases suddenly at $b \approx -0°.2$, reaching a very large positive value at $b \sim 0°$, and then decreases gradually to zero at $b \sim 0°.6$. There is a reversal of RM-sign at around $b \sim -0°.2$. This indicates that there exists a reversal of the field direction on the line of sight. Namely, the line-of-sight component of the field runs away from the observer at $b \lesssim -0°.3$, while toward us at $b \gtrsim 0°$ (Fig. 1).

We also plotted RM across the rotation axis along $b = 0°.3$ as a function of the longitude ℓ [4]. The RM varies with ℓ in an antisymmetric way with respect to the rotation axis. This indicates the existence of a toroidal (ring) field at this latitude [8]. The toroidal field runs away from us at $\ell < 0°$, while it is toward us at $\ell > 0°$ (Fig. 1).

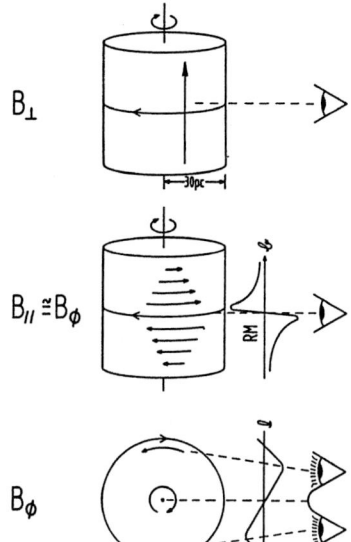

$B_\perp$

$B_{//} \cong B_\phi$

B_ϕ

Figure 1: Characteristic features of (a) observed transverse direction of magnetic field showing the existence of a poloidal field near the galactic center, (b) the RM variation across the galactic plane showing a reversal of the line-of-sight component (toroidal field), (c) the RM variation across the rotation axis showing the existence of a ring (toroidal) field

4. Twisted Poloidal (Semi-helical) Field

Combining these facts we propose the following simple model (Fig. 2). The magnetic field in the central ~ 50 pc is predominantly poloidal but twisted by the disk rotation near the galactic plane. The twist produces a toroidal field whose direction in the line of sight reverses from the lower to the upper side of the galactic plane. Consequently the field becomes helical, but the pitch angle changes its sense from the lower to the upper side of the galaxy.

The extremely large Faraday rotation ($|RM| \gtrsim$ several 10^3 rad m^{-2}) at $b = -0.4$ to -0.2 and at $b = -0.1$ to 0.1, where the toroidal field strength reaches the maximum, results in depolarization as the observations show [3,9]. However, at the reversing point of the toroidal component, the field lines become shortly perpendicular to the line of sight, where no depolarization takes place. This narrow "Faraday window" allows the point-like polarization emission be detected on the arc [1,5].

A question remains why the polarized features are bright only along the large scale twisted poloidal field. This may be answered, if we take the model that the radio bridge is a magnetized jet from the galactic center with relativistic electrons [9]. The jet encounters the ambient poloidal field injecting the electrons near the arc, where the synchrotron emission is most intense. Electrons leaking along a magnetic tube connected to the arc region emit the polarized radiation at high latitudes (Fig. 2).

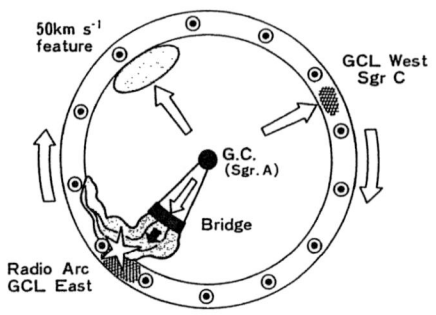

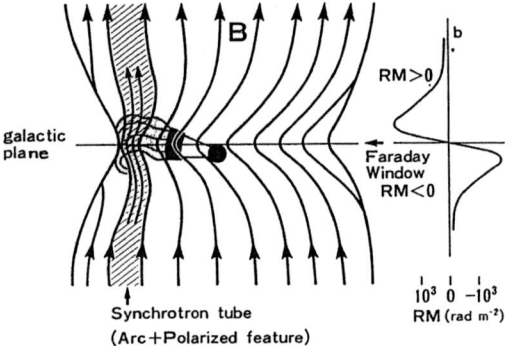

Figure 2: A twisted poloidal (semi-helical) magnetic field model proposed from the observed facts in Fig. 1. The radio bridge as a jet from the galactic center excites the synchtrotron emission along a certain flux tube on the poloidal field

References

1. M. Inoue, T. Takahashi, H. Tabara, T. Kato, M. Tsuboi: Proc. Astron. Soc. Japan <u>36</u>, 633 (1984)

2. J.H. Seiradakis, A.N. Lasenby, F. Yuseh−Zadeh, R. Wielebinski, U. Klein: Nature <u>317</u>, 69 (1985)

3. M. Tsuboi, M. Inoue, T. Handa, H. Tabara, T. Kato, Y. Sofue, N. Kaifu: Astron. J., in press (1986)

4. Y. Sofue, W. Reich, M. Inoue, J.H. Seiradakis: Proc. Astron. Soc. Japan, in press (1986)

5. W. Reich, Y. Sofue, M. Inoue, J.H. Seiradakis, this volume

6. F. Yusef−Zadeh, M. Morris, O.B. Slee, G.J. Nelson: Astrophys. J. Lett. in press (1986)

7. F. Yusef−Zadeh, M. Morris, D. Chance: Nature <u>310</u>, 557 (1984)

8. Y. Sofue, this volume

9. Y. Sofue, M. Fujimoto: Nature, submitted (1986)

Part III

Magnetic Fields in Star-Forming Regions and Supernova Remnants

The Strength of the Interstellar Magnetic Fields and Its Possible Role in HI Cloud Dynamics

G.L. Verschuur

Unaffiliated [1]

If interstellar HI regions contain even a small fraction of ionized material, the neutral gas will be coupled to the field. What would be the observational consequences of such coupling? It is suggested that the data already exist to demonstrate that the field may be the arbiter of motions in so-called HI "clouds". The possibility that the field may act to determine line broadening and constrain HI concentrations is considered.

1. The Strength of the Field in HI Clouds

The strength of the field in interstellar HI coulds has been determined through measurements of the Zeeman splitting. Various attempts to relate the observed field strength to cloud HI density have met with mixed results (VERSCHUUR [2], TROLAND and HEILES [3]). VERSCHUUR [4] has recently reported that the field strength appears to correlate well with z-distance (height above the Galactic disk) of the cloud. The distances to the absorbing clouds in which the Zeeman effect has been detected are derived either from a knowledge of the distance to the background radio sources or from an interpretation of the absorption line velocity in terms of a Galactic rotation law. Figure 1 shows the field strength (derived from published Zeeman effect measurements) as a function of a cloud's z-distance. The filled circles and error bars represent data available in the literature and include VERSCHUUR'S [5] claimed detections, two points measured by TROLAND and HEILES [6], and two points in the Orion region given by HEILES and TROLAND [7]. The latter did not give formal errors so it was assumed that these were of the same order as those given by Troland and Heiles. A recent Zeeman experiment has added more points to this plot, but until these are formally analyzed (VERSCHUUR and SCHMELZ [8]) they are not used in Fig. 1. One new detection suggests a 7 microgauss field in an HI emission feature in the direction of the North Polar Spur.

It should be pointed out that the two points at the left-hand end of the diagram represent the fields in two absorption clouds immediately adjacent to the Orion nebula. VERSCHUUR [9] has argued that the conditions close to the Orion nebula are not typical of the z-distance of the nebula (which is about 140 pc). It was found that the density and gas pressure in these clouds, when compared with data for several hundred other clouds, are consistent with an environment near the nebula

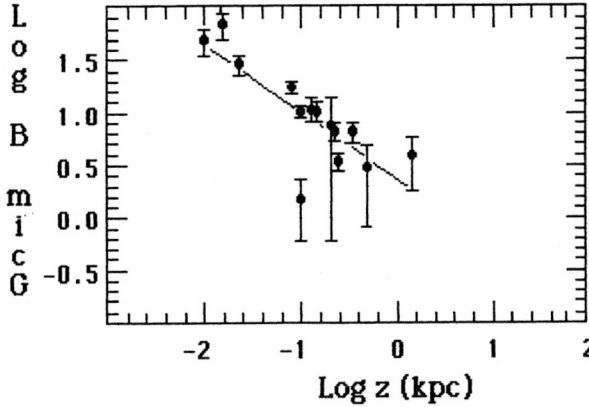

Figure 1. The magnetic field strength in HI clouds, as determined from Zeeman effect measurements, plotted against the z–distance of the clouds

that would be equivalent to a z–distance of the order of 10 to 20 pc. Upper limits to fields in several other directions are not plotted in Fig. 1.

2. Cloud Linewidths and Magnetic Field Strengths

This and the following section are meant to outline some avenues for further research and deal with largely unexplored territory. The work on HI cloud distances and cloud properties to be published by VERSCHUUR [9,10] shows that data for some 480 HI emission clouds exist in the literature which reveal interesting relationships between cloud parameters. At least half of these clouds have distances either esti-mated by the original authors or which can be derived from Galactic rotation. For the rest, a new method for finding distances has apparently been discovered (VERSCHUUR [2,9]). Thus HI cloud properties, especially mass and density, can be evaluated for the entire sample.

Data on HI emission clouds clearly indicate that profile linewidths are not due to thermal motions. It is hypothesized that Alfvén waves may cause the gas to move in a variety of ways so as to cause line broadening. If we allow this hypothesis it is possible to derive an expression which relates field strength to the product of cloud density (n, in cm^{-3}) and linewidth expressed as a temperature.

We begin with the usual expression for the kinetic temperature, T, given by,

$$T = 121 \ \sigma^2, \tag{1}$$

where σ is the dispersion of the emission profile in km/s. The Alfvén velocity, V_A, in a medium of density ρ and magnetic field strength B is given by,

$$V_A^2 = (B^2/4\pi\rho). \tag{2}$$

If we express density in terms of n and B in microgauss, and assume that to first order a "cloud" has a linewidth determined by Alfvén wave motion, then, if V_A is equated to the dispersion of the HI profile, we can derive an expression for the field, B. It is found to be given by,

$$B = 4.16 \times 10^{-2} \ (nT)^{1/2}.$$ (3)

For the large sample of HI clouds whose distances are either known or derived using the new technique (VERSCHUUR [9]), the values for nT for each fully mapped HI cloud was derived. Density was found by assuming that the clouds are as deep as they are wide and the equivalent values for T were found from the observed line-widths using Equation 1 above. This product is sometimes treated as a measure of the gas pressure in a cloud. However, if line broadening is produced by Alfvén wave motion, the term nT might be called a pseudo-gas-pressure because it now appears to be related to the field according to Equation 3.

Figure 2 shows the value of the "cloud field" (derived from Equation 3) plotted as a function of z-distance. Also shown are the Zeeman detections from Fig. 1. Two additional points, recently measured and still unpublished, are also included in this group. The derivation of the line shown in Fig. 2 is discussed in the next section. It is both very surprising and tantalizing that the Zeeman data lie in the same general domain as the points representing the data for the emission clouds, where field has been related to observed cloud properties according to Equation 3.

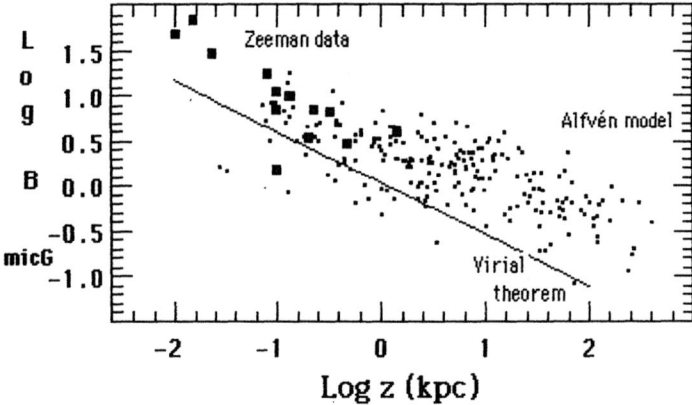

Figure 2.   The strength of the field versus z-distance for HI clouds. Small points: Field derived from emission cloud parameters assuming Alfvén wave line broadening. Squares: Zeeman data from Fig. 1 plus two unpublished points. Line: Derived from virial theorem, Section 3

3. The Magnetic Field and "Cloud" Stability

None of the HI emission clouds discussed in the literature appear to be near gravitational equilibrium. Thus we must ask what creates possible cloud control. Pressure balance with an intercloud medium may be important, and since nT in the clouds is found to be z-dependent (VERSCHUUR [2,9]) this may be reasonable. However, this simple view may be invalid since the "clouds" are probably not single entities. Whenever higher resolution studies are made of so-called "clouds" they are found to break into smaller structures, each with narrower linewidths than the overall structure. Thus the apparent gas pressure term, nT, based on observed linewidths and used in Section 2, is a function of resolution. This aspect of the data will be more carefully discussed in future.

We next explore the possibility that the magnetic field is the force that determines the *cloudiness* in so far as the field may keep regions of HI coherent. Even if Alfvén waves create line broadening, the very fact that the HI is coupled to the field also means that, provided the field is anchored somewhere outside the clouds, the coherence of the structures may be controlled by these fields.

The HI in all emission cloud structures is assumed to have a kinetic temperature of 80 K. Since the mass of an apparent cloud (as defined by the original researchers and assuming no molecular content) is known, because the distance is known, the thermal energy content can be evaluated. If we assume that HI structures are stable we must ask what controls them. Gravity is not a determinant even if the temperatures are assumed to be 80 K (VERSCHUUR [9]). Gas pressure may control the stability of 80 K clouds but does not account for the observed emission linewidths. If Alfvén waves are involved in determining line broadening, as hypothesized above, does the magnetic field also enter into the question of cloud stability? Why wouldn't the structures fly apart due to thermal pressure? The answer may lie in the possibility that the magnetic field acts to constrain thermal motion transverse to the field, and perhaps even along the field in a sufficiently complex morphology of fields and clouds.

Imagine that magnetic fields are anchored and invoke the classical image of field lines acting like elastic bands. They create a restoring force confining the gas which is coupled to the field. The gas would appear to be free to stream along the field lines and such streaming motion might also act to produce line broadening. However, it is not necessarily true that the HI can freely flow along the field lines either, if space around the "cloud" is filled with fields and other HI clouds. It has been known for some time that HI structures appear to be filamentary (see, for example, VERSCHUUR [11,12,4] and COLOMB et al. [13]), while the highest resolution data of VERSCHUUR [12] show that clouds appear to be strung together in filaments. This is strongly suggestive of the interaction between fields and gas and that there is control of HI morphology by the field.

At this point we ask what field strength would be required to balance the internal thermal motion in the clouds, assuming a kinetic temperature of 80 K and neglecting gravity. Using the virial theorem in its simplest form it can be shown that the field required to constrain the thermal motion is given by,

$$B = 0.1 \times (nT)^{1/2}. \tag{4}$$

This expression has the same general form as Equation 3, but the value for T to be used here is 80 K. This is unrelated to the observed profile linewidth, except for the coldest clouds where T < 80 K and is known. Using Equation 4, the value of B was found for each HI cloud and the results are shown in Fig. 3. The line fit to these data is indicated and was plotted in Fig. 2.

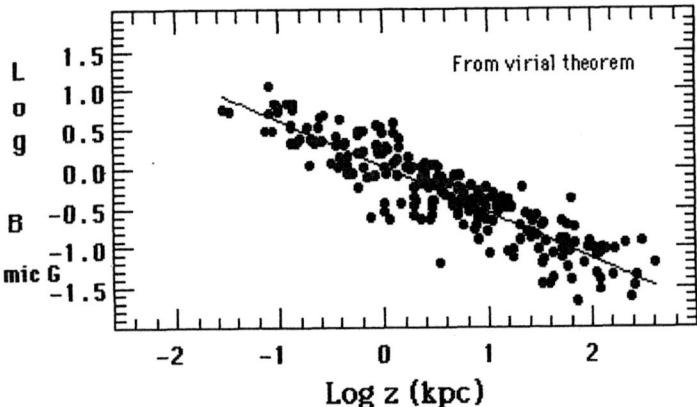

Figure 3.   The field strength in HI emission clouds found from application of the virial theorem, neglecting self-gravitation, and assuming balance between thermal and magnetic energies in the clouds

4. Discussion

We have considered three ways of getting at the strength of the interstellar mag-netic field in HI clouds and plotted the derived strength as a function of z–distance (Fig. 2). The agreement between the three methods was found to be remarkable, especially considering the diversity (and relative simplicity) of the approaches. However, it is possible that the apparent agreement is no more than a glorious coincidence, yet coincidences should be explored lest they contain information.

An interesting fact emerges from examination of Fig. 2, one that suggests a further avenue for exploration. Figure 2 contains two points clearly below and to the left of the line. This pair represents data for the cold clouds mapped by KNAPP and VERSCHUUR [14] which are significantly different from all the others in the sample in that they are the coldest ones with the narrowest linewidths. They are not broadened by non–thermal motion. In Section 2 the determination of field strength made use of the observed linewidth and therefore it is not surprising that the values of B derived for these cold clouds would be low. According to Fig. 2 the field in these clouds should be of the order of 2 microgauss. Recent attempts to measure a Zeeman effect in these two clouds (VERSCHUUR and SCHMELZ [8]) have produced very low limits (< 2.5 microgauss) consistent with this prediction.

It has been customary to plot the field strength versus cloud density where sufficient data were available to do so (VERSCHUUR [2]; TROLAND and HEILES [3]). This approach has produced mixed results. When the Zeeman detections are instead plotted versus the value of nT for the clouds, a relationship given by,

$$B = 0.15 \times (nT)^{0.51}, \tag{5}$$

is found, as might have been predicted by Equation 4. Again this might be a coinci-dence, yet it is significant that Equation 5 was derived for mostly absorption clouds whose field strengths, densities, and temperatures (from Gaussian line fitting) were taken from the literature. In a few cases where Gaussian fitting was not available

158

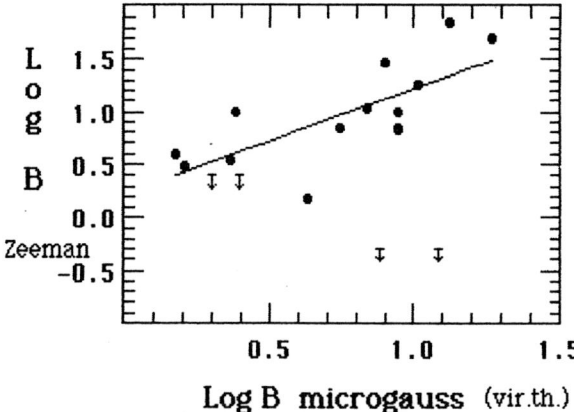

L o g B Zeeman

Log B microgauss (vir.th.)

Figure 4. The field strength in several HI clouds found from the 21-cm Zeeman data plotted against the field expected for these clouds if they were in virial equilibrium, assuming balance between magnetic and thermal energies

80 K was assumed. This analysis was taken a step further and is summarized in Fig. 4. Here the Zeeman field was plotted against the <u>predicted</u> field derived according to Equation 4. Four reliable limits are also shown, but not used to derive the best-fit line which is,

$$B(\text{Zeeman}) = 1.7 \times B(\text{virial theorem}). \tag{6}$$

It has been customary to assume that fields are frozen-in to a cloud which undergoes gravitational contraction and so the field may be expected to increase according to $n^{1/2}$ (TROLAND and HEILES [3]). Plotting the field versus cloud density for the same data used in Fig. 4 (considering detections only) suggests that the field could be defined by,

$$B = 1.7 \times n^{0.43}. \tag{7}$$

A term like $n^{1/2}$ has appeared in many contexts in our discussion and it is not evident which of the mechanisms describing the interaction between HI gas and field is most clearly distinguished by the data. It has been traditional to consider that the field is frozen-in to a contracting cloud, but few, if any, HI clouds are anywhere near gravitational equilibrium. So perhaps we should re-examine the alternative models suggested here.

<u>References</u>

1.   Temporary mailing address: Box 995, Arecibo, PR 00613, USA
2.   G.L. Verschuur: In <u>Interstellar Gas Dynamics</u>, IAU Symposium #40, ed. by H. Habing (Reidel, Dordrecht 1970), p. 150
3.   T.H. Troland, C. Heiles: Astrophys. J. <u>252</u>, 179 (1982)
4.   G.L. Verschuur: In <u>Workshop on Interstellar Processes</u>, Wyoming, in press
5.   G.L. Verschuur: Fund. of Cosmic Phys. <u>5</u>, 113 (1979)

6. T.H. Troland, C. Heiles: Astrophys. J. <u>260</u>, L19 (1982)

7. C. Heiles, T.H. Troland: Astrophys. J. <u>260</u>, L23 (1982)

8. G.L. Verschuur, J.T. Schmelz: in preparation (1986)

9. G.L. Verschuur: in preparation (1986)

10. G.L. Verschuur: in preparation (1986)

11. G.L. Verschuur: Astron. J. <u>78</u>, 573 (1973)

12. G.L. Verschuur: Astrophys. J. Suppl. <u>27</u>, 65 (1974)

13. F.R. Colomb, W.G.L. Pöppel, C. Heiles: Astron. Astrophys. Suppl. <u>27</u>, 65 (1980)

14. G.R. Knapp, G.L. Verschuur: Astron. J. <u>77</u>, 717 (1972)

Optical Polarisation Studies of Star Formation Regions

S.M. Scarrott, T.M. Gledhill, and R.F. Warren-Smith

Physics Department, The University, South Road, Durham DH1 3LE, UK

Linear optical polarisation maps are presented for nebulae associated with recently formed stars. The data suggest that in many cases there appears to be a circumstaller gas/dust disk that not only collimates any outflows but also contains a toroidal field which ultimately links up with the field in the surrounding dark clouds.

1. Introduction

Star formation is a complex process and appears to be influenced by several factors, primarily the conditions within the dark cloud out of which the star is or has formed. We have investigated several systems via the technique of imaging polarimetry to attempt to qualitatively assess the role played by local magnetic fields in the processes leading to star formation.

The data presented here are the result of observations with the Durham Imaging CCD Polarimeter [1] on the 1m telescope of the Wise Observatory, Israel during the period 1984 May to 1986 March.

2. The PV Cephei Nebula

PV Cep is a variable star with a small associated nebula which has undergone dramatic changes in its luminous structure in the last three decades [2]. Scarrottt et al. [3,4] discovered that the nebula is now biconical in form.

In fig 1 we show optical polarisation and intensity contour maps of the star and nebula. The intensity map clearly shows the characteristic biconical structure consisting of a central illuminating /exciting star (PV Cep), two colinear lobes of nebulosity and the narrow *waist* corresponding to the central gas/dust disk.

The polarisation map shows several features indicative of the physical conditions within the system. The faint S lobe (directed into the dark cloud out of which the system is forming) shows some characteristics of a reflection nebula illuminated by the central star but there are distortions to the expected pattern. The brighter N lobe does not show reflection properties which is surprising since this lobe has similar spectral characteristics to the central star. PV Cep is itself polarised, as is the central *waist* region, but again these polarisations do not result from simple scattering from a central source.

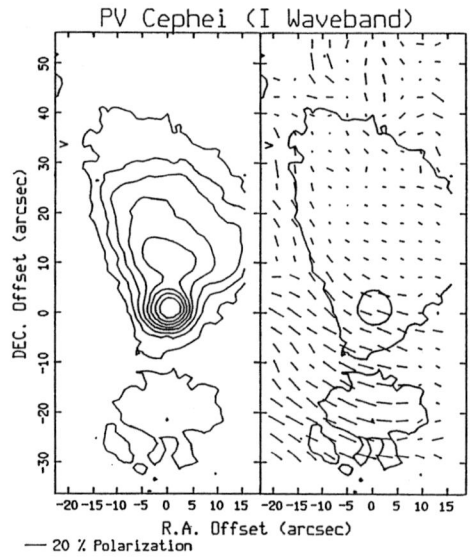

Fig 1. Optical and intensity contour maps of the PV Cep nebulosity. The R waveband was used for these measurements.

These polarisation features are not typical of those of a simple reflection nebula. We suggest that the observed polarisation arises not only from scattering but also from extinction by magnetically aligned grains and the two mechanisms complete to varying degrees in different parts of the nebula.

We believe that the central disk contains a toroidal magnetic field which aligns the non spherical dust grains therein.

162

The radiation from the central star and surrounding disk region is extinguished by these grains and in the process the light is linearly polarised parallel to the plane of the disk, as observed. To the N of the disk, in the region overlying the N lobe, the field leaves the disk in a radial-like fashion so that the polarisation induced by close-by and foreground extinction offsets and overcomes any polarisation induced by scattering. This would lead to our observed pattern and in particular to the polarisation orientations running parallel to the E edge of the N lobe.

3. NGC2261 and R Mon

This cometary nebula, illuminated by the variable and apical star R Mon, is a well-known reflection nebula. One of our earlier polarisation maps [5] revealed an anomalous band of polarisation running across the head of the nebula which we interpreted as an indirect consequence of a toroidal field within the plane of the circumstellar disk. This disk corresponds to the inner parts of the *interstellar* disk proposed by Canto et al [6] as part of their *nozzle model* for NGC2261 and associated HH objects.

In fig 2 we present a recent polarisation map of NGC2261 that again shows the central polarisation band and the apparent reflection nebula that constitutes the N lobe. However these new data reveal other more subtle effects within the polarisation pattern. Although the N lobe seems to exhibit the classic centro-symmetric pattern of polarisation orientations characteristic of reflection nebulae there is a systematic distortion running throughout the pattern. We suggest that the N lobe is seen by reflected central star light which is subsequently extinguished by aligned grains to yield the more complex pattern that is observed. The magnetic field configuration required to explain our observations would be toroidal in the disk and connected to the external dark cloud field in a helical fashion. To the SW of R Mon is the *jet* discovered by Malin & Walsh [7]. Oue polarisation map reveals that it is a structural feature (the rim of the counter cavity) illuminated directly by R Mon.

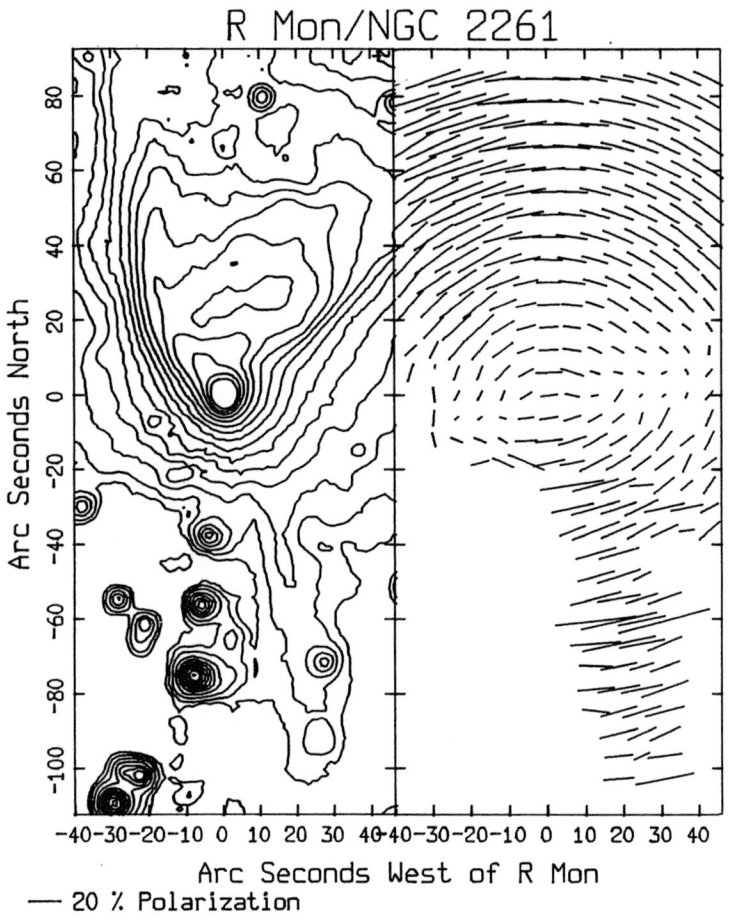

Fig 2. An optical polarisation map of the central regions of NGC2261 including the star R Mon and the newly discovered S jet.

4. **Other Systems**

We have investigated many other similar systems. Parsamyan 21, a small compact cometary nebula, shows the characteristic anomalous band we associate with central disks and toroidal fields.

The Serpens nebula, an object discussed elsewhere in this Workshop, has the central dust feature plus significant deviations from the typical reflection polarisation pattern.

The classic L1551-IRS5 bipolar outflow also shows evidence for magnetic effects, in fact the field may well be responsible for producing multiple parallel outflows in the same dark cloud.

5. Conclusion

Polarisation studies of nebulous objects in star formation region reveal anomalous bands of polarisation orientations about the central (proto)stellar objects which we believe correspond to circumstellar disks containing toroidal magnetic fields that appear to link up to the field in the surrounding dark clouds in a variety of ways.

Magnetic fields play an influential role in star formation.

6. References

1. Scarrott. S.M., Warren-Smith, R.F., Pallister, W.S., Axon, D.J. & Bingham, R.G., Mon.Not.R.astr.Soc., **204,** 1163, (1983).

2. Cohen M., Kuhi L.V., Harlan E.A. & Spinrad H., Astrophys. J., **245,** 920, (1981).

3. Scarrott,S.M., Warren-Smith,R.F., Draper,P.W. & Gledhill,T.M., Cosmical Gas Dynamics, Ed. Khan,F.D., VNU Science Press (1985).

4. Scarrott,S.M., Warren-Smith,R.F., Draper,P.W. & Gledhill,T.M., Can.J.Phys. **64**, 426, (1986).

5. Gething,M.R., Warren-Smith,R.F., Scarrott,S.M. & Bingham,R.G. Mon.Not.R.astr.Soc., **198,** 881, (1982).

6. Canto,J,, Rodriguez,L.F., Barral,J.F. & Carral,P., Ap.J., **244,** 102, (1981).

7. Walsh,J.R. & Malin,D.F., Mon.Not.R.astr.Soc., **217**, 31. (1985).

Observation of Magnetic Field Structure Around Newly Formed Stars

R.F. Warren-Smith, P.W. Draper, and S.M. Scarrott

Physics Department, The University, South Road, Durham DH1 3LE, UK

1. Introduction

The Serpens Object (RA=18:27:24, Dec=1°12'40") is an optical bipolar reflection nebula situated in a molecular cloud in Serpens which is undergoing an active phase of low-mass star formation [1]. The optical nebulosity results from light escaping into the surrounding dusty medium through cavities produced by a bipolar outflow from the illuminating star SVS2 [1,2]. SVS2 appears to be a very young object [3] and is still surrounded by recognisable remnants of the protostellar condensation from which it has formed.

Deep CCD imaging (Fig. 1) shows the bipolar nature of this object and reveals details of the density structure in the remnant protostellar material. Of particular note is the dark spiral filament which encircles the nebula, connecting from a nearby faint reflection nebula to the NE (GGD29 [4]), into the inner regions of the Serpens Object. This structure is surprising because it is believed that other similar young objects are surrounded by rotating disks [5]. This represents the first clear identification of spiral structure in protostellar material.

2. Polarimetry

High levels of linear polarization are common in reflection nebulae such as this, and arise from the scattering of light by dust grains. The orientation of the resulting polarization is perpendicular to the scattering plane; giving a characteristic polarization map with a circular pattern surrounding the source of illumination (in this case

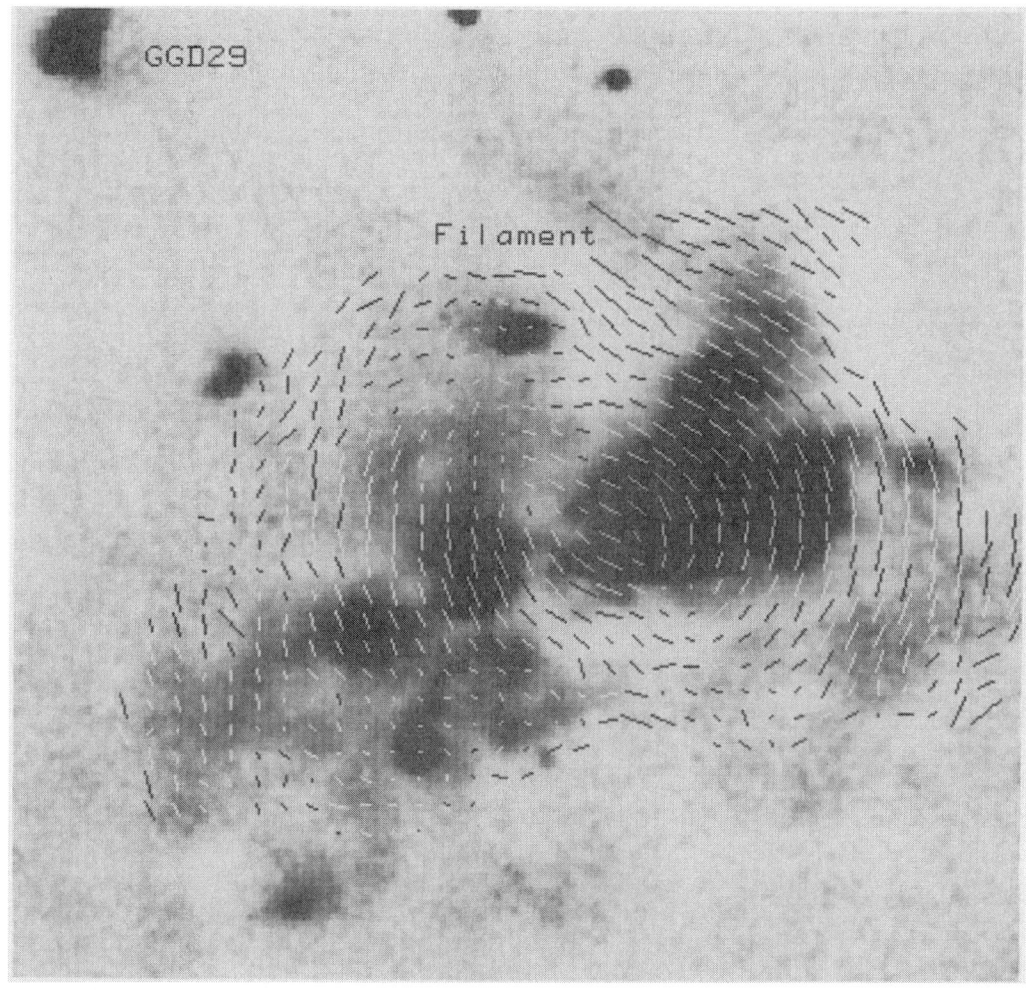

Figure 1. An I band (0.78μm) linear polarization map of the Serpens Object, superimposed on an R band (0.67μm) CCD image. The polarization of the illuminating star is 15.8%

SVS2). Non-spherical dust grains may also be aligned by the magnetic field in the cloud surrounding the nebula (e.g. by the Davis-Greenstein mechanism [6]) and light from the nebula will be extinguished by them. These aligned grains impose an additional polarizing effect on the light, distorting the circular appearance of the underlying polarization map so that the polarization tends to align (to some extent) with the foreground field direction. Details of the surrounding magnetic field structure can therefore be deduced by examining the (sometimes small) departures from a circular polarization pattern.

We have used the Durham polarimeter [7] to obtain polarization maps of the Serpens Object. These show considerable distortion from a circular pattern (Fig. 1), mainly in the form of a band of parallel polarization lying along the minor axis of the nebula. This may be interpreted as indicating an approximately toroidal magnetic field trapped in a disk-like configuration close to the star. In addition, polarization throughout the nebula shows a small but significant "spiral" distortion, in which the polarization deviates away from a circular pattern, towards parallelism with the filamentary spiral nebular structure. This indicates that the foreground magnetic field also posesses similar spiral structure.

3. Magnetic field structure

We propose a model to explain both the density and magnetic field configuration around the nebula as the result of a non-axisymmetric magnetically braked protostellar collapse. We suggest that the protostellar cloud initially rotated about an axis which was considerably inclined to the magnetic field direction. Ultimately, a disk-like structure formed at the centre, but magnetic braking during the initial phases of collapse caused a gradual re-alignment of the rotation axis, so that regions outside the disk have no single axis of symmetry. The magnetic field in these outer regions therefore adopted an approximately spiral structure (Fig. 2), while in the inner regions it was wound into the disk, becoming toroidal.

4. Binary Star Formation

Near the centre of the nebula, there is an approximately symmetric polarization structure, which includes two null points where the disk and scattering polarizations cancel. Surprisingly, there is no distinct

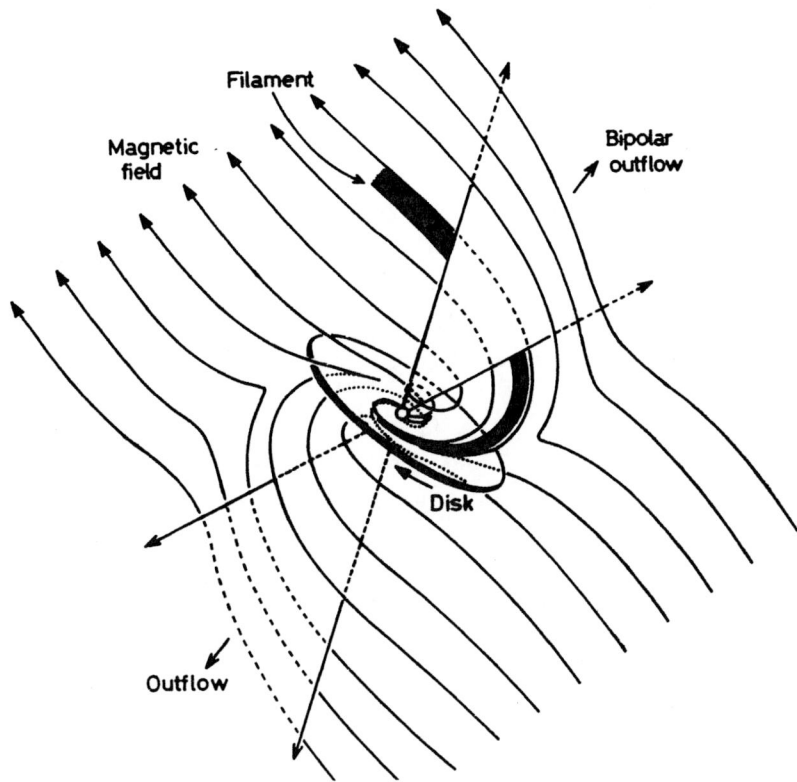

Figure 2. The proposed magnetic field configuration around the Serpens object as deduced from polarization mapping

polarization feature associated with the illuminating star SVS2 and the centre-of-symmetry of the central polarization pattern actually lies to the NE of SVS2, displaced towards the nearby faint red star HL3 [8]. The central polarization pattern appears to enclose both these stars.

This indicates that the central polarization is dominated by aligned grains (rather than scattering) and that the magnetic field in the centre of the nebula is not symmetric about SVS2, but encircles both SVS2 and HL3. We conclude that these two stars are physically associated in a binary system which has resulted from fragmentation of the core (or inner disk) of the collapsing cloud. Both stars are now enclosed in a common magnetic envelope which renders the surrounding nebular medium strongly polarizing, but because SVS2 is the more massive star, it supplies the only significant source of optical illumination for the nebula.

References

1. Strom, S.E., Vrba, F.J., Strom, K.M., 1976. Astr. J., **81,** 638.

2. Worden, S.P., Grasdalen, G.L., 1974. Astr. Ap., **34,** 37.

3. Churchwell, E., Koornneef, J., 1986. Ap. J., **300,** 729.

4. Gyulbudaghian, A.L., Glushkov, V.I., Denisyuk, E.K., 1978. Ap. J. Lett., **224,** L137.

5. Torrelles, J.M., Rodriguez, L.F., Canto, J., Carral, P., Marcaide, J., Moran, J.M., Ho, P.T.P., 1983. Ap. J., **274,** 214.

6. Davis, L., Greenstein, J.L., 1951. Ap. J., **114,** 206.

7. Scarrott, S.M., Warren-Smith, R.F., Pallister, W.S., Axon, D.J., Bingham, R.G., 1983. Mon. Not. R. astr. Soc., **204,** 1163.

8. Hartigan, P., Lada, C., 1985. Ap. J. Suppl., **59,** 383.

Stochastic Star Formation, Magnetic Fields and the Fine Structure of Spiral Arms

J.V. Feitzinger and J. Spicker

Astronomisches Institut, Ruhr-Universität Bochum, Postfach 10 21 48,
D-4630 Bochum, Fed. Rep. of Germany

1. Introduction

Spiral arms are not smooth features. The assumption of regularity of spiral arm structures is much more introduced by theoretical concepts and the need for simplification
than by observational facts. Star forming activity disturbs and changes any regular
spatial structure and the radial and vertical velocity fields. The intermediate scale
dynamics (1-2 kpc) organizes locally well defined deviations in velocity space and
in the spatial locations of star forming sites. These deviations, i.e. the fine structure of spiral arms, are the result of the interaction of stellar dynamical forces,
star forming activity and magnetic fields. The interstellar medium is reshuffeled
and stirred by these influences. Here we present some building blocks on the route
towards a unifying picture concerning the z-structure of spiral arms and magnetic
field configurations.

2. The Dark Clouds

Discussing the fine structure of spiral arms means first of all to look at the local
arm. We performed a statistical analysis of the distribution and orientation of 2622
dark clouds (FEITZINGER and STÜWE /1/). The mean distance of the clouds from the sun
is 400-500 pc, but there is only some marginal evidence for a connection of the dark
cloud distribution with other local spiral arm indicators. The investigation of the
relation of the dark clouds with respect to the local magnetic field orientation and
to the galactic plane reveals the following: For clouds of opacity classes 5 and 6
there is only a slight tendency to allign parallel to the galactic plane, whereas
clouds of opacity classes 1, 2, 3, 4 are clearly aligned. Elongation and alignment
with respect to each other and to the galactic plane is a common property of the dark
clouds on all scales. The distribution of the orientation angles with respect to the
magnetic field shows no tendency towards prependicular orientations. Cloud orientation and magnetic field direction are predominantly parallel. In Fig. 1 we present
a computer processed picture of the local dark cloud distribution in galactic coordinates. The individual clouds are folded with a Gaussian filter (FEITZINGER and
STÜWE /2/) of half width $2^{\circ}.5$. This clearly reveals the elongation of the low opacity
clouds along the galactic plane. The spatial distribution of the dust clouds and the
incremental polarization (Ellis and Axon /3/) show very little correlation with the

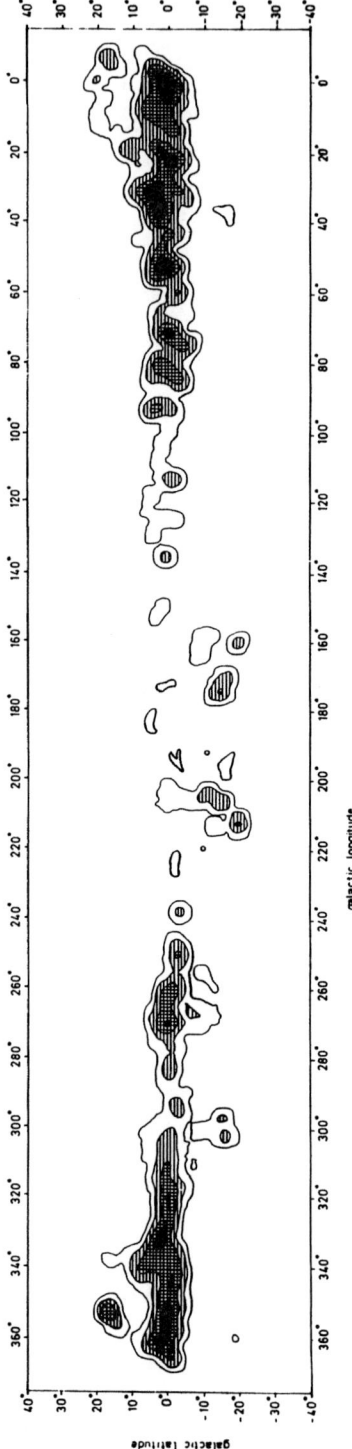

Fig. 1 Dark cloud distribution in galactic coordinates. 2622 dark clouds are folded with a Gaussian filter ($\theta_0 = 2^\circ5$). The opacity classes are coded: 10C = white, 20C = line hatched, 3 to 40C = cross hatched, 5 to 60C = solid black.

local spiral structure. The same is true for the direction of the minimum polarization at l = 50° and l = 230° and the dark cloud distribution.

3. Velocity Active Regions and Magnetic Field Configurations

Another piece of information on the fine structure of spiral arms can be deduced from the so called "rolling motion" phenomenon (FEITZINGER and SPICKER /4/). This phenomenon is visible on (b, v_r)-maps as a pronounced variation of the HI radial velocity with galactic latitude; a linear gradient dv/db is observed. Such a gradient can only partly be explained by geometric effects. In nearly all spiral arms of the Milky Way velocity gradients perpendicular to the galactic plane are found with typical values of ±20 km/s/kpc. In Fig. 2 we show the projection on the galactic plane of the HI centroids for the spiral arms measured (SPICKER and FEITZINGER /5/). The regions with remarkable z-velocity gradients are denoted. We call the localized portions of spiral arms being characterized by pronounced velocity gradients Velocity Active Regions

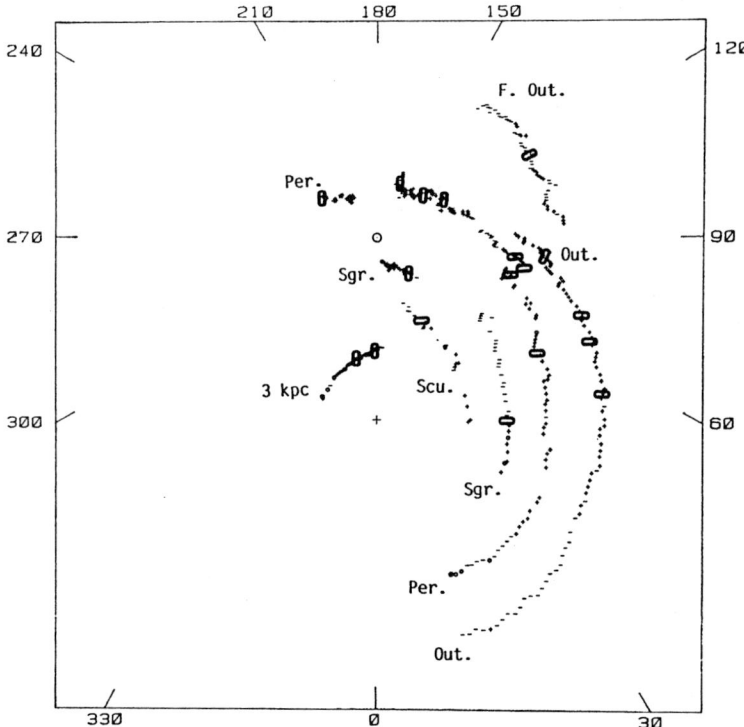

Fig. 2 Composite view of the HI centroid distribution projected onto the galactic plane. The galactic warp has been subtracted. +: $b_o > 0°$, −: $b_o < 0°$, ∘: $b_o = 0°$, ⊙ = sun, + = galactic center, **0** = velocity active regions (VARs)

(VARs, FEITZINGER and SPICKER /6/). The VARs are correlated with deviations of the arms radial velocity ($<\Delta z = 300$ pc$>$), a velocity gradient perpendicular to the galactic plane ($<dv/dz = \pm20$ km/s/kpc$>$) and star forming regions (HII, OB, WR stars). The VARs are examples of the intermediate scale dynamics of our Galaxy. These non-circular motions are primarily powered by stellar energy sources, leading to rising and falling motions of HI gas above star forming regions.

There is a striking preference for scale length of 300 - 600 pc and time scales of 20 - $40 \cdot 10^6$ yrs for objects or properties associated with the magnetic field structure of galactic spiral arms (SPICKER and FEITZINGER /7/). In Tab. 1 these properties are summarized. The first column refers to the observed constituents, the second refers to mean separations between these objects or in the case of corrugation waves to $\lambda/2$. The third column gives typical dynamical evolution time scales.

Table 1. Scale properties of various constituents of galactic spiral arms

object	length scale pc	time scale 10^6 yr
turbulent eddies	400	100
helical field	500	
correlation length of electron & magn. field	200	
rotating eddies	300	50
turbulent eddies	200-500	30-70
Parker instab. clouds	500-600	20-40
corrugation waves	500-1000	
HII regions	300	
gal. fountain cycle		30
OB- and WR stars		20-40

To quantify these phenomena we propose a simple model (FEITZINGER and SPICKER /6/) for the gaseous component of the VARs in spiral arms. We suppose a tumbling motion about the longitudinal axis of the arm; in an ellipsoidal shell structure the gas moves in elliptical stream lines (Fig. 3). For reasonable values of $l = 400$ pc, $a_1 = 500$ pc, $a_1/b_1 = a_2/b_2 = 4$, $a_1/a_2 = b_1/b_2 = 1.25$ and 40 km/s/kpc the mean energy involved in the tumbling motion is $5 \cdot 10^{51}$ erg.

Guiding the gas motion needs the additional influence of magnetic field configurations. The tumbling motion can be incorporated into a bisymmetric spiral configuration of magnetic fields. Using the model of SAWA and FUJIMOTO /8/ (see also SOFUE et al.

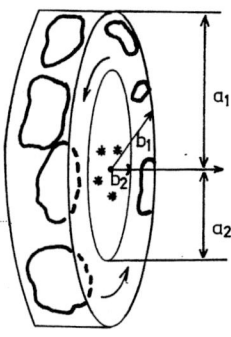

Fig. 3. Geometry of the tumbling
gas motion. The centre is occu-
pied by a star forming region.

/9/), the magnetic field z-structure follows naturally the ellipsoidal stream lines
(Fig. 4). The B_z-component of the field above the disk takes an arch line configuration. The direction of the gas z-motion, which shows no reflection symmetry, could
be produced by such a mechanism.

4. Stochastic Star Formation and Magnetic Fields

It is well known that star forming activity and magnetic fields are intimately connected. The correlation found recently between the content of molecular hydrogen gas
in galaxies and the magnetic field strength (SOFUE et al. /9/) corroborates such a
correlation. Since molecular clouds are bearing the star forming activity, there
should be a connection between magnetic field configurations and the propagation of
star formation. CHIANG and ELMEGREEN /10/ were the first to take this into consideration. In section 2 we have shown that dark clouds (= molecular clouds) are elongated
along the magnetic field lines. Such an alignment produces an additional magnetic
pressure on the gas perpendicular to the field lines. If a percolating star forming
process is ignited, an anisotropic motion and gas reshuffeling results; as a consequence, an anisotropic propagating star forming wave results. To account for such
an anisotropy we have incorporated in the models of selfpropagating stochastic star
formation an anisotropic stimulated probability distribution for the percolating star
forming process. The ratio of the anisotropy is chosen to be 5, similar to the axis
ratio of dark (molecular) clouds. The percolation probability becomes predominantly
greater along preexisting molecular cloud complexes. The models with a small shear
(the maximum rotational velocity of flat rotation curves must be less than 160 km/s)
produce longer and better defined spiral arm filaments. For higher rotational velocities the anisotropy is compensated by shear. In Fig. 5 a comparison is shown for a
model with and without a simulated magnetic field structure. The parameters are chosen to match the M 101 galaxy (Sc (s) I). The anisotropy enlarges the coherence scale
of the spiral arms by a factor 2 with a preponderance of coeval star forming events
on 2-3 kpc scales.

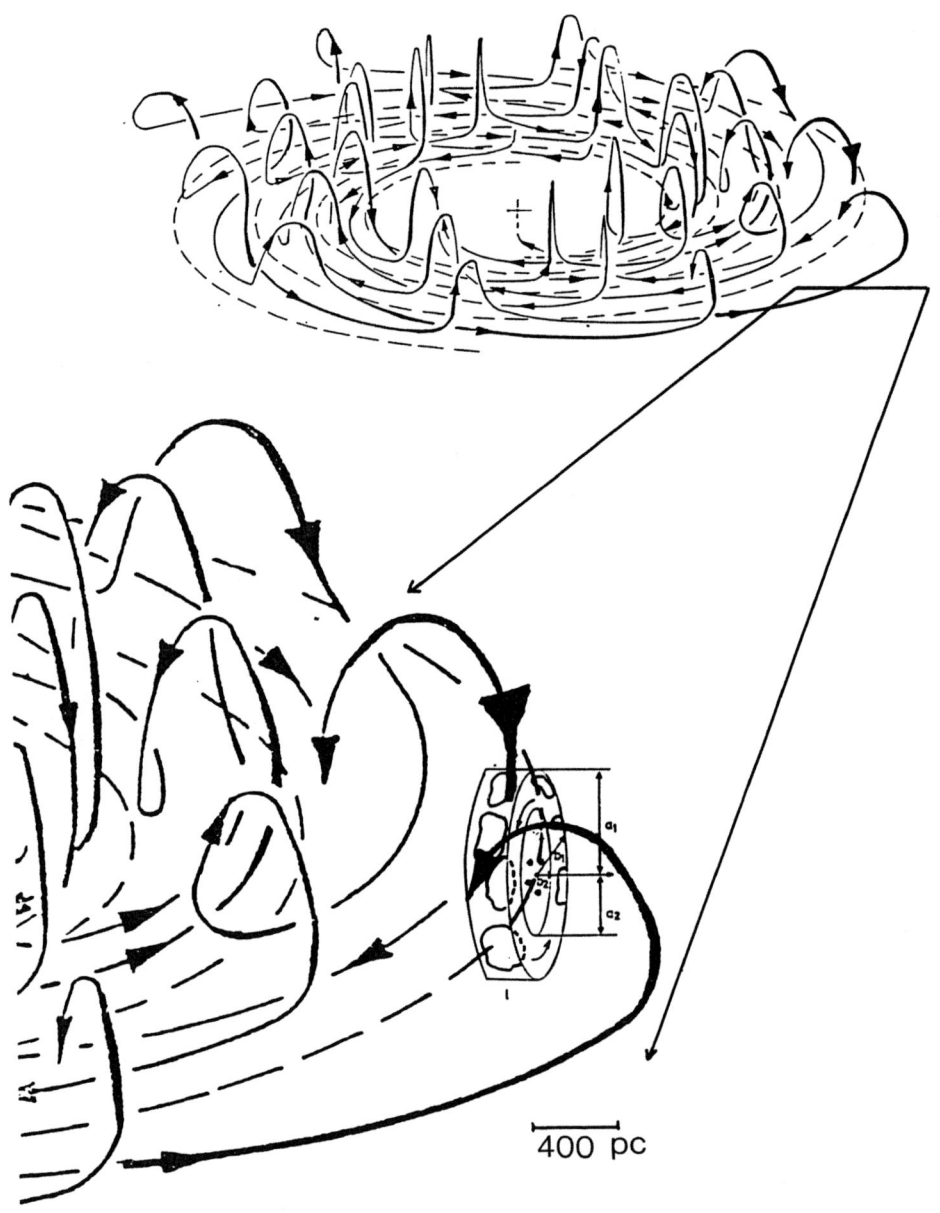

400 pc

Fig. 4 Global distribution of the bisymmetric spiral field
according to Fujimoto and Sawa /11/. The tumbling motion
phenomenon is incorporated.

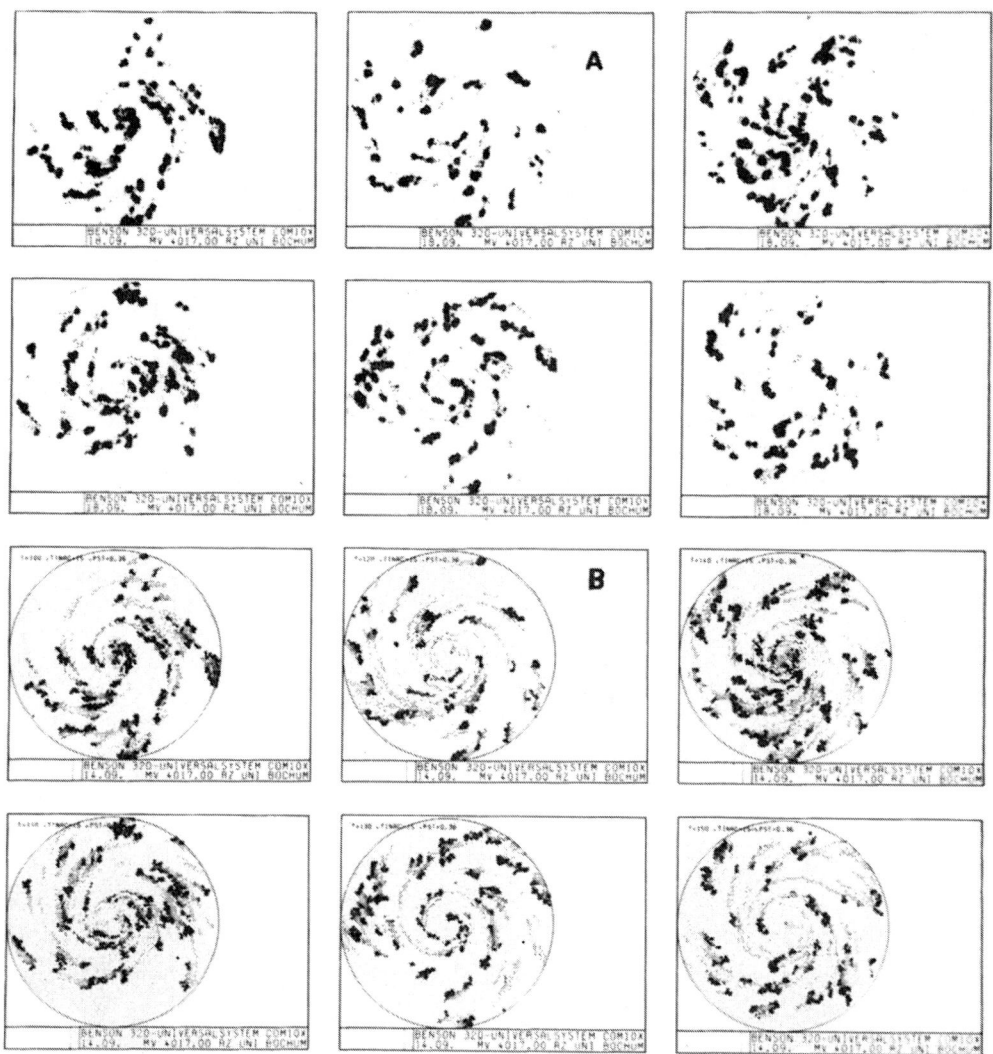

Fig. 5 Model galaxies with anisotropic (magnetic field A) and
with isotropic (without magnetic field B) percolating
star formation. The disk radius is 1o kpc, the flat
rotation curve has v_{max} = 15o km/s. Symbol size is in-
versely proportional to the age of the star forming
sites. All star forming events younger 10^8 yrs (1o time
steps) are plotted; every tenth time step is given.

References

1. Feitzinger, J.V., Stüwe, J.A., 1986, Ap. J. 305, 534
2. Feitzinger, J.V., Stüwe, J.A., 1986, Vistas in Astronomy, in press
3. Ellis, R.S., Axon, D.J., 1978, Ap. Spc. Sci. 54, 425
4. Feitzinger, J.V., Spicker, J., 1985, Mon. Not. Roy. Astr. Soc. 214, 539
5. Spicker, J., Feitzinger, J.V., 1986, Astron. Astrophys. 163, 43
6. Feitzinger, J.V., Spicker, J., 1986, Pub. Astr. Soc. Japan 38, 485
7. Spicker, J., Feitzinger, J.V., 1984, in Plasma Astrophysics, eds. T.D. Guyenne and J.J. Hunt, ESA SP 207 (European Space Agency, Paris), p. 225
8. Sawa, T., Fujimoto, M., 1986, Pub. Astr. Soc. Japan 38, 133
9. Sofue, Y., Fujimoto, M., Wielebinski, R., 1986, Ann. Rev. Astron. Astrophys. 23
10. Chiang, W.H., Elmegreen, B.G., 1982, Bull. Am. Astr. Soc. 14, 655
11. Fujimoto, M., Sawa, T., 1986, Pub. Astr. Soc. Japan, in press

Magnetic Fields in Supernova Remnants - Results from Radio Continuum Observations

E. Fürst

Max-Planck-Institut für Radioastronomie, Auf dem Hügel 69,
D-5300 Bonn 1, Fed. Rep. of Germany

1. Introduction

The observation of magnetic fields in supernova remnants (SNRs) may help to under-stand evolutionary effects, to study a possible coupling between the magnetic field of SNRs and the general galactic magnetic field, and to investigate the influence of galactic magnetic field variations on the morphology of SNRs.

The exploration of the magnetic field in SNRs relies on radio observations of the linearly polarized synchrotron emission. The orientation of the magnetic field component perpendicular to the line of sight can be derived from observed polariza-tion vectors obtained at several radio frequencies, after removing the Faraday rota-tion caused by free thermal electrons.

In case of galactic SNRs the observed Faraday rotation measure RM (rad m^{-2}) = $8.1 \cdot 10^5 \int N_e(cm^{-3}) \cdot \vec{B}(G) \cdot d\vec{r}(pc)$ is typically $|RM| < 200$ rad m^{-2}. At high radio fre-quencies the Faraday rotation angle $\tau(rad) = RM \cdot \lambda^2$ (cm) is rather small ($\tau(10$ GHz) $< 10°$, $\tau(30$ GHz) $< 1°$). In this case the detected polarization vectors are close to the intrinsic field vectors. However, the majority of polarization measurements on SNRs have been made at lower frequencies, particularly at ≈ 2.7 and ≈ 5 GHz. Such two-frequency measurements involve the problem of $\pm n \cdot \pi$ ambiguity corresponding to an ambiguity ΔRM in the estimation of RM. At the two mentioned frequencies the ambiguity is $|\Delta RM| \approx 360$ rad m^{-2}, larger than the typical values of RM. In this case it is often possible to obtain the intrinsic magnetic field vectors from a set of two measurements.

The strength of the magnetic field cannot be obtained directly from the obser-vations. In SNRs relativistic electrons coexist with comparatively dense thermal gas. Equipartition between the energy densities of the magnetic field and the relativistic electrons is therefore doubtful. Based on theoretical assumptions the magnetic field strength in thin sheets or filaments may reach 100 μG or more.

In the following we concentrate on polarization properties and the magnetic field orientations in SNRs.

2. The percentage of polarization in SNRs

About 50 percent of ≈ 150 known galactic SNRs are also detected in linear polariza-tion. Most of them are weakly polarized (integrated polarization 10% at 5 GHz). According to WEILER and SHAVER [1] SNRs form two main classes: filled-center and shell-type SNRs. A catalogue of filled-center or plerionic SNRs has recently been published by WEILER [2]. This type of SNRs is normally considered to be highly polarized (Crab Nebula, 3C58, G21.5-0.9 and G74.9+1.2, integrated polarization at 5 GHz between 15 and 20%). However, there are exceptions, in particular G24.7+0.6 (integrated polarization < 5% at 5 GHz).

Similarly, while most shell-type SNRs show weak polarization, there are several objects with high polarization, for example G30.7+1.0 (integrated polarization 32% at 5 GHz [3]), G93.3+6.9 (polarization up to 55% at 5 GHz [4]), G116.5+1.1 (polarization up to 60% at 2.7 GHz [5]).

In these cases a large fraction of ordered magnetic field and a very low Faraday depolarization is demonstrated.

In case of weakly polarized SNRs a large fraction of depolarization may be caused by interstellar clouds [6]. Polarization measurements of SNRs may, therefore, also be used to probe the interstellar medium.

3. The magnetic field orientation in SNRs

(a) Filled-center SNRs

Filled-center SNRs are sometimes considered to be sources with a uniform magnetic field structure. An example (3C58) is shown in Fig. 1. The uniform field is found to be aligned to the main axis of the source [7]. A similar structure is found for the Crab Nebula [8], where the uniform field follows the minor axis. There is, however, some doubt that a uniform field can be considered as common for the class of filled-center SNRs. The magnetic field orientation in G21.5-0.9 has been found to be very likely radial [9]. Vela X, the plerionic component of Vela XYZ shows a complicated cell structure of magnetic field orientation [10].

The classification of a SNR as filled-center implies several common properties, such as filled-center radio morphology, flat radio spectrum, central energy source, etc. However, the magnetic field orientation and the linear polarization percentage vary within this class of SNRs and can hardly serve as an identification.

(b) Shell-type SNRs

A typical example of young shell-type SNRs is shown in Fig. 2 [11]. The radial orientation of the magnetic field is also found in other historical shell-remnants. This field configuration demonstrates the stretching of the magnetic field lines by the fast expanding ejected matter.

Old SNRs often show a tangential magnetic field orientation (see for example G127.1+0.5 ([12], Fig. 3), demonstrating the interaction of the ejected matter and the ambient interstellar medium. Objects like G127.1+0.5 may easily be interpreted in terms of adiabatic shock compression of the interstellar medium as proposed by VAN DER LAAN [13].

In most cases, however, the magnetic field orientation shows a confusing picture and cannot be classified as either radial or tangential across the whole object [14].

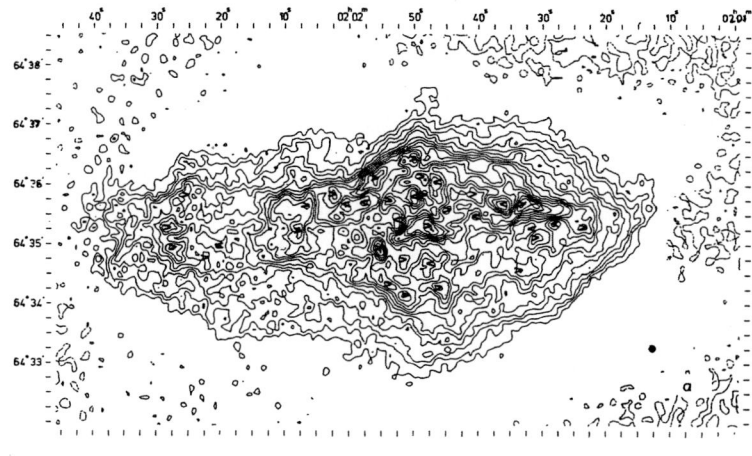

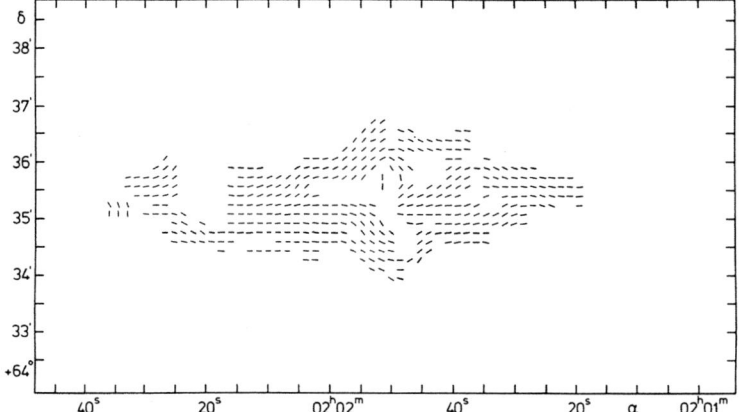

Figure 1: Total intensity and magnetic field distribution over 3C58 [7]
(a)   Distribution of the total intensity at 21 cm. The first positive contour is 10
*   mJy/beam area and the following three contours are 20, 35 and 50 mJy/beam*
*   area. After that the contours increase in equal steps of 25 mJy/beam area. No*
*   zero contour is plotted*
(b)   Direction of the component of magnetic field in a plane perpendicular to the
*   line of sight at a resolution of 24"×27". Line length has no significance*

Very complicated distributions are evident in supernova remnants presenting a well developed filamentary system (see for example HB21 [15]).

Regardless this complexity the original conclusion by DICKEL and MILNE [14] is still principally valid: Young shell-type SNRs present radial magnetic field orientation, while in old SNRs the orientation tends to be tangential.

A possible exception concerning the restriction of radial magnetic fields to young SNRs is possibly given by G179.0+2.7, which has recently been detected by FÜRST and REICH [16]. This source shows a radial magnetic field orientation, although it is probably an old SNR.

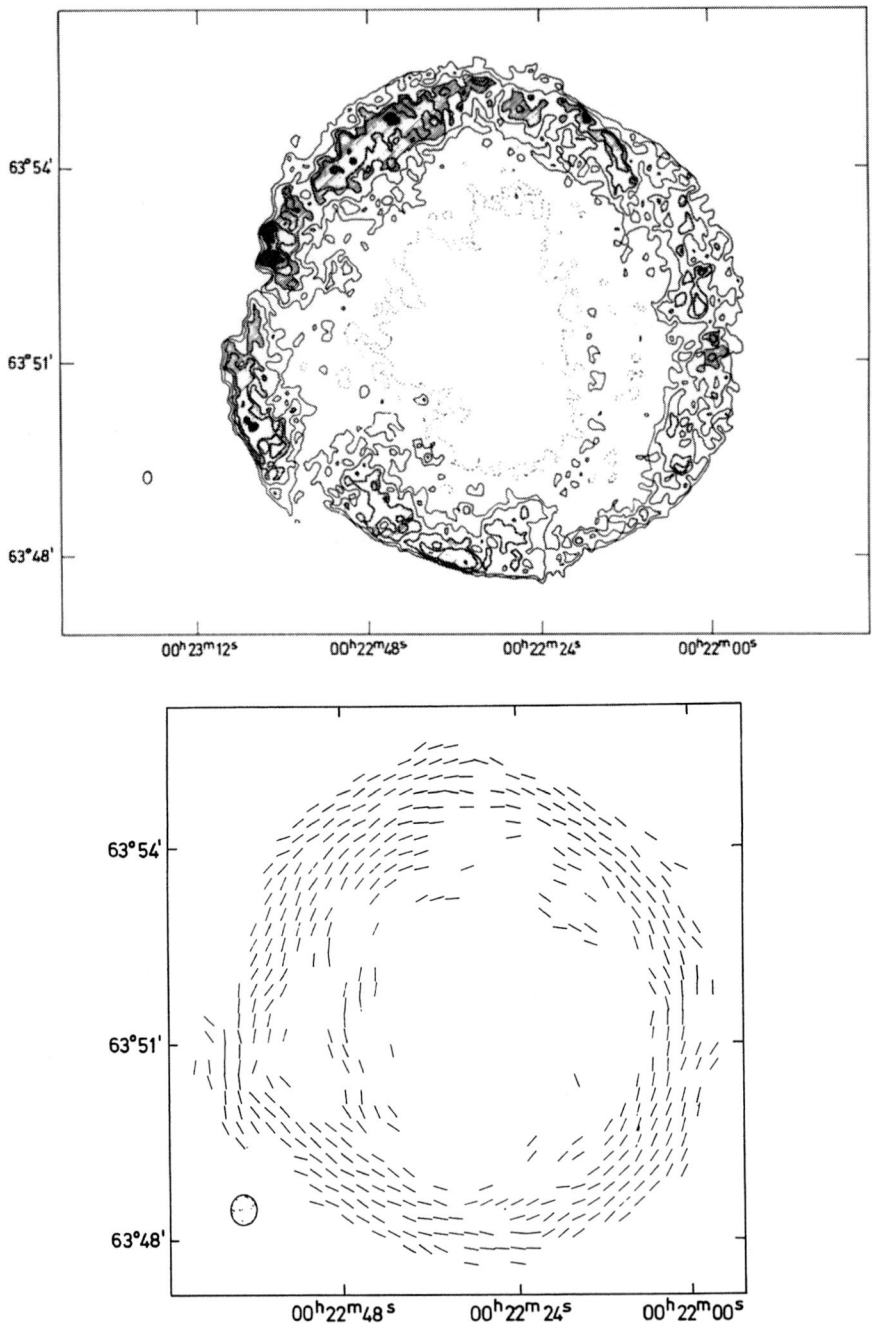

Figure 2: Total intensity and intrinsic polarization angle distribution of 3C10 (Tycho's SNR) [11]
(a) Full resolution contour map of the total intensity distribution. A detailed description of this figure is given by DUIN and STROM [11]
(b) Intrinsic polarization E-vector distribution as derived from 6 and 21 cm data. The resolution used is 24"×27" and the separation between points is 14". The length of the vectors has no meaning

PROJECTED MAGNETIC FIELD STRUCTURE OF G 127.1+0.5

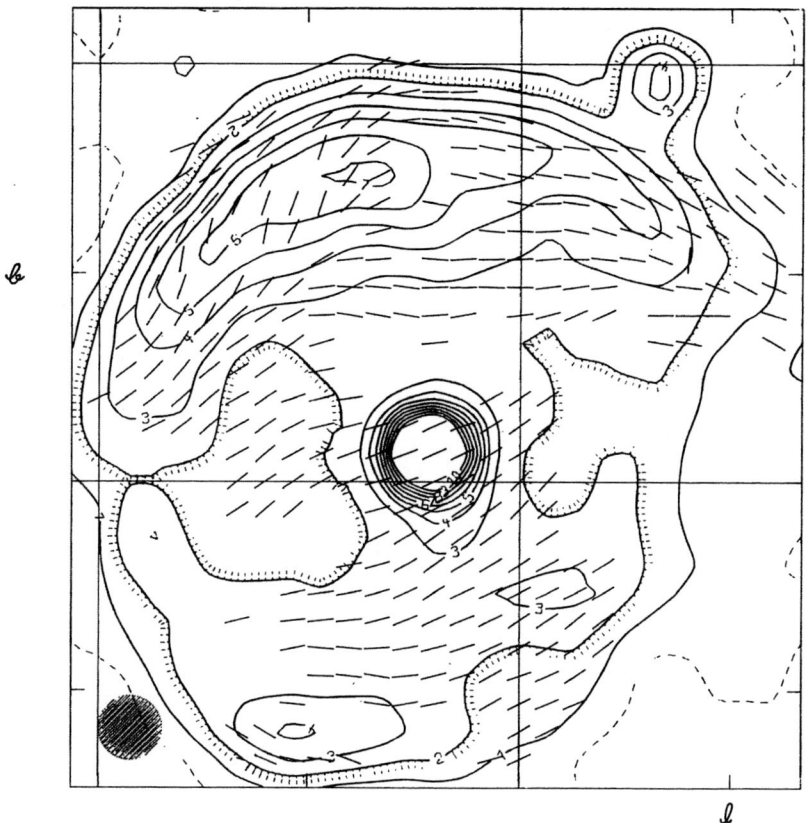

Figure 3: Total intensity and magnetic field distribution over G127.1+0.5 [12]. Vectors of unit length indicate the magnetic field structure, they are superposed on convolved 4750 MHz total intensity contours. The angular resolution is 4'5

Summary

The general results of the observations of magnetic fields in SNRs still present a confusing picture. There is little common structure neither for filled-center SNRs nor for shells. It is difficult to establish by which amount this result is caused by a limited sensitivity of the polarization measurements. Common properties are found only for young shells (radial field), or for tangential field orientations, which are found only in old SNRs. However, the radial field in the filled-center SNR G21.5-0.9 and the radial field in the old SNR G179.0+2.7 demonstrate that possibly the structure of the ambient medium and/or the initial conditions of the supernova explosion have a strong influence on the magnetic field orientation. The interpretation of the magnetic fields in SNRs is therefore most often limited to individual objects.

References

1. K.W. Weiler, P.A. Shaver: Astron. Astrophys. <u>70</u>, 389 (1978)

2. K.W. Weiler: In <u>The Crab Nebula and Related Supernova Remnants</u>, ed. by M.C. Kafatos and P.B.C. Henry (Cambridge University Press, Cambridge 1985), p. 265

3. W. Reich, E. Fürst, P. Reich, Y. Sofue, T. Handa: Astron. Astrophys. <u>155</u>, 185 (1986)

4. P. Lalitha, F. Mantovani, C.J. Salter, P. Tomasi: Astron. Astrophys. <u>131</u>, 196 (1984)

5. W. Reich, E. Braunsfurth: Astron. Astrophys. <u>99</u>, 17 (1981)

6. B.J. Burn: Monthly Notices Roy. Astron. Soc. <u>133</u>, 67 (1966)

7. A.S. Wilson, K.W. Weiler: Astron. Astrophys. <u>49</u>, 357 (1976)

8. A.S. Wilson: Monthly Notices Roy. Astron. Soc. <u>157</u>, 229 (1972)

9. R.H. Becker, A.E. Szymkowiak: Astrophys. J. <u>248</u>, L23 (1981)

10. D.K. Milne: Astron. Astrophys. <u>81</u>, 293 (1980)

11. R.M. Duin, R.G. Strom: Astron. Astrophys. <u>39</u>, 33 (1975)

12. E. Fürst, W. Reich, R. Steube: Astron. Astrophys. <u>133</u>, 11 (1984)

13. H. van der Laan: Monthly Notices Roy. Astron. Soc. <u>124</u>, 125 (1962)

14. J.R. Dickel, D.K. Milne: Australian J. Phys. <u>29</u>, 435 (1976)

15. W. Reich, E. Fürst, W. Sieber: 1983, In <u>Supernova Remnants and their X-ray Emission</u>, IAU Symp. 101, ed. by J. Danziger and P. Gorenstein (Reidel Publ. Co., Dordrecht 1983), p. 377

16. E. Fürst, W. Reich: Astron. Astrophys. <u>154</u>, 303 (1986)

Magnetic Fields in Supernovae and Supernova Shells

W. Kundt

Institut für Astrophysik, Auf dem Hügel 71, D-5300 Bonn 1, Fed. Rep. of Germany

There are indications that supernova explosions are driven by magnetic springs, i.e. that the spin energy of the collapsing core is transferred to the envelope via magnetic stresses which subsequently convert to relativistic $e^{\pm}$-pair plasma. The expanding pair plasma decomposes the star's shell into magnetised thermal filaments. Supernova shells are 3-dim networks of such filaments, of high mass density and low filling factor, whose explosion velocity decreases exponentially with time. Their surrounding space is filled with relativistic pair plasma and/or ambient matter of interstellar composition. The various (polarized) morphologies of SN shells can be understood in terms of the relative motion of these three components.

1. Introduction

The presence of magnetic fields in SN shells (remnants) - in approximate local pressure equilibrium with warm plasma (cf. Singal [1]) - is revealed by their linear polarization. As clearly demonstrated by Ernst Fürst at this workshop, the morphology of the magnetic fields can be different from shell to shell, hence any viable model for their explanation must be complex.

More precisely, linearly polarized radio waves are emitted by a source whenever it contains both relativistic electrons (and/or positrons) and ordered magnetic fields. It is widely believed by the experts that the necessary relativistic electrons can be provided in situ via shock acceleration, a mechanism whose needed high efficiency I consider unrealistic (Kundt [2]). Quite generally, it is difficult to devise engines with efficiencies exceeding several percent. Even more surprising: SN shells appear to contain relativistic electrons (and/or positrons) at high overpressure, as is indicated by the two bursting shells shown in figs. 2,3 (which will be discussed in section 3).

The strongest indication against efficient in-situ acceleration is the fact that supernovae of type Ib are radio emitters within one week of the explosion, those of type II L within several months (only), and the others (types Ia, II P) not at all (within years; Weiler et al [3]). This distinction in radio-loudness cannot be explained by different absorption screens. It indicates that the supply of relativistic electrons to the radio-photosphere can be almost instantaneous or strongly retarded, depending on the conditions near the exploding star (e.g. wind of binary companion, density of progenitor's windzone).

In order to explain why young SN shells tend to have radial magnetic fields and old shells circumferential or parallel fields, we must therefore look into their detailed structure, both in view of the supply of relativistic electrons and of the amplification of magnetic fields. As will be seen, the Sedov-Taylor description proposed by Shklovskiy in 1964 [4] turns out to be inadequate.

2. Supernova Explosions

It is often maintained that the outcome of an explosion (in air or plasma) is independent of the detailed explosion mechanism: at large distances from the center, most of the energy has been transferred to the ambient medium which is swept outward in the form of a thin, dense shell, a Sedov-Taylor wave. Hydrogen bomb explosions have verified this expectation. But in general the conclusion cannot be drawn.

It is well-known that there are two types of bombs: pressure bombs and splinter bombs, see fig. 1. Pressure bombs are thin-walled; their energy is transferred to the ambient medium as predicted by the Sedov solution. Splinter bombs transfer their energy to small-filling-factor, heavy fragments which are poor sweepers and thereby have a much larger destruction range. I maintain that SN explosions are of the splinter type.

In order to predict the outcome of a SN explosion, one should know the detailed explosion mechanism. Most of the literature of the 70s and 80s prefers neutrinos as a driving agent because rotation and magnetic fields are too difficult for present-day computers. Exceptions are Kardashev [5], Bisnovatyi-Kogan et al [6], Kundt [7], and Srinivasan [8] who envisage the transfer of the collapsing (degenerate) core's rotational energy to the surrounding shell via a magnetic spring. This latter mechanism seems to be indicated by the filamentary morphology of SN shells, reminiscent of magnetic Rayleigh-Taylor instabilities - as opposed to the cumulus-type morphologies of volcanic eruptions. At the same time it guarantees an approximately universal explosion energy and universal mass of the neutron-star remnant, independent of the progenitor star's mass, be-

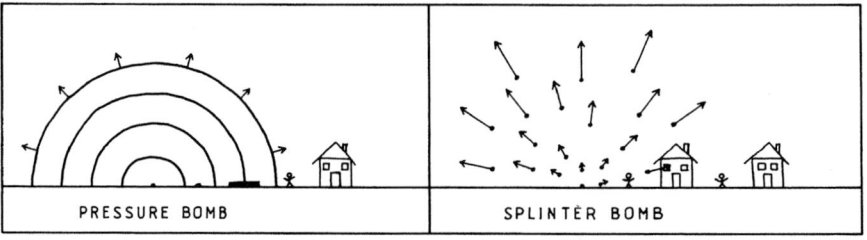

| PRESSURE BOMB | SPLINTER BOMB |

Figure 1: Pressure bombs give rise to Sedov waves, splinter bombs do not. The latter have a larger undecelerated range.

cause magnetic fields enhance the sweeping efficiency (cf. Kundt [9]). More-over, if a magnetic spring transfers the core's spin energy, of order 10^{52} erg, it must itself take an energy of this order which will afterwards be converted into high-energy particles, via reconnections. The high-energy particles can transfer their energy to relativistic $e^{\pm}$-pairs, on collision with thermal photons, so that the explosion may be driven by relativistic pair plasma (instead of by neutrinos). Once the pairs have pierced the stellar shell, they turn into a radio source.

The piston of a supernova may therefore be magnetised relativistic $e^{\pm}$-pair plasma. Even if this conclusion were premature, it is clear that the piston's (terminal) sound speed must exceed the terminal speed v of the exploding shell, of order 10^4 km/s; i.e. its (terminal) temperature must exceed 10^{10} K if it consisted of hydrogen. The piston is therefore lighter than its load, a Rayleigh-Taylor unstable setup during the switchon phase (of increasing pres-sure). During this phase, the stellar shell is likely to be torn and squashed into filamentary fragments. Details of supernova light curves and spectra confirm this expectation (to be published).

3. Inadequacy of the Sedov-Taylor Model

Supernova shells are not well described by a Sedov-Taylor wave (Kundt [10], [11]; Hamilton [12]). They consist of 3-dim networks of thermal filaments ('ropes': Kirshner & Arnold [13]), of small volume-filling factor $f \lesssim 10^{-3}$, pressure-confined by relativistic pair plasma, or ram-pressure confined by the ambient medium. Ram-pressure confinement is not a stable form of existence because ϱv^2 fluctuates strongly; but sound speed at temperature $T = 10^4$ K is some 10^{-3} times slower than the expansion speed v so that the squeezed filaments spread into conical domains of small opening angle $\approx 10^{-3}$ ($\leftrightarrow$3 arcmin). Their densities exceed those of the confining hydrogen plasma by a factor $m_p v^2/2kT \lesssim 10^6 \cdot v_9{}^2/T_4$, or confining pair plasma by a factor $m_p c^2/6kT \gtrsim 10^8/T_4$; i.e. the filaments move like tanks.

Not only the (spatial) density profile of a filamentary SN shell is vastly different from that of a Sedov wave. Also its (spatial) temperature profile is vastly different: instead of the predicted temperature $T \gtrsim T_{shock} \approx 0.1\ m_p v^2/k = = 10^9 K\ v_9{}^2$, the observed temperatures are $\gtrsim 10^4$ K (optical emission lines), $\lesssim 10^7$K (X-rays), and $\gtrsim 10^{13}$ K (radio synchrotron), revealing at least three different components.

Here are some further difficulties of the Sedov interpretation:

(3) The filaments of the Crab nebula are 8% post-accelerated, i.e. move faster now than on average (cf. Kundt & Krotscheck [14]).

(4) The dispersion measure and rotation measure of the Crab pulsar have shown linear increases throughout 2 years, corresponding to a plasma column length of

$b \lesssim 10^{14.5}$ cm/n_3 and parallel magnetic field srength $<B_{\parallel}> \gtrsim 10^{-4}$ G. (n_3:=number density/10^3 cm^{-3}; Rankin et al [15]).

(5) Cas A shows three spatially coincident shells of 'knots', 'flocculi' and 'radio peaks' with vastly different velocities and divergence centers (Kundt [10]).

(6) The Crab, Cas A, Cygnus Loop and others have 'forerunner' filaments of vastly higher speed. Their shell thicknesses are comparable to the radius.

(7) Even old (large) shells, like the Cygnus Loop or Puppis A, show strong nuclear-chemical gradients (Winkler & Kirshner [16]).

(8) The Sedov age of MSH 15-52 = G 320.4-1.2 = RCW 89 exceeds that of its neutron star remnant by a factor >7 (van den Bergh & Kamper [17]).

(9) The number of all known Galactic supernova remnants exceeds that of all $\leq 10^3$ yr old ones by only a factor of 10, implying an average age of 10^4 yr, whereas the Sedov description allows for ages >10^5 yr.

(10) Estimates of the expansion velocity as a function of shell radius, and (initial) explosion energy based on the Sedov model grow with the shell diameter (Dopita [18]; Mathewson et al [19]; Berkhuijsen [20]; Mills et al [21]; Braun [22]; Blair & Kirshner [23]).

(11) The log(cumulative number)-versus-log(diameter) plot suggests undecelerated expansion to diameters of order 30 pc for the SN shells of both the Magellanic Clouds, M31, M33, and the Galaxy (references under (10)). Green [24] and others have argued that this defect can be understood as due to a clumpy ISM; the difficulty is thereby swept under the carpet of our insufficient knowledge of the structure of the ISM and its interaction with SN shells.

(12) The X-ray map of RCW 86 shows a bright 'bump' in the southwest, of equal (X-ray) temperature to the rest of the shell. A bump means higher velocity, corresponding to a lower ambient density in the Sedov interpretation. But a lower ambient density should imply a lower X-ray brightness, contrary to the observation (Claas [25]).

(13) The Crab, Cas A, and Kes 32 show an 'outlet' (= 'chimney', 'spur', 'jet') of the relativistic plasma, see figs. 2,3 (Crab: Fesen & Gull [26]; Kes 32: Roger et al [27]). To me, these outlets look like overpressure reliefs of relativistic pair plasma (Kundt [28]), left over from the SN explosion and/or ejected by the central neutron star. If the X-ray 'bumps' of RCW 86 and the Cygnus Loop are milder versions of the same 'outlet' phenomenon, this phenomenon may be common to all SN shells.

4. Filamentary Supernova Shells

When compared with a Sedov wave, a filamentary shell has a low sweeping efficiency, long free-expansion stage, no reverse shock, and shorter (exponential) lifetime (Kundt [11]). What are its expected magnetic fields?

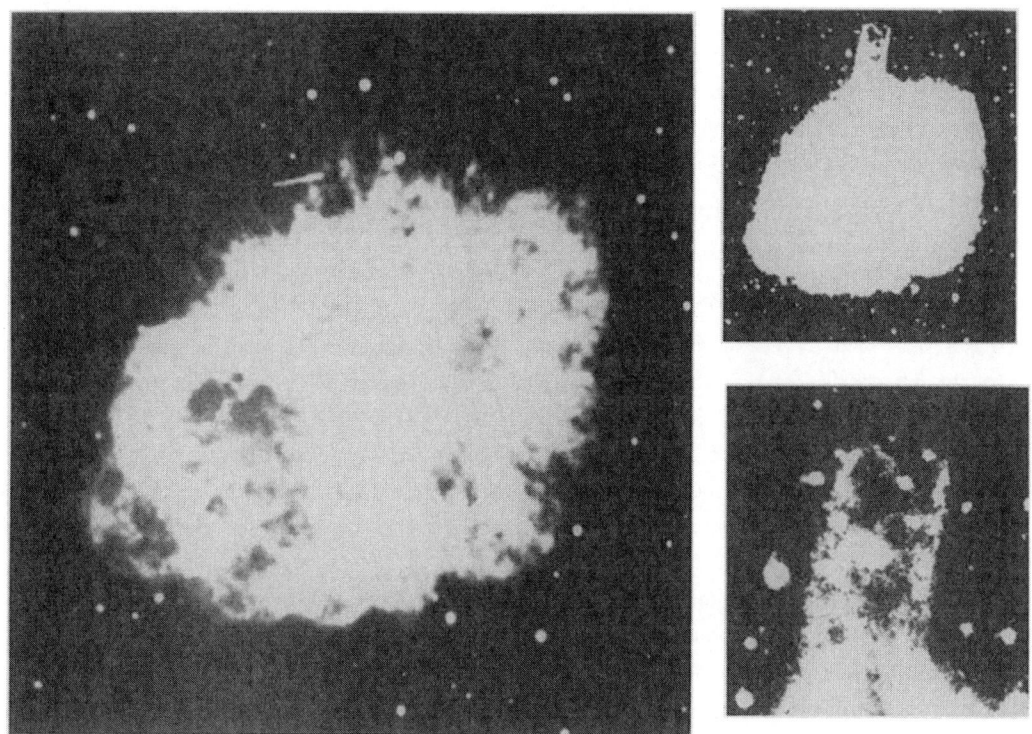

Figure 2: Emission-line map of the Crab nebula and its 'chimney', from Gull &
Fesen [29].

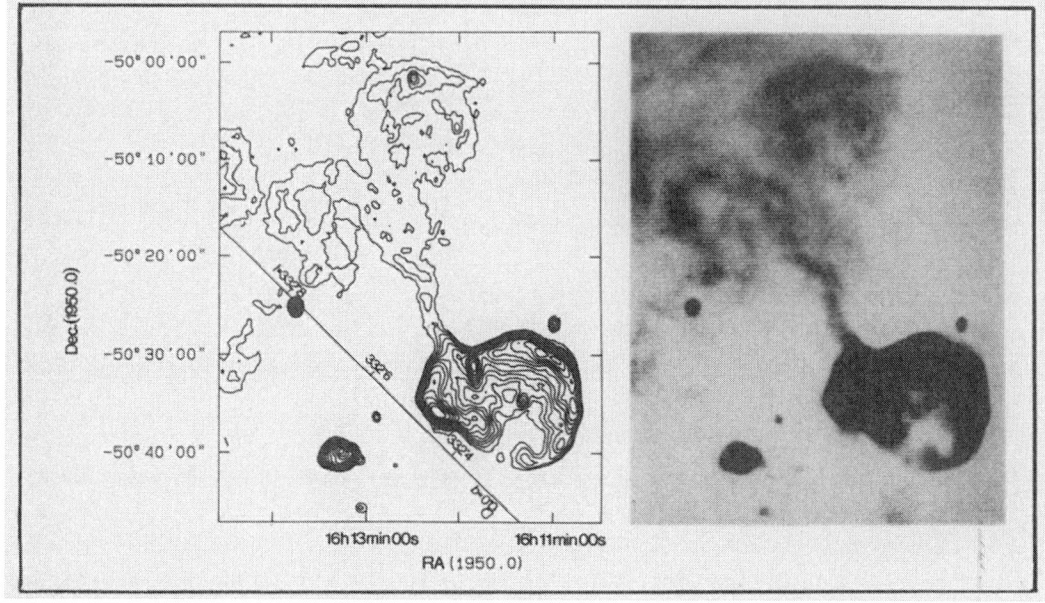

Figure 3: Radio map of the 'outgassing' SN shell Kes 32, from Roger et al [27].

Filaments in relative motion to their (weakly) magnetised environment tend to get (strongly) magnetised because they are good conductors (cf. Kundt [10]). Once their motion changes from subsonic - through relativistic pair plasma, like in the Crab - to supersonic - through interstellar plasma, like in Cas A - the strong supersonic shear is expected to stretch all available magnetic fields in radial direction. In this stage, SN shells are expected to show radial magnetic polarization.

At a later time, when the filaments have been decelerated to the sound speed of the (reheated) interior, of order $\lesssim 3 \cdot 10^2$ km/s, the interior will expand explosively and overtake the filamentary shell (Kundt [11]). The piston is again lighter than its load, but relaxing, hence the expansion is stable, the ambient matter is swept out. In this stage, magnetic fields are expected to be brushed towards the contact discontinuity and therefore to align peripherally, both from the inside and from the outside. The average field configuration may then appear circumferential or linear depending on whether the interior or exterior fields dominate.

Note, however, that only those field lines will contribute to the linear polarization which are populated by relativistic pair plasma, and that this pair plasma will be preferentially supplied by the interior, as a remnant of the supernova explosion and/or supply by the central neutron star. Only a multicomponent theory can hope to explain the large variety of different-looking SN shells.

Acknowledgements

A number of discussions with Sam Falle and Bob Kirshner are gratefully acknowledged.

5. References

1. A.K. Singal: Astron.Astrophys. 155, 242 (1986)
2. W. Kundt: J.Astroph.Astron. 5, 277 (1984)
3. K.W. Weiler, R.A. Sramek, N. Panagia, J.M. van der Hulst, M. Salvati: Astrophys.J. 301, 790 (1986)
4. I.S. Shklovskiy: A.Zh. 41, 176, 676 (1964)
5. N. Kardashev : Sovj.Astron. 14, 375 (1970)
6. G.S. Bisnovatyi-Kogan, Yu.P. Popov, A.A. Samochin: Astrophys.Sp.Sci. 41, 287 (1976)
7. W. Kundt : Nature 261, 673 (1976)
8. G. Srinivasan: invited lecture at XIXth General Assembly of IAU, New Delhi (1985)
9. W. Kundt: Bull.Astron.Soc.India 13, 12 (1985)
10. W. Kundt: Ann.N.Y.Acad.Sci. 336, 429 (1980)
11. W. Kundt, in: The Crab nebula and related SNRs, eds. Kafatos & Henry, Cambridge Univ. press, p. 151 (1985)

12. A.J.S. Hamilton: Astroph.J. 291, 523 (1985)

13. R.P. Kirshner, C.N. Arnold: Astroph.J. 229, 147 (1979)

14. W. Kundt, E. Krotscheck: Astron.Astrophys. 83, 1 (1980)

15. J.M. Rankin, D.B. Campbell, R.B. Isaacman, R.R. Payne: submitted to Astron. Astroph. (1986)

16. P.F. Winkler, R.P. Kirshner: Astroph.J. 299, 981 (1985)

17. S. van den Bergh, K.W. Kamper: Astroph.J. 280, L51 (1984)

18. M.A. Dopita: Astroph.J.Suppl. 40, 455 (1979)

19. D.S. Mathewson, V.L. Ford, M.A. Dopita, I.R. Tuohy, K.S. Long, D.J. Helfand: Astroph.J.Suppl. 51, 345 (1983)

20. E.M. Berkhuijsen: Astron.Astrophys. 140, 431 (1984)

21. B.Y. Mills, A.J. Turtle, A.G. Little, J.M. Durdin: Austr.J.Phys. 37, 321 (1984)

22. R. Braun: The interaction of supernovae with the ISM, Proefschrift Leiden (1985)

23. W.P. Blair, R.P. Kirshner: Astroph.J. 289, 582 (1985)

24. D.A. Green: Mon.Not.R.Astron.Soc. 205, 449 (1984)

25. J. Claas: contribution to Irsee ASI on 'Physical Processes in Interstellar Clouds' (1986)

26. R.A. Fesen, T.R. Gull: Astroph.J. 306, 259 (1986)

27. R.S. Roger, D.K. Milne, M.J. Kesteven, R.F. Haynes, K.J. Wellington: Nature 316, 44 (1985)

28. W. Kundt: Astron.Astrophys. 121, L15 (1983)

29. T.R. Gull, R.A. Fesen: Astroph.J. 260, L75 (1982)

Part IV

Magnetic Fields in Accretion Disks and Jets

Magnetic Fields, Black Holes and Accretion Disks

E. Asseo

Ecole Polytechnique, F-91128 Palaiseau, France

Abstract: We here briefly review the essential features associated to the presence of a magnetic field around Black Holes and Accretion Disks.

1. Introduction

While intensive future search for Black Holes hereafter BHs is one of the proposed aim for the Hubble Space Telescope, their existence has been suspected from observations a long time ago / i.e 1,2/.Cygnus X1 and A0620 - 00 are among the BH candidates. Determination of the mass for both of these objects ($M > 3M_\odot$) excludes any association with a NS or a WD. Besides, the existence of a supermassive central BH in the center of our own Galaxy (with $M \lesssim 10^6 M_\odot$) has been proposed but is still very controversial /3/. The source LMC X 3 (with $M < 4 - 9\ M_\odot$) in the Large Magellanic Cloud (LMC) belongs to extragalactic BH candidates as well as the two radio - sources Cygnus A and M87 and of course as all the observed Active Galactic Nuclei (A G Ns), Quasars, BL Lacs.......In these different contexts,a large scale magnetic field has been observed or inferred from observations: a magnetic field of about 10^{-6} Gauss pervades our own Galaxy while stronger fields have been measured in some regions of the LMC or associated to the radio emissions of Cygnus A and M 87 /4/.

BHs are also known to be very important from a theoretical point of view. Due to the huge work done during the last three decades, it became clear that "BHs are dynamical, evolving, energy - storing and energy - releasing objects" /5/. From detailled studies in BHs physics it has been shown possible to extract rotational and/or electromagnetic energy from a BH. On the other hand, recent developments of accretion disk models, necessary to interpret the observations from A G Ns and quasars, have revealed the importance of systems formed by a BH and an accretion disk in which electromagnetic effects play an essential role. Therefore a good knowledge of physical phenomena in the vicinity of BHs in the presence of a magnetic field is of great astrophysical significance. In particular, a study of magnetospheres around BHs and of magnetic effects in accretion disks and a correct evaluation of the electromagnetic energy extractable from such systems is inevitable (For a detailed and refined discussion see /2/ and /5/ for example).

2. Electrodynamical Properties of Black Holes

2.1. Isolated Black Hole

It is a very classical result that the electrodynamical properties of BHs are uniquely determined by the knowledge of their total mass - energy M_{BH}, of their charge Q_{BH} and of their angular momentum J (i.e.6). The horizon of a rotating Kerr BH is defined as the surface where $r = r_+ = M + (M^2 - Q^2 - J_*^2/M^2)^{1/2}$. This relation describes a true BH if and only if $M^2 \geqslant Q^2 + J_*^2/M^2$. Here $M = GM_{BH}/c^2$, $Q = \sqrt{G} Q_{BH}/c^2$ and $J_* = GJ/c^3$, c is the velocity of light and G the gravitational constant. This constraint links the three independent characteristic BH quantities. It implies that the BH can be endowed with a maximum charge $Q_{BH}^{max} = \sqrt{G} M_{BH}$. A crude estimation of the ratio of the electromagnetic to gravitational forces suffered by a particle near the horizon of the BH yields $F_E/F_G = Q_{BH}e/M_{BH} m_p G \leqslant e/m_p G^{1/2} \simeq 10^{18}$. Consequently when $Q_{BH} < 10^{-18} Q_{BH}^{max}$, F_G is dominant over F_E, the BH remains charged. In the reverse case F_E dominates and there is selective accretion of electric species according to the sign of the electric field. This leads to neutralization of the charged BH /7/. For this reason essentially uncharged BHs have been considered in the literature. However, BHs immersed in a magnetized medium can maintain a non-zero charge more easily.

2.2. Electromagnetic Structure Around Isolated BHs

The electromagnetic configuration around a BH rotating in vacuum depends on the magnetic moment, μ, associated to the BH, $\mu = Q_{BH} J/M_{BH}c$. Very far from the BH, the electromagnetic fields are quite analogous to the fields created by a rotating magnetic star /8,9/. As measured by an observer at rest in a local inertial system of reference, up to terms $O(\frac{1}{r^4})$, the electric field components contain /1/ a Coulomb term, Q_{BH}/r^2, /2/ a quadrupole term χ $\mu J M_{BH}^{-1}/r^4$ and /3/ an additional term χ $Q_{BH}J^2M_{BH}^2/r^4$, special to the choice of the system of reference. The magnetic field $B_r \simeq 2\mu\cos\theta/r^3$, $B_\theta \simeq \mu\sin\theta/r^3$, $B_\varphi = 0$, is dipolar. Later improvements and more detailed calculations concerning the multipolar electromagnetic structure around BHs can be found in /10, 11/.

2.3 Black Holes Immersed in a Magnetized Medium

The electromagnetic field configuration around a rotating BH immersed in a uniform magnetic field B_o is such that the BH attracts positive charges until $Q_{BH} = 2B_o JG/c^3$ /12/. A comparison between the isolated BH and immersed BH in B_o cases shows that it is easier to maintain charges on supermassive BHs when a magnetic field is present. Indeed, an isolated BH maintains its charge if $(Q_{BH}/M_{BH}) < 10^{-18}G^{1/2}$ while a Kerr BH in B_o keeps its charge when $(Q_{BH}/M_{BH}) < 2B_o M_{BH}G^{3/2}/c^2 \simeq 4 \cdot 10^3$ (with $M_{BH} = 10^8$ M , $B_o \simeq 10^{-6}$G). Clearly the presence of a magnetic field enhances the possibility for BHs to remain charged.

2.4. The Surface of the Black Hole Considered as an Electrically Conducting Bubble for the Horizon of a BH.

Damour /13, 14/ has generalized the classical notions of surface charge σ_H, surface current $\vec{K}$, electric ($\vec{E}$) and magnetic ($\vec{B}$) fields and has derived the equivalent classical laws of electrodynamics by means of a careful study of the interaction between a BH and an electromagnetic field (for a different approach see also /15/). There is conservation of charge for a system constituted by the horizon and the exterior of the BH, namely: $\frac{1}{\sqrt{\gamma}} \frac{\partial}{\partial t} (\sqrt{\gamma} \; \sigma_H) + \text{div} \; \vec{K} = J$ where $\sqrt{\gamma}$ is characteristic of the geometrical properties and J is the flux of charges entering the horizon of the BH. Integrating over the horizon the total charge of the BH, defined from Gauss theorem, is: $Q_H = \oint \sigma_H \, dS$. At the horizon, the tangential electric field and the normal magnetic field verify Faraday's law $\vec{\nabla} \wedge \vec{E} = -\frac{1}{\sqrt{\gamma}} \frac{\partial}{\partial t} (\sqrt{\gamma} \vec{B})$ and Ohm's law $\vec{E} + \vec{V} \wedge \vec{B} = 4\pi \; (\vec{K} - \sigma_H \vec{V})$. From the simultaneous validity of these laws it is possible to consider the surface of the BH as a bubble endowed with a surface electrical resistivity equal to 4π or equivalently to 377 ohms.

3. Energy Extraction from Black Holes

3.1. Possibility for an Electromagnetic Energy Extraction from Charged Black Holes

Three different contributions combine in the expression of the total mass-energy of an isolated BH: 1) the irreducible mass-energy of the BH, M_{ir}, represents the total energy of the BH after all its rotational and electromagnetic energy has been extracted, 2) an electromagnetic mass-energy, 3) a rotational energy. $M^2 = (M_{ir} + Q^2/4M_{ir})^2 + J_*^2/4 \; M_{ir}^2$, together with the constraint $M^2 \geqslant Q^2 + J_*^2/M^2$. There is no way to decrease M_{ir}: it is left unchanged by reversible transformations, it increases through irreversible ones. However, Q_{BH} and/or J, and thus M_{BH} can be increased or decreased at will through reversible or irreversible transformations i.e. by the addition of particles into the BH /16/. One deduces from this that an important fraction of the energy of the BH can become available. If J = 0, up to 50% of the total mass-energy of the BH stored in electromagnetic form can be drawn out, whereas, if Q_{BH} = 0, up to 29% of the energy stored in rotational form can be extracted. Thus BHs appear as the "largest storehouse of energy in the Universe". /16/.

3.2. The Penrose Process

The possibility to pull out energy from a rotating BH has been first realized by Penrose /17/. He discovered the existence of negative total mass-energy orbits in the "ergosphere" of a BH. The ergosphere is the region located between the horizon of the BH and a unique characteristic surface on which observers can remain at rest and through which the time like Killing vector $\partial/\partial t$ turns spacelike. A particle (p_1) with positive energy E_1, which enters the ergosphere, may be splitted into two particles (p_2) and (p_3) which show different behaviors. Particle

196

(p_2) may belong to a negative total mass - energy orbit and therefore will fall into the BH. Due to energy conservation, particle (p_3) will then recede from the horizon of the BH with positive mass - energy E_3 greater than E_1. Such an energy extraction from the region close to the BH may occur only if the angular momentum of particle (p_2) is negative. However, the huge astrophysical consequences of such a particle disintegration inside the ergosphere first appeared very difficult to exploit due to the necessity for a relativistic particle disintegration.

3.3. Influence of a Magnetic Field on the Penrose Process

It has been shown recently /18, 19, 20, 21/ that the Penrose process is more easy to achieve if the BH is immersed in a magnetic field. Matching a dipolar magnetic field and a Petterson vacuum solution /11/ at a surface located farther than the horizon it can be shown that negative energy states for particles p_2 exist with energy E_2: $V < E_2 < 0$, V being an effective potential. Near the horizon $V = (J_*/2r_+M^2)(J_* - 3\lambda/4(1 - J_*/M^2))$, where λ is characteristic of magnetic effects. Thus V should be negative, not only for a wide choice of negative angular momenta J but also for positive J. Moreover, very large V can be obtained if J is large and negative while λ is large and positive. Also in this case, the energy extraction region extends very far from the BH, farther than the ergosphere. Consequently the Penrose process is no more restricted to negative J, to small values of the energy or to the region of the ergosphere. The possibility to extract energy from the BH greatly enhances in the presence of a magnetic field.

4. Accretion Disks and Magnetic Field

Many books and reviews describe the physics of accretion disks /i.e 2, 22, 23, 24/. Some refer more specifically to magnetized accretion disks /i.e, 4, 24, 25, 26, 42/. Besides molecular viscosity, negligible in accretion disks, turbulent viscosity may help to interpret the angular momentum transport. Part of the recent thin accretion disk models focus on the role of magnetic viscosity. The hypothesis is that the accretion disk suffers a strong Keplerian differential rotation together with a convective turbulence. A highly turbulent dynamo - resulting magnetic field triggers an anomalous viscous transport.

Three different regions are defined in the standard model for thin accretion disks around BHs /27/. They are characteristic of the relative values of the gas and radiative pressures, P_g and P_R, and also of the relative values of the free - free absorption coefficient due to bremsstrahlung, κ_{ff}, and of the Thomson diffusion coefficient on free electrons, κ_T "Fig.1". In the vertical direction, compensation between the pressure gradient and the gravity is assumed. At this time there is no definite answer to the difficult problem of knowing on which pressure, P_g or P_R or $B^2/8\pi$, the viscous stress tensor depends. Some of the results are reported in Table 1 and can be found in references /27 - 33/.

-Table 1 -

	Basic Process for Angular Momentum Transport	Stress Viscosity Tensor.
Standard Model 1973	Hydrodynamical and Magnetic Turbulence	ad - hoc law $T_{r\varphi} = \alpha P_g, \alpha \leqslant 1$ $\alpha = v_t/v_s + v_A^2/v_s^2$
Eardley, Lightman 1975	Chaotic Magnetic Field $= \Sigma$ magnetic cells. Shear and Reconnection	$T_{r\varphi} = \frac{B^2}{8\pi} \times \frac{1}{2}\left(\frac{RL}{H}\right)^2 P_g$
Coroniti 1981	MHD Turbulence $= \Sigma$ magnetic cells. Shear and Reconnection	$T_{r\varphi} \leqslant \frac{9}{4}\frac{P_g}{R^{2/3}}$
Sakimoto Coroniti 1981	Magnetic Turbulence	$T_{r\varphi} \sim \alpha\frac{B^2}{8\pi} \sim \alpha P_g$ for the equipartition model
Chageli shvili Lominadze 1984	Convective Turbulence and Magnetic Field	$T_{r\varphi} \times P_R$ in the inner region

v_t is the turbulent velocity, v_S the sound velocity, v_A the Alfven velocity. R is the reconnection rate, L the spatial scale length of a magnetic cell, H the scale height of the accretion disk.

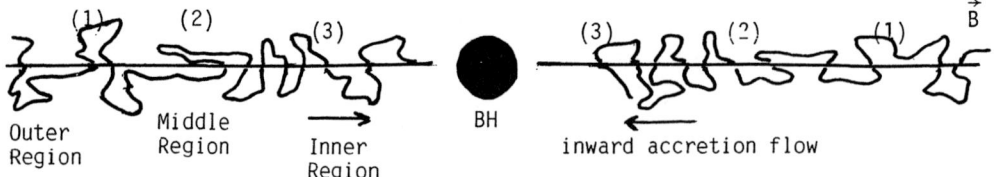

Outer Region   Middle Region   Inner Region   BH   inward accretion flow

Figure 1. The different regions in a thin accretion disk surrounding a black hole. (1) $P_g \gg P_r$, $\kappa_H \gg \kappa_T$, (2) $P_g \gg P_R$, $\kappa_H \ll \kappa_T$, (3) $P_g \ll P_R$, $\kappa_H \ll \kappa_T$.

5. Electromagnetic Energy Extraction from Black Holes and from Accretion Disk Systems

The problem is to know whether a rotating BH can energize relativistic jets from AGNs by liberating its rotational energy through electromagnetic processes. From

Penrose process /17/ and Christodoulou and Ruffini analysis /16/, part of the
total mass - energy of the BH stored in rotational or electromagnetic form can
be removed. When a magnetic field is present such processes would likely be
performant in an astrophysical situation and lead to an efficient extraction of
energy /20,21/. However in the models introduced by Lovelace /34/, Blandford /35/
or Blandford and Znajek /36/, the electromagnetic energy extraction process is
different. It is based on an analogy with the theory of pulsars /i.e 37/ assuming
that AGNs or quasars could be powered by BHs. The accretion disk and BH system is
considered as connected to a large - scale magnetic field, probably rather strong
near the surface of the hole (B $\approx 10^4$ Gauss) "Fig.2". The energy and angular
momentum lost by the accreted matter and by the BH itself can be dragged out by
electromagnetic torques. The idea is that the toroïdal component of the field, B_T,
created by the circulation of currents on the external surface of the accretion
disk, pulls out the rotational energy of the hole and disk and transfers it to
accelerated charged particles forming a jet. "Fig.3". Thus the possibility to
remove the gravitational energy of infalling matter and to collimate it in the
direction of the rotation axis is described.

In the Blandford and Znajek model /36/, combination between the rotation of
the disk and BH, and of the large magnetic field $\vec{B}$ frozen in the accreted matter,
induces a strong non-zero electric field parallel to $\vec{B}$ near the horizon of the
BH. Then pulsar - like mechanisms /i.e 37, 38/ lead to pair creation in a gap
located outside the BH horizon and to the existence of a force - free magnetosphere beyond this gap "Fig.4". A self-consistent model can be studied by means
of a single equation for the azimuthal potential A_ϕ. In particular $B_T = B_T(A_\phi)$
and the angular velocity $\omega = \omega(A_\phi)$. Thus for the total rates energy , $R_{\&}$, and
of angular momentum, R_J, electromagnetically extractable from the BH , one obtains

$$R_{\&} = - 4\pi\varepsilon_0 \int_{A_\phi(pole)}^{A_\phi(equator)} dA_\phi \, B_T(A_\phi) \, \omega \, (A_\phi),$$
and
$$R_J = - 4\pi\varepsilon_0 \int_{A_\phi(pole)}^{A_\phi(equator)} dA_\phi \, B_T(A_\phi).$$

Spinning up a paraboloïdal (radial) magnetic field and BH it can be shown that up
to 38% (50%) of the electromagnetic energy can be extracted from the rotating hole.

The structure of such force - free BH magnetospheres has been considered anew
recently by Thorne and Macdonald /5, 39, 40/ using a much more general viewpoint.
They have shown that there is a mathematical equivalence between the 4 - dimensional
description of General Relativity and a "3 + 1" formalism corresponding to a separation of the space-time into a curved 3 - dimensional space and a universal time t.
Then the stationary, axisymmetric BH space-time is characterized by a diagonal metric
$ds^2 = (\exp 2\mu_1) \, dr^2 + (\exp 2\mu_2) \, d\theta^2 + \tilde{\omega}^2 \, d\phi^2$ and by the universal time t
(μ_1, μ_2 and $\tilde{\omega}$ depend on r and θ only). The laws of electromagnetism, equivalent to

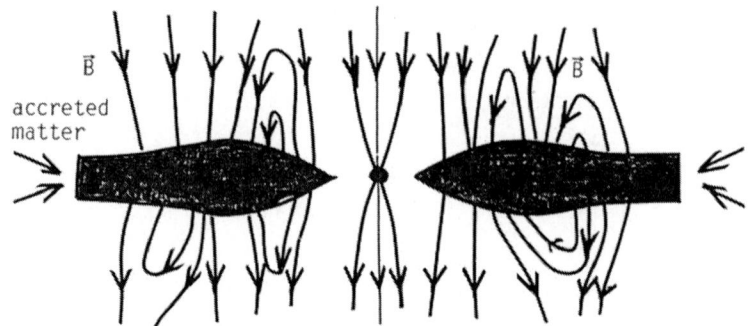

Figure 2.   Accretion disk and black hole system connected to a large-scale poloïdal magnetic field (After/5/).

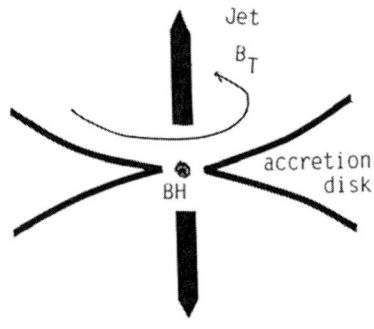

Figure 3.   The azimuthal magnetic field B_T is responsible for the extraction of energy and angular momentum from the black hole and accretion disk system which ultimately leads to the formation of jets.

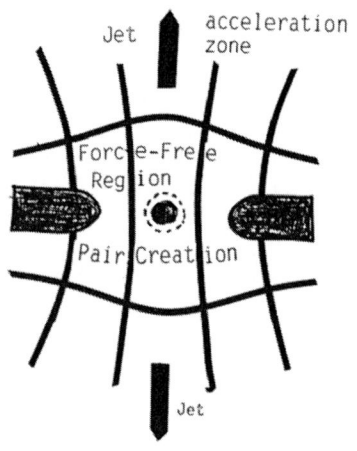

Figure 4.   The Blandford-Znajek model/36/.

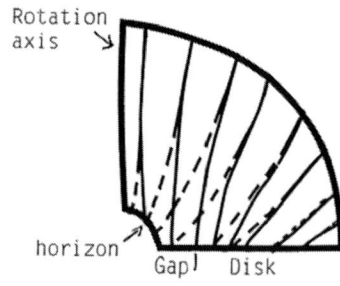

Figure 5. Deformation of paraboloïdal magnetic field lines after spinning up the BH (dotted lines are paraboloïdal lines, full lines are deformed lines).

the laws in classical electrodynamics, write in terms of the electric and magnetic fields and of the charge and current densities, as measured by a set of particular observers. The electromagnetic configuration of the magnetosphere derives from a single equation for the unique stream function $\Psi(r,\theta) = 2\pi A_\phi$, with additional boundary conditions. Ψ physically represents the total magnetic flux through an azimuthal loop. In Macdonald's numerical analysis /41/ the boundary conditions are given at the limits of the poloīdal force - free region. Starting from a static vacuum magnetic field solution in the force - free magnetosphere he shows that in the case of paraboloīdal field lines there is concentration of magnetic flux on the hole and the electromagnetic energy focusses along the rotation axis, as observed in astrophysical jets "Fig.5".

Summary

Magnetic fields are an essential ingredient in the study of the physics of black holes and accretion disks. In the presence of a magnetic field the possibility to extract energy from the rotating hole is greatly enhanced and the region where the process can occur extends much farther than the ergosphere. A BH and accretion disk system is the nowadays accepted model for the interpretation of observations from AGNs and quasars. From recent jet formation models it is possible to estimate how much electromagnetic energy can be extracted from such systems and to fulfill the energy requirements. Finally, the role of magnetic viscosity in accretion disks appears primordial, although yet unclear.

References

1. S.L. Shapiro, S.A. Teukolsky: Black Holes, White Dwarfs and Neutron Stars (John Wiley and Sons, Wiley Interscience Publication, New York 1983)
2. M.C. Begelman, R.D. Blandford, M.J. Rees: Rev. Mod. Phys. 56, 2, 255 (1984)
3. M.J. Rees: In The Milky Way Galaxy, ed. by H. Van Woerden et al., IAU Symposium No. 106 (Reidel Publ. Co., Dordrecht 1985), p. 379
4. E. Asseo, H. Sol: Physics Reports (to be published) (1987)
5. K.S. Thorne, R.H. Price, D.A. Macdonald (Eds.): Black Holes: The Membrane Paradigm (Yale University Press, New Haven and London 1986)
6. C.W. Misner, K.S. Thorne, J.A. Wheeler: Gravitation (W.H. Freeman and Company, San Francisco 1973)
7. D.M. Eardley, W.H. Press: Ann. Rev. Astron. Astrophys. 13, 381 (1975)
8. D. Christodoulou, R. Ruffini: In Black Holes, ed. by C. De Witt and B.S. De Witt (Gordon and Breach, New York 1973)
9. A. Deutsch: Annales d'Astrophysique 18, 1 (1955)
10. A.R. King, J.P. Lasota, W. Kundt: Phys. Rev. D. 12, 3037 (1975)
11. J.A. Petterson: Phys. Rev. D. 12, 2218 (1975)
12. R.M. Wald: Phys. Rev. D. 10, 1680 (1974)
13. T. Damour: Ph.D. thesis, University Paris VI (1979)
14. T. Damour: Phys. Rev. D. 18, 3598 (1978)
15. R.L. Znajek: Monthly Notices Roy. Astron. Soc. 185, 833 (1978)
16. D. Christodoulou, R. Ruffini: Phys. Rev. D. 4, 3552 (1971)
17. R. Penrose: Revista Nuovo Cimento 1, 252 (1969)

18. S.V. Dhurandar, N. Dadhich: Phys. Rev. D. $\underline{29}$, 27 (1984)

19. S.V. Dhurandar, N. Dadhich: Phys. Rev. D. $\underline{30}$, 1625 (1984)

20. S.V. Dhurandar, N. Dadhich: Astrophys. J. $\underline{307}$, 38 (1986)

21. S.M. Wagh, S.V. Dhurandar, N. Dadhich: Astrophys. J. $\underline{290}$, 12 (1985)

22. J.E. Pringle: Ann. Rev. Astron. Astrophys. $\underline{19}$, 137 (1981)

23. J. Frank, A.R. King, D.J. Raine: Accretion Power in Astrophysics (Cambridge University Press, Cambridge 1985)

24. F.V. Coroniti: In Unstable Current Systems and Plasma Instabilities in Astrophysics, ed. by M.R. Kundu and G.D. Holman, IAU Symposium No. 107 (Reidel Publ. Co., Dordrecht 1985), p. 453

25. J.J. Aly: In Magnetospheric Phenomena in Astrophysics, Mexico (1984)

26. R.V.E. Lovelace: Preprint (1984)

27. N.I. Shakura, R.A. Sunyaev: Astron. Astrophys. $\underline{24}$, 337 (1973)

28. D.M. Eardley, A.P. Lightman: Astrophys. J. $\underline{200}$, 187 (1975)

29. A.A. Galeev, R. Rösner, G.S. Vaiana: Astrophys. J. $\underline{229}$, 318 (1979)

30. F.V. Coroniti: Astrophys. J. $\underline{244}$, 587 (1981)

31. P.J. Sakimoto, F.V. Coroniti: Astrophys. J. $\underline{247}$, 19 (1981)

32. R.E. Pudritz: Monthly Notices Roy. Astron. Soc. $\underline{195}$, 881 and 897 (1981)

33. G.D. Chagelishvili, J.G. Lominadze: Proceedings of a Course and Workshop on Plasma Astrophysics, Varenna, Italy, ESA SP. $\underline{207}$, p. 131 (1984)

34. R.V.E. Lovelace: Nature $\underline{262}$, 649 (1976)

35. R.D. Blandford: Monthly Notices Roy. Astron. Soc. $\underline{176}$, 465 (1976)

36. R.D. Blandford, R.L. Znajek: Monthly Notices Roy. Astron. Soc. $\underline{179}$, 433 (1977)

37. P. Goldreich, W.H. Julian: Astrophys. J. $\underline{157}$, 869 (1969)

38. M.A. Ruderman, P.G. Sutherland: Astrophys. J. $\underline{196}$, 51 (1975)

39. K.S. Thorne, D. Macdonald: Monthly Notices Roy. Astron. Soc. $\underline{198}$, 339 (1982)

40. D. Macdonald, K.S. Thorne: Monthly Notices Roy. Astron. Soc. $\underline{198}$, 345 (1982)

41. D. Macdonald: Monthly Notices Roy. Astron. Soc. $\underline{211}$, 313 (1984)

42. Ya.B. Zeldovich, A.A. Ruzmaikin, D.D. Sokoloff: Magnetic Fields in Astrophysics (Gordon and Breach Science Publishers, New York 1983)

The Influence of External Magnetic Fields on the Structure of Thin Accretion Disks

U. Anzer, G. Börner, and E. Meyer-Hofmeister

Max-Planck-Institut für Physik und Astrophysik, Institut für Astrophysik,
Karl-Schwarzschild-Str. 1, D-8046 Garching, Fed. Rep. of Germany

1. Introduction

There is a generally accepted picture of many binary X-ray sources wherein a rotating magnetized neutron star accretes matter from a surrounding disk of plasma. As the accreting matter approaches the rotating neutron star it is more and more influenced by the stellar magnetic field until eventually its motion is dominated by the field. Finally the plasma will flow along the field lines towards the surface of the neutron star.

The description of the vertical structure of thin accretion disks without external magnetic fields has been achieved with the use of stellar evolution codes which allow one to calculate the vertical energy transport [Meyer & Meyer-Hofmeister 1982].

On the other hand studies of a simple magnetospheric configuration around a neutron star [Anzer & Börner 1983] using a greatly simplified disk structure have convinced us that many interesting observations can be explained from the interaction of the accretion disk and the magnetosphere.

It seems promising therefore to combine these two approaches. Thus the modest aim of this work is a description of the equilibrium structure of a thin accretion disk subjected to the pressure of an external magnetic field.

2. The Accretion Disk Model

The disk is considered as a radial structure of small vertical extent, $z \ll r$ ("thin disk"), as suggested from estimates of the scale height for hydrostatic equilibrium.

Conservation of mass and angular momentum require

$$\partial_t(2\pi r\Sigma) - \partial_r M = 0, \qquad (1)$$

$$\partial_t(2\pi r\Sigma r^2\Omega) - \partial_r(r^2\Omega\dot{M} + 2\pi r^3\partial_r\Omega \int \mu dz) = 0. \qquad (2)$$

These are the basic equations for the radial structure integrated over the vertical extent z. Here μ is the viscosity,

$$\Sigma \equiv \int dz\rho \qquad (3)$$

Σ the surface density, z the height above the midplane of the disk, M the inward mass flow rate, $\Omega \equiv r^{-1}v_k$.

Consider a stationary disk ($\partial_t = 0$) for which, according to eq. (1), $\dot{M}$ is constant in time and radius. The radial inflow is governed by equation (2) which then reads

$$\frac{d}{dr} (r \, v_k \, \dot{M} + 2\pi \, r^3 \, \frac{d}{dr} (\frac{v_k}{r}) \int \mu dz) = 0. \tag{4}$$

The viscosity is parametrized by a dimensionless number α ("α-disks" see e.g. Shakura & Sunyaev (1973)) such that

$$\mu r \mid \frac{d}{dr} (\frac{v_k}{r}) \mid = \alpha \, p, \tag{5}$$

where p is the pressure in the disk. Usually the parameter α is taken to be a constant of the order of 1.

The vertical structure of the disk [see Meyer & Meyer-Hofmeister 1982] in hydrostatic equilibrium follows from

$$\frac{dp}{dz} = -\rho \, g_z = -\rho \, \frac{GM}{r^2} \, z \tag{6}$$

g_z is the z-component of the gravitational field of the neutron star.

The vertical energy transport can be radiative or convective depending on the temperature gradient $d\ln T/d\ln\rho$. The temperature gradient is given as

$$\frac{dT}{dz} = \begin{cases} \dfrac{-3\kappa\rho}{4acT^3} \, F & \text{for} \quad \nabla \leqslant \nabla_{ad} \\[2em] \dfrac{\rho T}{p} \, g_z \, \nabla_{conv} & \text{for} \quad \nabla > \nabla_{ad} \end{cases} \tag{7}$$

where κ is the opacity. The flux gradient is given by

$$\frac{dF}{dz} = \frac{9}{4} \, \frac{GM}{r^3} \, \mu \, . \tag{8}$$

At each radius the heat produced by friction is transported vertically outwards, and radiated away from the surface. This leads to a relation between the effective surface temperature T_{eff}, and the mass-flow

$$\sigma \, T_{eff}^4 = \frac{3}{8\pi} \, \frac{GM \, \dot{M}}{r^3} \, . \tag{9}$$

The solution of the vertical structure equations (6) to (9) at a given radius r results in functions $p(z,r)$, $F(z,r)$, $T(z,r)$ with input parameters M and $\dot{M}$.

3. The Model of the Magnetosphere

We choose a configuration where the magnetic dipole field $\underline{B}$ has its moment $\underline{\mu}$ in the plane of the disk, perpendicular to the rotation axis $\underline{\Omega}$ of the neutron star. We assume that the magnetic field does not penetrate into the disk at all, that only an external magnetic pressure is exerted on the disk. In such a configuration the thin disk is in good approximation a potential surface of the field.

We assume that the external magnetic pressure depends only on the radius

$$p_B = \frac{B_0^2}{8\pi} \left(\frac{r}{R}\right)^{-6} . \tag{10}$$

The outer boundary for the disk is defined by the criterion that the gas pressure in the disk is equal to the external magnetic pressure

$$p_G = p_B . \tag{11}$$

In the nonmagnetic case the disk thickness z_0 is defined as the height, where the optical depth τ is 2/3.

4. Disk and External Magnetic Field

The vertical structure equations (6) to (9) for the disk are solved with the outer boundary condition (11), instead of the boundary condition for a free atmosphere.

The parameters were chosen from the X-ray source Her X-1:

mass $M = 1.4\ M_{\odot}$ period $P = 1.24$ sec

$\dot{M} = 10^{17}$ g/sec; $B_0 = 5 \times 10^{12}$ Gauss.

The corotation radius is $r_c \equiv \left(\frac{GM}{\Omega^2}\right)^{1/3} \simeq 2 \times 10^8$ cm.

In figure 1 the run of various quantities with z is displayed for $r = r_c$.

At this radius a pronounced difference of the cases with and without magnetic field is evident. There is a much more sharply defined outer edge of the disk located at $z_0 = 2 \times 10^6$ cm.

The temperature drops steeply near the boundary, and is below the non-field value by a factor ~5 at $z = 2 \times 10^6$ cm, whereas near $z \simeq 0$ both cases coincide.

The density stays almost constant, and then rises close to z_0. The disk has a lower density inside than at the boundary. This inversion of the density gradient may drive some additional mixing.

In figure 2 the thickness of the disk z_0 (determined by $p_B = p_G$ at $z = z_0$, or $\tau(z_0) = 2/3$ for $\underline{B} = 0$) is plotted versus the radius. The magnetic field compresses the disk, but at large radii this is a small effect only. Close to the inner edge, on the other hand, at $r = 2 \times 10^8$ cm the disk is about half as thick with an external magnetic field than without one.

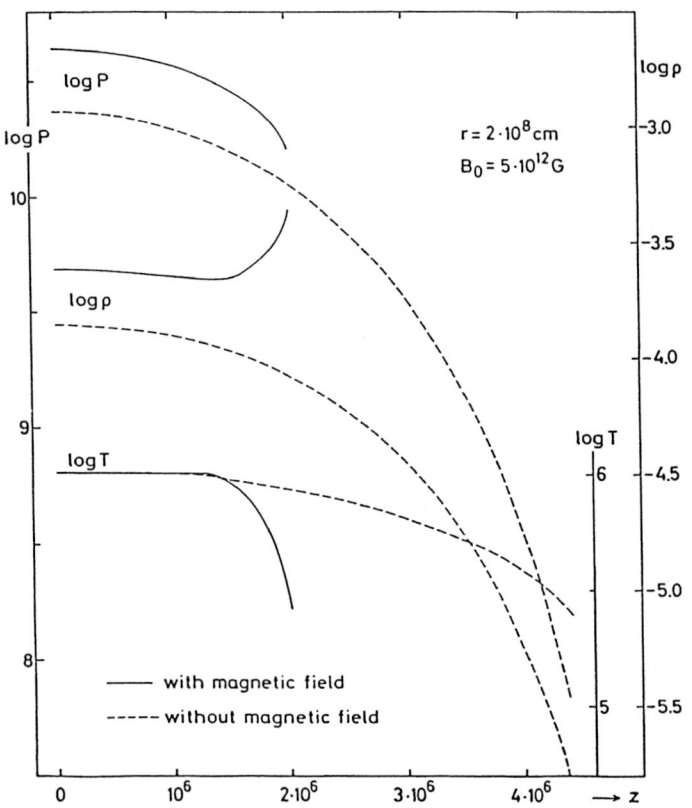

Fig. 1: Vertical disk structure at distance $r = 2 \cdot 10^8$ cm from the neutron star. B_0 magnetic dipole field; p pressure, ρ density, T temperature as functions of the height z above the midplane.

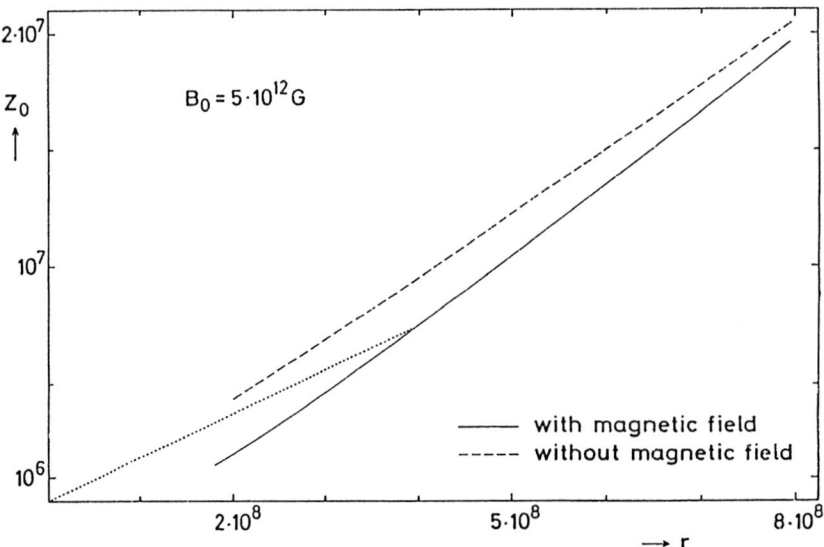

Fig. 2: Height of photosphere z_0 versus distance from the neutron star r, B_0 magnetic dipole field. Dotted line shows the direction of radiation from the neutron star.

5. Discussion

Our numerical computations show that the existence of a magnetosphere surrounding the disk produces basically two changes compared to the no-field case. One is the occurrence of a density inversion close to the disk surface. This certainly is a gravitationally unstable configuration and should thus increase the efficiency of the vertical mixing. It could also lead to a stronger interaction between the outer disk layers and the magnetosphere. The other is the steeper inclination of the disk surface which leads to an enhanced X-ray heating and thus a higher surface temperature. Both effects will be studied in more detail in a future paper.

References

Anzer, U., Börner, G.: 1983, Astron. Astrophys. 122, 73

Meyer, F., Meyer-Hofmeister, E.: 1982, Astron. Astrophys. 106, 34

Shakura, N.I., Sunyaev, R.A.: 1973, Astron. Astrophys. 24, 337.

Stellar and Extragalactic Jets: An Introduction

W. Hillebrandt

Max-Planck-Institut für Physik und Astrophysik, Institut für Astrophysik,
Karl-Schwarzschild-Str. 1, D-8046 Garching, Fed. Rep. of Germany

In recent years collimated outflows of gas, commonly called jets, have been discovered in various astrophysical objects. The most prominent ones are those in extragalactic radio sources including objects such as powerful and weak radio galaxies, quasars and Seyfert galaxies (see, e.g., the reviews by BRIDLE and PERLEY [1] and PERLEY [2]), but also jet-like structures near the center of our own galaxy have been found (LO et al. [3]; see also the contributions of REICH and of SOFUE in this volume). On stellar scales young stars show bipolar outflow and jets extending out to about 1 pc (MUNDT [4]). X-ray binaries such as Sco X-1 and Cyg X-3 possess jets (FOMALONT et al. [5], GELDZAHLER et al. [6]) as well as SS 433 (MARGON [7]). Finally, even supernova remnants show occasionally jets or jet-like structures as for example the Crab Nebula (VAN DEN BERGH [8], VELUSAMY [9], SHULL et al. [10]) or bilateral remnants (BECKER and HELFAND [11]). Despite the fact that jets seem to occur under a variety of rather different physical conditions all jet sources seem to have in common that they are related to a central compact object and very often are of bipolar nature. This has led to the suggestion that jets may be the result of non-spherical accretion into a strong gravitational potential well. Since the theoretical ideas are extensively discussed in recent reviews (see, e.g., BEGELMAN, BLANDFORD and REES [12] and BLANDFORD [13]) I will only briefly discuss a few aspects of the problem here which are directly related to the subject of the present workshop.

If we accept the basic hypothesis that jets form if matter is accreted non-spherically onto a compact object we have to answer the following questions:
1. How is the outflow accelerated and collimated?
2. Are magnetic fields important or do rotation and a density gradient suffice?
3. What is the physical state of the fluid-like plasma in the jets?
4. How do jets interact with the ambient medium?
Powerful radio galaxies may serve as an example how these questions can at least partially be answered by observations. Extended sources such as Cyg A show a central unresolved core and extended diffuse radio lobes with hot spots at their ends. These hot spots indicate that the gas flow is highly supersonic and that kinetic energy is dissipated into heat when the jets interact with the intergalactic medium. By assuming that the emission is mainly due to synchrotron

radiation the total energy in the lobes is estimated to be of the order of 10^{59} ergs and the magnetic field strength is around several tens of μG. Weak radio galaxies, on the other hand, show more diffuse and extended structures and no hot spots, as for example 3C 449 (PERLEY, WILLIS and SCOTT [14]). Here the gas flow is likely to be subsonic on large scales although the total energy in the jets may be significantly higher than in high-luminosity sources [2]. Estimates of the magnetic fields yield typical values around 1 μG [2]. In some cases the morphology of the jets is strongly influenced by the relative motion of the compact source with respect to the external medium and the jets are strongly bent (e.g., NGC 1265).

The presence of magnetic fields and the existence of a non-thermal plasma in extragalactic jet sources is manifested by the existence of polarized emission on all scales. It has been found [2] that in the radio lobes the projected magnetic field is mainly oriented parallel to the boundary of the emitting region and that the degree of polarization can go up to 70% in the hot spots. On arc sec scales, on the other hand, the field is generally oriented either parallel or perpendicular to the axis of the jet (BRIDLE [15]) showing that the jets contain a magnetized plasma.

VLBI observations of extended sources with strong core emission have shown that the jets are already existing and are well collimated on the milli arc sec (or pc) scale. In almost all cases these small scale jets are one-sided and are aligned with the large scale structures [2]. This one-sidedness as well as the apparent superluminal expansion velocities in several compact sources is commonly interpreted as being due to relativistic beaming (BLANDFORD, MCKEE and REES [16]) and requires Lorentz factors of the order of ten. On these small scales the magnetic fields seem to be generally parallel to the jet axis (RUSK and SEAQUIST [17]) indicating again that magnetic field may play an important role in both the formation and collimation of the jets. The fact that jets are already collimated on pc scales has been used as another argument in favour of magnetic rather than pressure confinement since otherwise the X-ray luminosity would exceed the observed values [13]. On the larger scales, finally, objects such as Virgo A show jets with internal pressure much larger than the external gas pressure [18] again indicating the need for magnetic confinement.

So we are left with the conclusion that for many of the observed extragalactic jets magnetic fields play a dominant role and similar arguments can be given for jets from young stars (MUNDT [19]). In all those cases, however, gas and/or radiation stresses may also be important and jet formation is very likely due to the existence of a strong gravitational field. Therefore, a complete theory of astrophysical jets has to include magnetohydrodynamics and radiation transport of a relativistic plasma in a strong gravitational field, and it is not surprising that the problems involved are far from being solved. In the following contributions to this volume by Browne, Kössl and Camenzind attempts to illustrate the importance of magnetic fields in the acceleration as well as the collimation process will be made, which, nevertheless, are based on various simplifying assumptions. We still have to go a long way before we really understand the physics of jets.

References

1. A.H. Bridle and R.A. Perley: Ann. Rev. Astron. Astrophys. 22, 319 (1984)

2. R.A. Perley: In Radiation Hydrodynamics in Stars and Compact Objects, ed. by D. Mihalas and K.-H.A. Winkler, Lecture Notes in Physics, Vol.255 (Springer Berlin, Heidelberg 1986) p.403

3. K.Y. Lo, D.C. Baker, R.D. Ekers, K.I. Kellermann, M. Reid, J.M. Moran: Nature 315, 124 (1985)

4. R. Mundt: In Nearby Molecular Clouds, ed. by G. Serra, Lecture Notes in Physics, Vol.237 (Springer Berlin, Heidelberg 1985) p.160

5. E.B. Fomalont, B.J. Geldzahler, R.M. Hjellming, C.M. Wade: Ap.J. 275, 802 (1983)

6. B.J. Geldzahler et al.: In "VLBI and Compact Radio Sources", IAU Symp. No.110, ed. by R. Fanti, K. Kellermann and G. Setti (Reidel, Dordrecht 1984) p.281

7. B. Margon: Ann. Rev. Astron. Astrophys. 22, 507 (1984)

8. S. Van den Bergh: Ap. J. Lett. 160, L27 (1970)

9. T. Velusamy: Nature 308, 251 (1984)

10. P. Shull, M. Carsentz, M. Sareandér, T. Neckel: Ap. J. Lett. 285, L75 (1984)

11. R.M. Becker, D.J. Helfand: Nature 313, 115 (1985)

12. M.C. Begelmann, R.D. Blandford, M.J. Rees: Rev. Mod. Phys. 56, 255 (1984)

13. R.D. Blandford: In Radiation Hydrodynamics in Stars and Compact Objects, l.c., p.387

14. R.A. Perley, A.G. Willis, J.S. Scott: Nature 281, 487 (1979)

15. A.H. Bridle: In Jets from Stars and Galaxies, ed. by R.N. Henriksen, Can. J. Phys. 64, 353 (1986)

16. R.D. Blandford, C.F. McKee, M.J. Rees: Nature 267, 211 (1977)

17. R. Rusk, E.R. Seaquist: Ap. J. 90, 30 (1985)

18. G.B. Benford: In Physics of Energy Transport in Extragalactic Radio Sources, ed. by A.H. Bridle and J.A. Eilek (NRAO, Green Bank, W. Virginia 1984) p.185

19. R. Mundt: In Radiation Hydrodynamics in Stars and Compact Objects, l.c., p.7

Magnetic Vortex Tubes and Charge Acceleration

P.F. Browne

Department of Pure and Applied Physics, University of Manchester,
Institute of Science and Technology, Manchester M601QD, UK

1. The Viscous Battery

In a differentially rotating plasma in which there is an electric field $\underline{E}$ and a
magnetic field $\underline{B}$, the equations of motion for a neutral atom (subscript n), a
positive ion (subscript i) and an electron (subscript e) are, neglecting gravity,

$$m_n Du_{-n}/Dt = -n_n^{-1}\nabla p - F_{-n} + K_{-n} \tag{1a}$$

$$m_i Du_{-i}/Dt = -n_i^{-1}\nabla p - F_{-i} + K_{-i} + e(\underline{E} + u_{-i} \times \underline{B}/c) \tag{1b}$$

$$m_e Du_{-e}/D = -n_e^{-1}\nabla p - F_{-e} + K_{-e} - e(\underline{E} + u_{-e} \times \underline{B}/c) \tag{1c}$$

where u is drift velocity, n is number density, p is pressure, $\underline{F}$ is viscous force,
$\underline{K}$ is drag force, m is particle mass and e electron charge.

For the viscous force on a particle of species r one writes [1,2]

$$F_{-r} = F_{-rn} + F_{-ri} + F_{-re}$$
$$\approx F_{-rn} = 6 n_n m_n \lambda_n v_{nr} Q_{nr} \nabla^2 u_{-n} \tag{2}$$

Here we neglect the viscous forces due to the ion and electron gases because mean
free paths in astrophysics become large compared to the gyroradii of electrons and
ions, and then only the neutral atom gas exerts a viscous force. In (2) λ is mean
free path, v_{nr} is the average relative thermal velocity between an n-particle and an
r-particle, and Q_{nr} is the cross section for momentum transfer from an n-particle
to an r-particle.

The drag force on an r-particle is K_{-r}, where [1,2]

$$K_{-r} = \sum_s K_{-rs} \quad ; \quad K_{-rr} = 0 \quad ; \quad n_r K_{-rs} = - n_s K_{-sr} \tag{3a}$$

$$K_{-rs} = - m_{r-rs} u_{rs} v_{rs} Q_{rs} n_s \tag{3b}$$

with $u_{-rs} = u_{-r} - u_{-s}$.

It will be assumed that radial drifts of charges are prevented by reactive
forces. The opposite radial drifts of electrons and ions are arrested by a
polarization electric field $\underline{E}$, and a pressure gradient prevents radial bulk motion.
Then the Lorentz forces in (1) each are zero.

Let $Q_{nn} \equiv Q_n$, $Q_{ni} = Q_{in} = k_i Q_n$, $Q_{en} = k_e Q_n$, $Q_{ne} = \mu k_e Q_n$, $Q_{ee} = Q_{ii} = Q_{ei} \equiv Q_o$
and $Q_{ie} = \mu Q_o$, where $\mu = m_e/m_i$. Also assume that $v_{rr} = \sqrt{2}v_r$ and $v_{en} = v_{ei} = v_e$.
Then, noting that $u_{-ni} + u_{-ie} + u_{-en} = 0$, a solution for u_{-ie} is obtained from any two

of (1), and by eliminating ∇p by use of $-3\nabla p = n_n F_{n-n} + n_i F_{i-i} + n_e F_{e-e}$, and noting $\underline{j} = e n_e \underline{u}_{ie}$, one obtains with the approximation $\mu \ll 1$,

$$\underline{j} = \frac{e \nabla^2 \underline{u}}{18(2\mu)^{\frac{1}{2}}(1 + k_i \eta_i)(n_i k_e n_e Q_{en} + n_i Q_{io})Q_n}$$ (4)

where $\eta_i = n_i/n_n$. The magnetic field due to $\underline{j}$ is obtained by taking the curl of (4) and use of $\nabla \times \underline{j} = -(c/4\pi)\nabla^2\underline{B}$. The result is

$$\underline{B} = \frac{2\pi e \xi}{9(2\mu)^{\frac{1}{2}}c(1 + k_i \eta_i)(n_i k_e n_e Q_{en} + n_i Q_{io})Q_n} \quad ; \quad \xi = \nabla \times \underline{u}$$ (5)

An order of magnitude estimate for B can be obtained by assuming that $n_i < n_n$, and $k_i = k_e = 1$ and $Q_o = Q_n = 10^{-15}\text{cm}^2$. Writing $\xi \simeq u/L$, one finds

$$B = 10^{10}u/Ln_n \quad \text{gauss}$$ (6)

In table 1 the predictions for B are compared to the observed field strengths, using the quoted values of u, L and n_n in (6). It will be seen that the viscous battery is capable of providing fields of the observed strengths in all systems, but the values required for n_n may be thought low in some cases. Amplification of the field by magnetohydrodynamical compression or by a dynamo process in response to compression should be considered. The time scale for diffusion of plasma across field lines do not cause difficulties if there is hierarchical vorticity.

Table 1. Predicted magnetic field strengths

object	u [km/s]	L [cm]	n_n [cm^{-3}]	B [G] (pred)	B [G] (obs)
galaxy	100	10^{21}	1	10^{-4}	$10^{-4} - 10^{-6}$
solar flux tube	10	10^{7}	10^{6}	10^{3}	10^{3}
A_p star	30	10^{7}	10^{6}	3×10^{3}	3×10^{3}
white dwarf	300	10^{6}	10^{5}	3×10^{6}	$10^{6} - 10^{8}$
γ-ray burster	10^{5}	10^{5}	10^{3}	10^{12}	10^{12}

Two familiar equations may be derived from (1), one for the evolution of ξ and the other for the evolution of B. The former follows by taking the curl of the equation of fluid motion, and the latter follows by taking the curl of Ohm's law after reduction to the simple form $\underline{j} = \sigma(\underline{E} + \underline{u} \times \underline{B}/c)$. Thus

$$\delta\xi/\delta t = \nabla \times (\underline{u} \times \xi) + (\eta/\rho)\nabla^2\xi$$ (7a)

$$\delta B/\delta t = \nabla \times (\underline{u} \times \underline{B}) + (c^2/4\pi\sigma)\nabla^2\underline{B}$$ (7b)

where

$$\underset{r}{\Sigma} \underset{r}{n} \underset{r}{F} = \eta \nabla^2 \underline{u} \quad ; \quad \rho = \underset{n}{n} \underset{n}{m} \quad ; \quad \sigma = e^2/m_e v_e Q_o \qquad (8)$$

The formal analogy between (7a) and (7b) means that the proportionality $\underline{B}$ to $\underline{\xi}$ is maintained during dissipationless evolution of the fields. The lines of both $\underline{\xi}$ and $\underline{B}$ remain frozen into the fluid, on neglecting the diffusion terms. By writing $\delta / \delta t \simeq 1/t$ and $\nabla \simeq 1/L$, one obtains a time scale for diffusion t in terms of a scale length L:

$$t_\xi = 6(2^{\frac{1}{2}}) Q_n n_n L_\xi^2/v_n \quad ; \quad t_B \simeq 4\pi a L_B^2/v_e Q_o \qquad (9)$$

where $a = e^2/m_e c^2$. If one requires $t_B < 10^{10}y$, then $L_B < 10^{13}cm$, which is a severe constraint.

It is widely thought that turbulence resolves this time-scale problem. Turbulent motions may stretch and wind field lines on a small scale, and (9) permits the dissipation of such small scale fluctuations fairly rapidly. But the energy is not necessarily transferred from a large scale magnetic field into small scale fluctuations of field. The latter obtain their energy from fluid motions, rather than the large scale field - for comment see reference [3]. In the present theory the difficulty is avoided. The scale length of $\underline{B}$ never becomes large. What one observes is ordering of a multitude of small scale fields brought about by the ordering of small scale vorticity in relation to large scale vorticity. The only other way to increase the scale length of $\underline{B}$ is by magnetohydrodynamical expansion - which need not weaken the field if there is continual generation of new field lines on the small scale.

The energy source for astrophysical magnetic fields is large scale differential rotation. In order to persist for $10^{10}y$ its scale length must exceed about 1 pc. The energy is continually fed down through the hierarchy of eddies to small scale vorticity, from which it can be converted into magnetic field energy but also is viscously damped on a short time scale. Dissipation as well as creation of astrophysical magnetic fields must occur on a small scale. The synchrotron sources represent regions where magnetic field energy is being dissipated.

2. The Magnetic Vortex Tube

At an early stage during formation of a star or galaxy from a gas cloud a vortex is likely to form along the rotation axis, and also a disc structure normal to this axis [2]. The reduced pressure in the core of the vortex tube (due to centrifuging) has an axial gradient which dives a bipolar outflow of gas. The outflow is maintained by an inflow in the equatorial plane, and the angular momentum which is sucked in with the gas maintains the vorticity much as an updraught maintains a terrestial tornado.

In the case of a partially ionized plasma the vortex will acquire a magnetic field. Since $\underline{B} \propto \underline{\xi}$, the magnetic field for a classical vortex would be that of a laboratory solenoid, the field being uniform for $r < R$ and zero for $r > R$, and the current density being azimuthal and confined to the surface r = R. Introducing cylindrical polar coordinates (r, ϕ, z), one has $\underline{B} = (0, 0, B_z)$ and $\underline{j} = (0, j_\phi, 0)$. For stability as well as other reasons a vortex tube is more likely than the classical vortex. Then both $\underline{\xi}$ and $\underline{B}$ are nonzero only for $R_1 < r < R_2$, and there are equal and opposite azimuthal currents per unit length on the surfaces $r = R_1$ and $r = R_2$.

An axial magnetic field B_z tends to expand, and if the expansion is uneven with respect to z the differential rotation will wind field lines into elongated loops which become increasingly coiled around the axis. Babcock [4] has envisaged such a process for solar field lines. The lines of both $\underline{B}$ and $\underline{j}$ become helical, but the longer the twisting persists the more nearly do the fields approach the $(0, B_\phi, 0)$ and $(0, 0, j_z)$ configurations. In other words, $(j_\phi, B_z) \rightarrow (j_z, B_\phi)$.

If the radial components of $\underline{j}$ and $\underline{B}$ vanish

$$\underline{j} \times \underline{B} = (j_\phi B_z - j_z B_\phi)\hat{r} \simeq (1/8\pi r^2)d(r^2 B_\phi^2)/dr \qquad (10)$$

the last step requiring that $B_\phi \gg B_z$. Thus, when $j_z B_\phi > j_\phi B_z$ the magnetic vortex tube (MVT) contracts due to magnetic pinch,

Contraction of the B_ϕ field (i.e. the core "wall") causes Fermi acceleration of charged particles which are trapped in the core. These particles acquire suprathermal velocities. In fact it can be shown that contraction from raduius R(0) to radius R(t) increases $\gamma\beta$ by a factor R(0)/R(t), where βc is the radial velocity of a trapped charge and γ the Lorentz factor [2]. Pressure changes as a step function across the core boundary r = R. Because the inward $\underline{j} \times \underline{B}$ force acts on electrons, whereas the outward pressure is exerted primarily by positive ions, the plasma will polarize in the radial direction, introducing a radial electric field E_r of limited extent Δr_E given by

$$eE_r \Delta r_E = (\gamma_i - 1)m_i c^2 \qquad (11)$$

Pressure is due to ions of momentum $\gamma_i \beta_i m_i c$ with flux density $n_i \beta_i c$. Hence

$$E_r^2/8\pi = 2\beta_i^2 \gamma_i n_i m_i c^2 \simeq B^2(R)/8\pi \qquad (12)$$

yielding $E_r \simeq B_\phi$. The radial electric field E_r permits electrons and ions to drift across the transverse magnetic field B_ϕ, which is essential for an axial current. The electric drift velocity is $c\underline{E} \times \underline{B}/B^2 = cE_r/B_\phi \simeq c$. But whilst electrons can take up the full electric drift because $\lambda_e < g_e$ ions cannot because $\lambda_i > g_i$, where g is radius of gyration. Specifically,

$$\frac{g_i}{\Delta r_E} = \left(\frac{\gamma_i + 1}{\gamma_i - 1}\right)^{\frac{1}{2}} \simeq \frac{2}{\beta_i} \qquad \frac{g_e}{\Delta r_E} = \frac{\mu\beta_e\gamma_e}{\gamma_i - 1} \simeq \frac{\beta_e\gamma_e}{918\beta_i^2} \qquad (13)$$

where $\beta_i \gamma_i \ll 1$ is assumed for the final values. In order that $\Delta r_E > g_e$ one requires that $\beta_e\gamma_e < 918\beta_i^2$. On the other hand $\Delta r_E < g_i$ for all β_i.

The current J flowing through the thin cylindrical channel of radius R and thickness Δr_E is the source of the B_ϕ field. Hence

$$B_\phi(R) = 2J/Rc \quad ; \quad J = 2\pi R \Delta r_E j_z \quad ; \quad j_z = e u_e n_e \quad (14)$$

yielding,

$$u_e/c \simeq 8 n_i/n_e \quad (15)$$

where again we assume $\beta_i \gamma_i < 1$. Equation (15) implies that $8n_i < n_e$, so that charge neutrality in the conduction channel is not achieved.

3 Acceleration of Charges

A mechanism which is capable of accelerating charges to the highest cosmic ray energies now becomes available. The tendency toward charge neutrality in the current channel implies $u_e \rightarrow c$, and also $E_r \simeq B_\phi$ implies electric drift velocity $\simeq$ c for electrons. Thus the limiting current density $e n_e c$ due to free charges can be reached, and if induction requires additional current it must be a displacement current.

In addition to this mechanism an instability can denude a region $\mathcal{R}$ of current channel of free carriers. The mechanism [2] is best understood by supposing, in the first instance, that the positive ions are positrons. Because induction maintains J constant a small decrease of electron density δn_e will be compensated by a small incrase of drift velocity δu_e. Then charges exit from $\mathcal{R}$ with increased velocity. At the boundary surfaces of $\mathcal{R}$, where n_e is average, the constancy of J demands a reduced rate of entry of charges of the opposite sign. This further decreases n_e in $\mathcal{R}$, and it is only a matter of time until $\mathcal{R}$ is denuded of free charges. The reduced rate of entry into $\mathcal{R}$ implies a build up of opposite charges at the boundary surfaces, giving rise to an axial electric field E_z in $\mathcal{R}$. The current through $\mathcal{R}$ is a displacement current, $(S/4\pi)\delta E_z/\delta t$, where $S = 2\pi R \Delta r_E$. The denudation is equivalent to inserting a capacitor C in the circuit. If one replaces the positrons by protons the current ceases to be divided equally between positive and negative charges, but the same argument applies.

As every electrical engineer knows, very large voltages can develop across a break in a circuit which carries an inductively maintained current. Specifically, a displacement current flows through the break in the circuit, and the growing axial electric field E_z is limited when all of the energy in the field B_ϕ is transferred to the field E_z. If $\mathcal{R}$ has length Δz and the current flows through a channel of cross sectional area S and length z, the field energies are:

$$\mathcal{E}_B = z \int_R^{r_1} (B_\phi^2/8\pi) 2\pi r \, dr = z \ln(r_1/R) J^2/c^2 \equiv LJ^2/2 \quad (16a)$$

$$\mathcal{E}_E = (E_z^2/8\pi) S \Delta z \equiv C(E_z \Delta z)^2/2 \quad (16b)$$

Here one uses (14) and introduces an upper limit r_1 which is determined by the distribution of return current. The expressions (16) define a self-inductance L and a capacitance C for the equivalent circuit. The oscillation of the energy between the fields $\underline{B}$ and $\underline{E}$ has a period P_e given by,

$$P_e = 2\pi(LC)^{\frac{1}{2}} = 2\pi kR/c \tag{17}$$

where

$$k^2 = (\Delta r_E/R)(z/\Delta z)\ln(r_1/R) \tag{18}$$

The maximum value of E_z ocurs when all of the energy in B_ϕ has been transferred to E_z and is obtained from $\mathcal{E}_B = \mathcal{E}_E$. The energy W gained by a charge e which passes through $\mathcal{R}$ when E_z is a maximum is therefore given by

$$W = eE_z\Delta z = KezB_\phi(R) \tag{19}$$

where

$$K^2 = (R/\Delta r_E)(\Delta z/z)\ln(r_1/R) \tag{20}$$

Ultrarelativistic energies are easily obtained from (19). The accelerated charges constitute a beam since they have collimated motions, and moreover the beam is pulsed with the period (17).

4. Cosmic Rays and Pulsars

4.1 Origin of Cosmic Rays

The maximum observed cosmic ray energies of order 10^{20} ev are obtainable from (19) for a value of $zB_\phi(R)$ which is reasonable in stellar sources with cyclotron lines in the 30 - 70 kev range [5,22]. Assuming $\Delta z/z \simeq \Delta r_E/R$, one obtains $K \simeq 1$ from (20), and adopting $B_\phi(R) = 3 \times 10^{12}$ G which corresponds to a 40 kev cyclotron line, one finds from (19) that $W = 10^{20}$ ev if z = 1 km. This is a modest requirement for the length of the current channel associated with a pinch. The length Δz of denuded region will be smaller.

4.2 X-Ray Pulsars

The electromagnetic oscillation with period (17) is likely to be coupled to an acoustic mode of the star via the oscillatory $\underline{j} \times \underline{B}$ force. Again assuming $\Delta z/z \simeq \Delta r_E/R$, (18) yields $k \simeq 1$, and then (17) yields $P_e \simeq 2\pi R/c$. Thus $P_e \simeq 0.2$ s if R = 10^9 cm. Such a period is characteristic of pulsars. X-ray pulsars have periods in the range 3.3 ms to 835 s [5,6].

A train of radiation pulses may be generated by bunches of ultrarelativistic electrons as they pass through a gas cloud (relativistic bremsstrahlung) or through a transverse magnetic field (synchrotron radiation). In both cases the emission is forward-beamed due to relativistic aberration. One may show that the ratio of bremsstrahlung to synchrotron power radiated per electron with Lorentz factor γ is $2\pi n_i m_e c^2/B_\phi^2$, where n_e is proton density and B_ϕ is the transverse magnetic field. It is proposed to explain X-ray pulsars by the former mechanism, and radio pulsars by the latter.

216

For pulsed emission the length of the cloud must be less than the separation of the bunches. More extended gas will give a steady component of flux. The large duty cycle of X-ray pulses and the tendency for emission to bridge the interval between pulses is thus explained. By contrast, radio pulsars provide narrow pulses and small duty cycles since the transverse field is that of an outer pinch. Moreover, interpulses could be attributed to positively charged particles accelerated during phases when E_z is reversed.

The gas with which the electron bunches interact is likely to be outflowing through the core of the MVT in which they are accelerated. The radiation pulses will therefore have a Doppler shift. Secondly, changes in the distance between the source of the electron bunches and the cloud where the radiation is generated will cause phase shifts of the radiation pulses. Together these two effects can convert a perfectly periodic oscillation into a quasi-periodic train of radiation pulses. In the case of dwarf novae the exflux of matter is guaranteed during outburst, and the oscillations then seen are indeed quasi-periodic [7,8]. In old novae, where the outflow has ceased, the oscillations are more coherent [9].

The belief that X-ray pulsars, and also dwarf novae, are close binary systems with periods in the range 1 - 8 hour may be founded on a misinterpretation of a sinusoidal Doppler shift. The detailed evidence for several systems seems best explained by a precessing jet model, as will be discussed elsewhere. The variation of Doppler shift then becomes a mild case of the phenomenon seen in SS 433, although possibly without the counterjet.

4.3 Radio Pulsars

Radio pulsar periods range from 1.56 ms to 4.3 s, with a most common value near 0.5 s. Synchrotron radiation generated when pulsed beam of electrons enters a transverse magnetic field (for example that associated with a second pinch farther out along the MVT) may be responsible for radio pulsars.

The trajectory of a bunch of electrons through the pinch field pattern will determine the pulse profile. Linear polarization may approach 100 %, and its position angle often rotates by a large amount during a single pulse [10]. A helical trajectory for the electron bunch through a magnetic field which is predominantly B_ϕ could explain the rotation, the line of sight being parallel to the axis so that the lines of B_ϕ appear circular.

Individual pulse profiles change from pulse to pulse. Sometimes there is a sequence of pulse profiles which repeats with period P_3 [11,12], suggesting a sequence of trajectories through the pinch field pattern. During this sequence a peak (sub-pulse) often is seen to drift to progressively earlier phases, an effect which could be attributed to a diminishing transit time of electrons between their source and where they radiate. The period P_3 presumably is that of fluid rotation, suggesting precession of the injector of electrons which might lead to a variation of pitch angle of the spiral trajectory of the electron bunches. On the other hand, sometimes the pulsar switches between two profiles each with its characteristic

polarization position angle and sense of circular polarization, the switches
occurring between pulses and being random [13,14]. This effect would be produced if
electrons followed two possible trajectories through the pinch field.

"Nulls" refer to periods when no pulses are received. When the pulses return the
phase of sub-pulse drift is preserved [12], so that whatever produces the null is
associated with an integral number of periods P_3 which has been interpreted as an
integral number of fluid revolutions.

5. Bursters and Cataclysmic Outbursts

5.1 Window to Degenerate Interior

A polar MVT may play another role in the production of stellar X-rays. Expulsion of
gas through the core of a polar MVT can open a "window" to a sub-layer of the star
where gas is degenerate, and hence very hot. The "window" is a low density path
which is sustained by magnetic pressure until such time as a new plug of cool gas
forms in the core of the MVT. Due to its high thermal conductivity, degenerate gas
cannot sustain an appreciable temperature gradient, and most of the temperature
fall from the centre to the surface occurs across the skin of normal gas
surrounding the degenerate interior.

The threshold for degeneracy is set by $E_F > kT$, where E_F is the Fermi energy
(the level up to which the states available in unit volume are filled at zero
temperature). The condition $E_F > kT$ implies $n_e > n_e^*$, where

$$n_e^* = (2mkT)^{3/2}(3\pi^2 \hbar^3)^{-1} \tag{21}$$

The pressure corresponding to n_e^* can be balanced by magnetic pressure if $B > B^*$,
where

$$B^{*2}/8\pi = 2n_e^* kT = 2(2m)^{3/2}(kT)^{5/2}(3\pi^2\hbar^3)^{-1} \tag{22}$$

Taking $kT = 3$ kev, one finds $B^* = 10^{10}$ G, and taking $kT = 400$ kev one finds $B^* = 4.5$
x 10^{12} G. The former situation will apply to X-ray bursters, and the latter to γ-
ray bursters.

The high thermal conductivity of degenerate gas is a consequence of the increase
of the mean free path λ of electrons due to the strong dependence of the cross
section for Coulomb collisions on energy. One has $\lambda \propto E_F^2$, and hence the rate of
heat diffusion varies as E_F^4. (Only electrons within kT of E_F can collide and exert
pressure in the sense of the kinetic theory of gases, electrons lower in the
distribution having ordered motions like the electrons orbiting a high Z atom.) As
E_F increases the sharing of the thermal energy among the given electrons becomes
increasingly unequal. Relatively few electrons at the top of the distribution have
all the energy, and are extremely hot.

Presumably the "blow out" which creates the window is due to radiation pressure
which has been increasing in the degenerate interior for some time, and at the time

218

of the outburst has reached the threshold value for lift-off. The "blow out" occurs in the core of a polar MVT because here there is less mass to lift off.

5.2 X-Ray Bursters

X-ray bursters deliver bursts of duration ~10 s and luminosity ~3 x 10^{38} erg/s. The bursts repeat semi-regularly after intervals of 10^4 - 10^5 s. The spectrum of the X-rays fits that of a blackbody whose temperature rises quickly to a maximum of 3 x 10^7 K and then declines slowly. The radius may remain constant as the temperature varies [15,16], or it may increase and then decrease with a counter variation of temperature [17,18].

It is usually thought that the small size of the source implies a neutron star. However, a more likely explanation is that 10 km represents the core radius R for an MVT which penetrates through the relatively opaque skin layer around a degenerate star. One supposes that a "window" is created in the core of the MVT to the very hot degenerate interior of the star, from which comes the X-rays. The "window" is suddenly opened when a plug of cool gas in the MVT core is blown out, probably by radiation pressure which has been steadily increasing in the interior of the star. Reformation of such a plug of cool gas rapidly closes the "window", resulting in a burst of only 10 s duration.

The observation of two peaks in the burst profile for hard X-rays, but not soft X-rays [17,18], suggests that the optical depth of the escaping gas for hard flux first increases and then decreases. Precursors [19] may have a similar explanation. There is little doubt that optical bursts, delayed by 2 - 3 s, are reprocessed flux [20]. In the present model the escaping gas does the reprocessing.

X-ray bursters do not show oscillations because of the small value of R. Thus, on substitution of R = 10 km into (17) one finds with k = 1 that P_e = 0.2 ms, which is too small for detection.

5.3 γ-Ray Bursters

γ-ray bursts have durations in the range 0.1 - 100 s, preferred values being 0.3 s and 10 s [21]. Whilst recurrences have been observed for one source [22] they are long term and unlike the recurrences of X-ray bursts.

The spectra of γ-ray bursts usually can be fitted to a blackbody curve with kT in the range 100 - 1000 kev [23]. Since mc^2 = 500 kev such a plasma is relativistic with a large density of positrons. This is confirmed by the presence of the positronium annihilation line in most spectra [23]. The line is seen in the range 420 - 460 kev, which represents a substantial redshift from the laboratory energy 511 kev. Although a gravitational redshift in a neutron star has been suggested, this seems unlikely in view of the line width and the variation of the redshift from source to source. A more acceptable possibility would be transverse Doppler effect due to relativistic rotation of gas in the core of a vortex tube, viewed axially. The rotational velocity would be ~0.5 c for a line at 440 kev. Relativistic rotation of gas in the core of the MVT might be expected since cyclotron lines in most sources in the 30 - 70 kev [23] range indicate magnetic fields of order 5 x 10^{12} G, and presumably $\underline{B} \propto \underline{\xi}$.

One unusually intense and unusually soft γ-ray burst on 5 March 1979 came from a direction which can be located to within an error box of 0.1 min , having a rapid rise time of < 0.25 ms and being detected by nine spacecraft [24]. Only the supernova remnant N49 in the LMC at distance 55 kpc is seen in this error box, and if this identification is correct one knows that $4\pi d^2 = 3.6 \times 10^{47}$ cm^2 which implies burst luminosity of 4×10^{44} erg/s. This exceeds the luminosity of a typical X-ray burst by a factor of 10^6 . Hence $R^2 T^4$ is greater by a factor of 10^6 . But one knows that T is up by a factor of about 100, and hence R must be down by a factor 1/10, yielding R = 1 km.

The explanation for the relativistic electron-positron plasma is degeneracy. In the γ-ray bursters the Fermi level is so high that kT ~ 400 kev.

5.4 Dwarf Novae

One is now in a position to offer a new unified theory for cataclysmic outbursts in general. The transition from the X-ray and γ-ray bursts to optical outbursts is a small step, requiring only that the X-rays be reprocessed into optical flux by the expanding surface layer. The observation of ocillations during outburst with P_e ≃ 9 - 39 s [6,7] suggests $R \simeq 10^{10} - 10^{11}$ cm, taking k ≃ 5 in (17).

One deals again with a basic instability in that internal radiation pressure increases slowly until eventually a "blow out" occurs. The "blow out" is most likely to happen in the core of a polar MVT because here there is less gas to lift off. After the outburst a plug of cool gas reforms in the MVT core, and conditions are then created for a subsequent outburst. One expects the time for radiation pressure to attain threshold to be greatest when the outbursts are greatest, because then there is more gas to expel and also cooling is greatest. This effect is known, being expressed by an empirical relationship, $\Delta m = 2 + 1.78 \log \Delta t$, between magnitude change Δm and interval between outbursts Δt (days).

5.5 Novae

The mechanism can be extended to novae. Luminosity will increase by a factor $f = (R'/R)^2 (T'/T)^4$, where R' and T' refer to a layer of the expanding gas which reprocesses the X-ray flux. Thus if T'/T = 100 and R'/R = 1/30 one obtains $f = 10^5$, corresponding to Δm = 12.5.

The shells surrounding old novae often have a conical shape, confirming the anisotropy of the outburst.

5.6 Supernovae

There appears to be no reason why supernova should not also be attributed to the same basic mechanism. Probably a large portion of the opaque skin is removed in the supernova outburst. One notes that the onset of degeneracy is a runaway process in which a substantial fraction of the trapped thermal energy of the star can be released. As E_F increases the rate of diffusion of heat increases, being proportional to E_F^4. The rate of cooling of the interior therefore increases, which promotes further contraction and hence further increases E_F . The outcome is rapid release of the thermal energy of the stellar core which was trapped by normal gas. Radiation pressure at the base of the skin of normal gas increases until the threshold for lift-off is reached, and then the supernova is seen.

Tests for such a model might be sought in the morphologies of supernova remnants. Do they, like nova shells, imply anisotropy of outburst? Certainly in some cases they do, notably those remnants in which jets have been found [25] and vorticity suspected [26]. One notes that a conical outburst viewed off-axis would give elliptical-shell morphology, which sometimes is seen.

References

1, P.F. Browne: Astrophys. Lett. 2, 217 (1968)

2. P.F. Browne: Astron. Astrophys. 144, 298 (1985).

3. T.G. Cowling: Ann. Rev. Astron. Astrophys. 19, 115 (1981).

4. H.W. Babcock: Astrophys. J. 133, 572 (1961).

5. H.V.D. Bradt, J.E. McClintock: Ann, Rev. Astron. Astrophys. 21, 13 (1983).

6. P.C. Joss and S.A. Rappaport: Ann. Rev. Astron. Astrophys. 22, 537 (1984).

7 J. Patterson: Astrophys. J. Suppl. Ser. 45, 517 (1981).

8. F.A. Cordova, T.J. Chester, K.O. Mason, S.M. Kahn, S.P. Garmire: Astrophys. J.
9   278, 739 (1984).

8. J. Patterson: Astrophys. J. 233, L13 (1979).

10. J.M. Rankin, D.B. Campbell, D.C. Backer: Astrophys. J. 188, 609 (1974),

11. L. Oster, D.A. Hilton, W. Sieber: Astron. Astrophys. 57, 1 (1977).

12. A. Filippenko, V. Radhakrishnan: Astrophys. J. 263, 828 (1982).

13. L Oster, W. Sieber: Astrophys. J. 58, 303 (1977).

14. D.C. Backer, J.M. Rankin: Astrophys. J. 42, 143 (1980).

15. J.S. Swank, R.H. Becker, E.A. Boldt, S.S. Holt, S.H. Pravdo, R.J. Serlemitsos:
    Astrophys. J. 212, L73 (1977).

16. J.A. Hoffman, W.H.G. Lewin, J.Doty: Astrophys. J. 217, L23 (1977).

17. J.A. Hoffman, L. Cominsky, W.H.G. Lewin: Astrophys. J. 240, L27 (1980).

18. J.E. Grindlay, H.L. Marshall, P. Hertz, A. Soltan, M.C. Weisskopf, R.F.
    Elsner, P. Ghosh, W. Darbro, P.G. Sutherland: Astrophys. J. 240, L121 (1980).

19. W.H.G. Lewin, W.D. Vacca, E.M. Basinska: Astrophys. J. 277, L57 (1984).

20. A. Lawrence et al.: Astrophys. J. 271, 793 (1983).

21. J.P. Norris, T.L. Cline, U.D. Desai, B.J. Teegarden: Nature 308, 434 (1984).

22. R.E. Rothschild, R.E. Lingenfelter: Nature 312, 737 (1984).

23. E.P. Mazets, S.V. Golenetskii, R.L. Aptekar, Y.U. Gur'yan, V.N. Il'inskii:
    Nature 290, 378 (1981).

24. T.L. Cline et al.: Astrophys. J. 255, L45 (1982).

25. R.S. Roger, D.K. Milne, M.J. Kesteven, R.F. Haynes, K.J. Wellington:
    Nature 316, 44 (1985).

26. P.A Shaver, C.J. Salter, A.R. Patnaik, J.H. van Gorkom, G.C. Hunt:
    Nature 313, 113 (1985).

Collimation of Supersonic Gas Flows by Toroidal Magnetic Fields

D. Kössl

Max-Planck-Institut für Physik und Astrophysik, Institut für Astrophysik, Karl-Schwarzschild-Str. 1, D-8046 Garching, Fed. Rep. of Germany

A numerical ideal magnetohydrodynamic (MHD) model is described that allows examination of magnetic confinement of supersonic gas flows. Some test calculations and a few preliminary results are discussed.

1. Introduction

Jets are a common phenomenon in the universe. Observations indicate that magnetic fields play an important role in the formation and collimation of jets (see, e.g., the reviews by BRIDLE nad PERLEY [1] , BEGELMAN, BLANDFORD and REES [2] and see also the contribution of HILLEBRANDT in this volume). At least for the most powerful radio galaxies it seems that the internal pressure of the ejected plasma is much higher than the pressure of the surrounding medium [2]. Thus pressure confinement is ruled out as collimation process at least in some cases. Toroidal magnetic fields are a promising way to solve the confinement problem of these jets.

2. Basic Equations

The equations appropriate for modelling the magnetic collimation of supersonic gas flows are the ideal magnetohydrodynamic (MHD) equations (LANDAU and LIFSHITZ [3]). These equations relate the magnetic field $\vec{B}$ to the velocity $\vec{v}$ and the thermodynamic variables (the mass density ρ, the pressure p and the internal energy e). They may be written in the form of a set of conservation laws for mass density ρ, total energy density $u = \frac{\rho v^2}{2} + \rho e + \frac{B^2}{8\pi}$, momentum density $\vec{g} = \rho\vec{v}$ and magnetic flux density $\vec{B}$. These conservation laws relate the rate of change of the density of a conserved quantity to the divergence of the corresponding flux.

Mass conservation yields the continuity equation

$$\dot{\rho} = -\mathrm{div}(\rho\vec{v})$$

and conservation of total anergy the heat transfer equation

$$\dot{u} = -\mathrm{div}\vec{q} \quad ,$$

where $\vec{q}$ is the energy flux density

$$q_i = \left(u + p + \frac{B^2}{8\pi}\right) \cdot v_i - \frac{1}{4\pi}\left(\vec{v} \cdot \vec{B}\right) \cdot B_i \quad,$$

containing the Poynting flux in MHD approximation as second term. Conservation of momentum gives the MHD equivalence of the Euler equation

$$\dot{\vec{g}} = -\mathrm{Div}\Pi \quad,$$

where Div is the tensor divergence. The momentum flux density tensor takes the form

$$\Pi_{ik} = \rho v_i v_k + \left(p + \frac{B^2}{8\pi}\right) \cdot \delta_{ik} - \frac{B_i B_k}{4\pi} \quad,$$

where the last term is the Maxwellian stress tensor. The induction equation in MHD form reads
$$\dot{\vec{B}} = -\mathrm{Div}F \quad,$$
where

$$F_{ik} = B_i v_k - B_k v_i$$

is an antisymmetric flux tensor that guarantees $\mathrm{div}\vec{B} = 0$ due to the relation

$$\mathrm{div}\mathrm{Div}F = 0 \quad.$$

To complete our set of equations, we specify the equation of state to be

$$p = (\gamma - 1)\rho e \quad,$$

i.e. we assume a simple ideal gas law with $\gamma = \frac{5}{3}$ for a non-relativistic monoatomic gas.

The above set of equations are based on a few approximations. They are non-relativistic and electromagnetic waves cannot occur. Furthermore viscosity and finite resistivity have been omitted, because for the problem of supersonic jet propagation both dynamical and magnetic Reynolds numbers are large. Today's computer facilites do not allow to solve the jet propagation problem in MHD approximation in three dimensions. But nevertheless as jets are well collimated on scales of a few hundreths of parsecs it seems to be sufficient to reduce the problem to an axisymmetric one. In this geometry, however, kinck instabilities of a pinched plasma column cannot be treated. But as the gas flow is supersonic one would expect that the growth rate of this instability is small.

Computing the tensor divergence in cylindrical coordinates yields the coriolis forces

$$f_r = +\frac{1}{r}\left(p + \frac{B^2}{8\pi} + \rho v_\varphi^2 - \frac{B_\varphi^2}{4\pi}\right)$$

$$f_\varphi = -\frac{1}{r}\left(\rho v_r v_\varphi - \frac{B_r B_\varphi}{4\pi}\right)$$

in r- and φ-direction respectively. The "centrifugal" force due to the toroidal field is responsible for the pinch-effect and is expected to be able to collimate the gas flow along the symmetry axis, which coincides with the z-axis.

3. Numerical Methods and Code Verification

There are a few basic requirements a finite difference algorithm should satisfy to be an appropriate tool for solving the problem of supersonic MHD gas flows.

(1) The conservation properties of the physical equations should be mirrored in their numerical counterparts. This is simply achieved by adopting the special form of the MHD equations described above.

(2) The scheme must not produce unphysical results as negative densities or pressure.

(3) Furthermore it should resolve discontinuities (shocks, etc.) within a few grid points without generating dispersive ripples (monotonicity).

(4) Finally it should not generate numerical monopoles (e.g. div$\vec{B} = 0$).

A method that satisfies these demands is the so-called "Flux-Corrected-Transport" (BORRIS and BOOK [4], BORRIS, BOOK and HAIN [5], BORRIS and BOOK [6], ZALESAK [7] and MORROW and CRAM [8]).

The correctness of the code has been tested in various ways. One of the most difficult tests is the nonlinear interaction of two colliding shock waves (see Fig. 1 and the review of WOODWARD and COLELLA [9]). This test has shown that the method is indeed capable of handling strong shocks and their interaction.

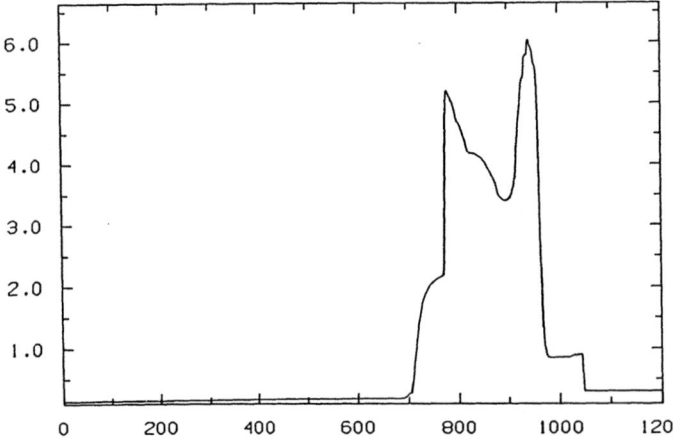

Fig.1 Density distribution after collision of a Mach 200 and a
Mach 60 shock (time is $t = 0.038$).

To verify the 2-D-MHD capacity of the code the time evolution of a blast wave running into a strong homogeneous magnetic field $\vec{B} = B\vec{e}_z$ has been computed (see Fig. 2). The initial

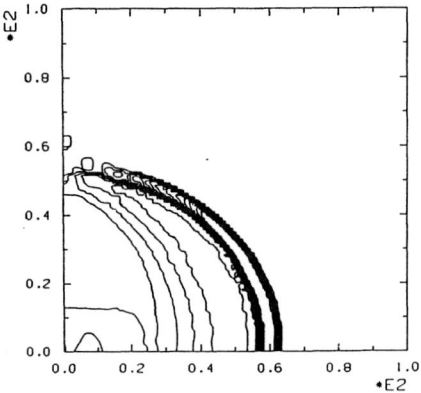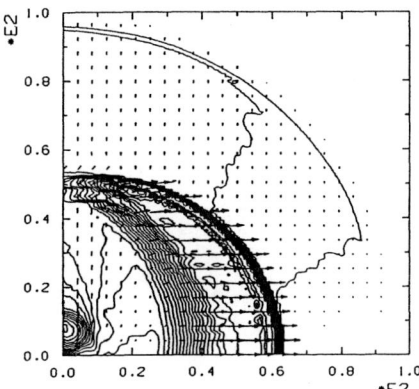

Fig. 2a Density distribution of a blast wave running into a homogeneous z-field. 20 equally spaced contour lines are shown.

Fig. 2b Velocity contours and vectors of the blast wave. Contour
lines are equally spaced.

conditions are as follows. A sphere of radius 50 contains a hot gas. The ratio of the inner
to the outer gas pressure is 100, the density ratio is 1 and the velocity is zero everywhere.
The result is a weak purely magnetic compression wave running in r-direction and a strong
gasdynamical shock in z-direction. The field lines remain almost unchanged. The solution
is correct within 1%. Nevertheless $\mathrm{div}\vec{B}$ is not identical to zero due to truncation errors of
the adopted algorithm, but the relative error is small (less then 1%).

4. Preliminary Results

We have chosen the following numerical setup. The grid has 75×180 uniform zones. The
jet enters the grid at the bottom part of the left boundary and has a radius of 10 grid
points. The basic variables are normalized, with the jet radius R_B being the unit of length.
The sound velocity c_M and density ρ_M of the medium are used as units of velocity and
density respectively. Thus the unit of time is $t = R_B/c_M$.

(a) Pressure-matched Hydrodynamic Jets: If the pressure of the medium p_M and of the
jet p_B are equal we have the following relation for densities and sound velocities

$$\rho_M c_M^2 = \rho_B c_B^2 \quad .$$

If γ is fixed (to a value of $\frac{5}{3}$), the problem has only two free parameters, namely the Mach
number of the beam and the density ratio

$$Ma_B = \frac{v_B}{c_B}$$

$$\eta = \frac{\rho_B}{\rho_M} \quad .$$

For $Ma_B = 6$ and $\eta = 1$ we get a naked beam with a terminal and a bow shock and a
distorted contact surface at the beam boundary. A light beam $(Ma_B = 6$ and $\eta = 0.1$,

225

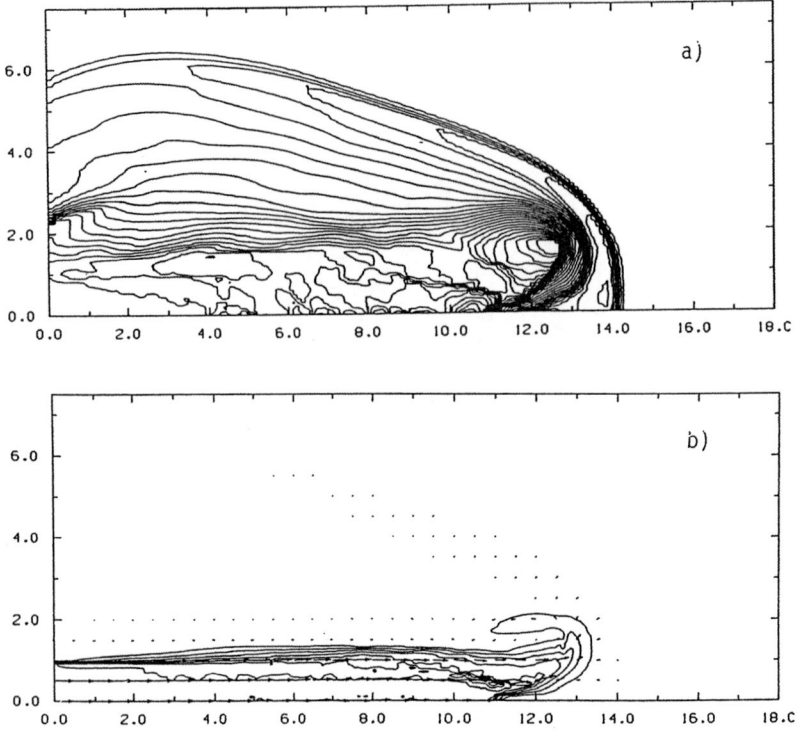

Fig. 3a Density distribution for a light supersonic jet ($Ma_B = 6$,
$\eta = 0.1$). 30 contour lines are shown spaced by a factor of
1.2 (time is $t = 3$).

Fig. 3b Velocity contours and vectors for a light supersonic jet. 10
equally spaced contour lines are shown.

see Fig. 3) shows a more complicated structure. There is an additional contact surface
between the terminal and bow shock and a cocoon with backflow is generated (NORMAN,
SMARR and WINKLER [10]). Furthermore the presence of a vortex flow above the head
of the beam is indicated by a minimum in the density and by the flow direction (see
Fig. 3). This vortex acts like an obstacle and is responsible for the comparatively small
backflow velocities (c.f. NORMAN, SMARR, WINKLER and SMITH [11] and ARNOLD
and ARNETT [12]).

Yet Fig. 3 reveals some of the difficulties that still have to be solved. The numerical
resolution is too low by approximately a factor of three. This leads to numerical noise in
the beam and to a distorted structure of the terminal shock. Furthermore there are some
difficulties with the boundary conditions near the nozzle.

(b) MHD Jets: The setup for this numerical experiment consists of an unmagnetized
jet boring into a magnetized medium. The magnetic field is purely toroidal and the field

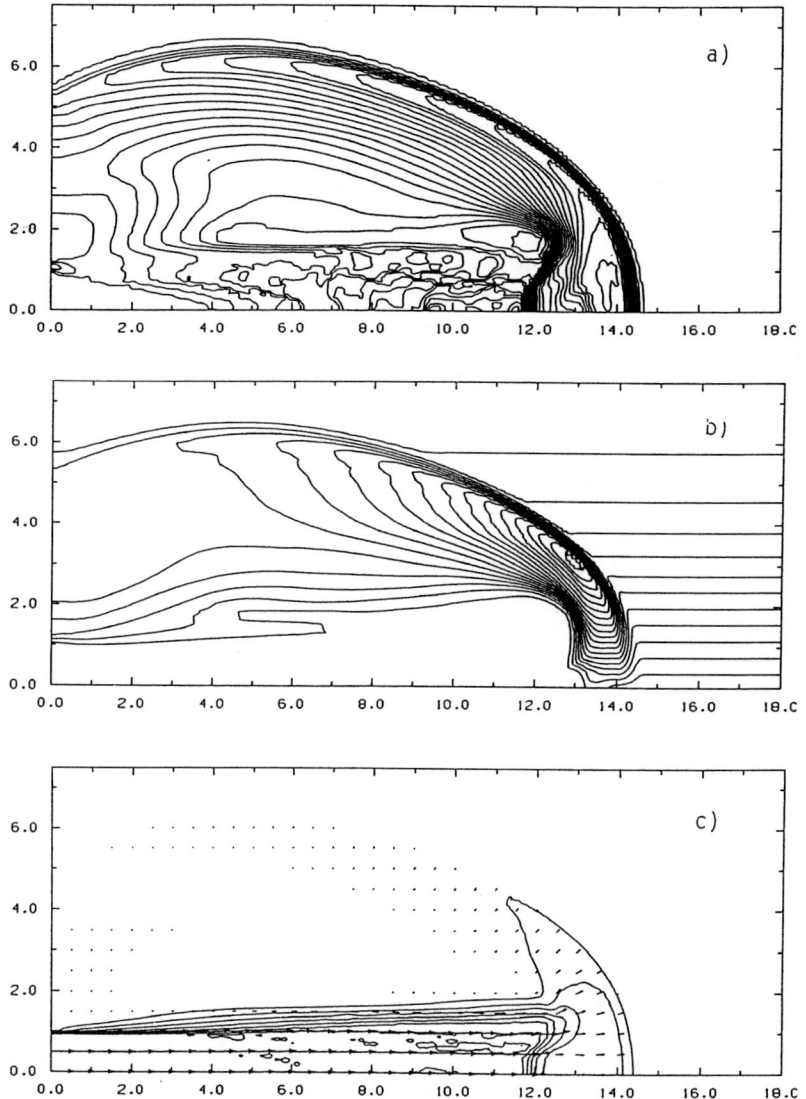

Fig. 4a Density distribution for a supersonic jet boring into a magnetized medium ($Ma_B = 6$, $\eta = 1$). 30 contour lines are shown spaced by a factor of 1.2 (time is $t = 2.9$).

Fig. 4b Toroidal magnetic flux density for a supersonic jet boring into a magnetized medium. 10 equally spaced contour lines are shown.

Fig. 4c Velocity contours and vectors for a supersonic jet boring into a magnetized medium. 10 equally spaced contour lines are shown.

strength is linearly growing in r-direction in the inner part of the grid and is proportional to $\frac{1}{r}$ in the outer part. There is no z-dependence and the peak value at $r = 3$ is slightly higher than the equipartition value. The pinch of the linear part of the field is counterbalanced by a pressure gradient. The beam parameters are $Ma_B = 6$, $\eta = 1$ and the jet is slightly overpressured (by a factor of 1.6). The result is a naked beam pushing the field out of its way, thereby compressing it by a factor of three (see Fig. 4). A comparison with the purely gasdynamical experiment ($\eta = 1$) shows that the propagation speed of the shocks is smaller by a factor of two. Furthermore there are some indications that vortices develop in a cocoon-like structure. But again the resolution is too low to obtain unequivocal results.

5. Conclusions

The results obtained show some interesting new features (vortices and small backflow velocities). To confirm these results the numerical experiments have to be repeated with a better grid resolution and with improved boundary conditions. Morever, magnetized jets boring into magnetized or unmagnetized media have to be studied according to observations (see, e.g., the review of MILEY [13]), and more realistic models with poloidal fields and strongly overpressured beams have to be taken into account. Finally models with other sets of beam parameters (Mach number Ma_B, density ratio $\eta = \frac{\rho_B}{\rho_M}$ and pressure ratio $\varsigma = \frac{p_B}{p_M}$) have to be investigated.

References

1. A.H. Bridle and R.A. Perley: Ann. Rev. Astron. Astrophys. **22**, 319 (1984)
2. M.C. Begelmann, R.D. Blandford and M.J. Rees: Rev. Mod. Phys. **56**, 255 (1984)
3. L.D. Landau and E.M. Lifshitz: *Electrodynamics of Continous Media* (Pergamon Press, Oxford, 1984, 2nd ed.) p. 225
4. J.P. Borris and D.L. Book: J. Computational Phys. **11**, 38 (1973)
5. D.L. Book, J.P. Boris and K. Hain: J. Computational Phys. **18**, 248 (1975)
6. J.P. Borris and D.L. Book: J. Computational Phys. **20**, 397 (1976)
7. S.T. Zalesak: J. Computational Phys. **31**, 335 (1979)
8. R. Morrow and L.E. Cram J. Computational Phys. **57**, 129 (1984)
9. P. Woodward and P. Colella: J. Computational Phys. **54**, 115 (1984)
10. M.L. Norman, L. Smarr and K.H.A. Winkler: In *Numerical Astrophysics* ed. by J.M. Centrella, J.M. LeBlanc and R.L. Bowers (Jones and Bartlett Publishers, Boston, Portola Valley, 1984) p. 88
11. M.L. Norman, L. Smarr, K.H.A. Winkler and M.D. Smith: Astron. Astrophys. **113**, 285 (1982)
12. C.N. Arnold and W.D. Arnett: Astropys. J. **305**, L57 (1986)
13. G. Miley: Ann. Rev. Astron. Astrophys. **18**, 165 (1980)

Jet Formation in Rapidly Rotating Magnetospheres

M. Camenzind

Landessternwarte Königstuhl, D-6900 Heidelberg 1, Fed. Rep. of Germany

1. Physics of Jets

The occurrence of jets is a common phenomenon in modern astrophysics.
Jets are observed in radio galaxies, they emerge from the nuclei of
active galaxies (AGNs) such as quasars, Seyfert galaxies and BL Lac
objects, but they are also produced by galactic objects like SS 433
and young stellar objects. Extragalactic jets are narrow streams of
plasma that appear to squirt out of the center of a galaxy, emitting
thereby radio waves, IR- and optical photons as well as X-rays. One
of the best studied example is the jet associated with the elliptical galaxy NGC 6251 (BRIDLE and PERLEY /1/). The narrow end of this
gigantic jet coincides with the very center of the galaxy, and VLBI
observations have revealed there a pointlike radio source with a
narrow extension of a few light years in the direction of the outer
jet. This example strongly indicates that jet formation already occurs on scales smaller than a few light years.

Jets have now been found in all types of AGNs. VLA observations
also indicate that extended radio emission is linked up with radio-
-quiet QSOs, though not on the large scales as in the radio-loud
quasars, but on a similar scale as found in Seyfert galaxies (RUDNICK
et al. /2/). These observations strongly suggest that plasma outflow
from the center of an active galaxy is a generic phenomenon and that
certainly many radiation features of AGNs are associated with this
outflow. This is definitely true for the broad emission lines. When
the nucleus in the Seyfert galaxy NGC 4151 is at a minimum, two narrow and variable emission lines appear on either side of the CIV 1550
line. The regions emitting these lines could be associated with a
two-sided jet which is the inner portion of the radio jet (ULRICH
et al. /3/).

Despite extensive observations of approximately 200 jets in extragalactic objects (BRIDLE /4/), their physical properties remain
poorly understood. In particular, the flow speeds in the jets are
not known. In superluminal sources, the jet speed has to be close to

the speed of light with minimal Lorentz factors between 2 and 10,
and for the galactic object SS 433 it is exactly known to be one-
-fourth of the speed of light. Apart from this direct evidence for
relativistic motion, the lack of inverse Compton X-rays which in
some quasars would be expected on the basis of canonical synchrotron
theory is indirect evidence for relativistic motion in the cores of
quasars. All this only refers to a relatively small number of sources. In a sample of 135 flat-spectrum quasars BROWNE & PERLEY /5/
find strong support for the ORR & BROWNE view /6/ that the parent
population of core-dominated quasars are the lobe-dominated (steep-
-spectrum) quasars. These strong quasars with extended radio-struc-
ture are consistent with the beaming model when the mean Lorentz
factor is about 5. But these results are certainly biased towards
strong objects, and the inclusion of the more numerous weaker sources will definitely broaden the Lorentz factor distribution towards
lower values. In addition, the structure of the emission regions in
the core of the quasars will influence the form of the luminosity
functions with the result that source counts are of limited use in
testing the beaming hypothesis /7/.

Motivated by many observations which require in situ replenish-
ment of the electron energy in radio sources, the basic idea underlying jet models is that there be continuous, collimated outflow of
energy from the nucleus to the radio source. The initial acceleration and collimation of the jets remain however a major unsolved
problem. There are essentially only two mechanisms for the initial
acceleration, radiation pressure around a compact object (see e. g.
EGGUM et al. /8/) and the magnetic sling effect in a rapidly rotating magnetosphere (CAMENZIND /9/). In the former case, the collimation of the outflow must be achieved with external "pressure walls"
such as the funnel of a geometrically thick accretion disk. In the
second case, the rotating magnetosphere also provides the collimation force in form of the pinching effect of the toroidal magnetic
field. This would also resolve the puzzle that at least some of the
jets are not freely expanding and that an agent is necessary to confine them. In addition, the strong polarisation of the radiation
from radio and optical jets also indicates the presence of organised
magnetic fields in these jets, which are then also important for
their dynamical behaviour.

The propagation of magnetized jets in the interstellar and intergalactic medium is a difficult question, since it involves relativistic MHD simulations, which are beyond the scope of the presently
available numerical techniques. The dynamical effects of magnetic

fields in the jets range from confinement, refocussing and knot formation to the possibility of electromagnetic interaction of the jets with their surroundings. In 1978, BENFORD suggested /10/ that if beams of particles originate in rotating magnetospheres, they would carry electric currents far into distant galactic and extragalactic regions. He also showed that spinning beams would have better stability properties than beams without currents. Ohmic dissipation will occur at the tips of the beams, thereby accelerating particles to relativistic energies. In the following we discuss numerical solutions of jet formation in rapidly rotating magnetospheres. These jets include a current system of the above type, and confirm therefore the suggestions made by BENFORD 8 years ago.

2. Plasma Injection

The magnetic sling effect is the most powerful way to produce high energetic plasma flows. When plasma is injected into a rotating magnetosphere of a compact object beyond the corotation radius, this plasma cannot accrete onto the central object, but it will immediately be flung out of the system by the strong centrifugal forces. As a result, this plasma flow opens up the magnetosphere. In the past few years, essentially 3 different situations have been discussed, where this process is operating in the vicinity of compact objects. We list these examples with increasing complexity of the physical problem.

First, let us consider a magnetosphere of a non-collapsed rapidly rotating compact object in interaction with an accretion disk. Let me illustrate the physical process for a rotating magnetized neutron star. As long as the object is slowly rotating (with periods P larger than about one second), the corotation radius R_{cor} is beyond the inner edge of the accretion disk, and plasma from the disk will be accreted by the central star leading in this way to a galactic X-ray pulsar. When the rotation period is however in the millisecond regime, the corotation radius

$$R_{cor} = 6 \ R_* \ (P/12 \ ms)^{2/3} \tag{1}$$

is now inside the inner edge of the disk

$$R_{in} \simeq 34 \ R_* \ B_{11}^{4/7} \ R_{*6}^{4/7} \ \dot{M}_{19}^{-2/7} \ (R_G/R_*)^{1/7} \ \alpha^{2/7} \ h^{2/7} \tag{2}$$

for surface magnetic fields B of the order of 10^{11} Gauss and typical galactic accretion rates $\dot{M}$ in units of 10^{19} g/s (α is the viscosity parameter and h the relative height of standard accretion disks). Plasma kept up by the rotating magnetosphere will be driven outwards and will flow through the light cylinder R_L

$$R_L = 60 \; R_* \; (P/12 \; ms) \tag{3}$$

The same situation occurs when an accretion disk forms around a magnetised supermassive rotator with the following characteristic radii /9/

$$R_{cor} \geq R_* \tag{4}$$

$$R_{in} \simeq 2.0 \; R_* \; (\frac{B}{10 \; kG})^{4/7} \; (\frac{R_*}{1.5 \cdot 10^{15} \; cm})^{4/7} \; (\frac{50 \; R_G}{R_*})^{1/7}$$

$$\times \; \dot{M}^{-2/7} \; \alpha^{2/7} \; h^{2/7} \tag{5}$$

$$R_L \simeq 10 \; R_* \; M_8^{-17/64} \; \omega \tag{6}$$

for an object of $10^8 \; M_\odot$ rotating approximately Keplerian at its surface ($\omega \simeq 1$) and having a surface magnetic field of typically 10 kGauss. The accretion rate $\dot{M}$ is here given in units of $M_\odot/yr$. This situation is sketched in Fig. 1. The accretion disk acts as a neutral sheet, which divides the magnetosphere into an upper and a lower hemisphere. For a large range of parameters, the inner edge of the disk is sandwiched between the corotation radius and the light cylinder. In a stationary state, all the plasma accreted must leave the system as a wind, which can be collimated beyond the light cylinder to form two oppositely directed jets. Energy and angular momentum in the plasma and the magnetic fields is ultimately extracted from the rotation of the central object. The energy released in the accretion disk is not important, since the inner edge is too far away from the Schwarzschild radius R_G of the central object.

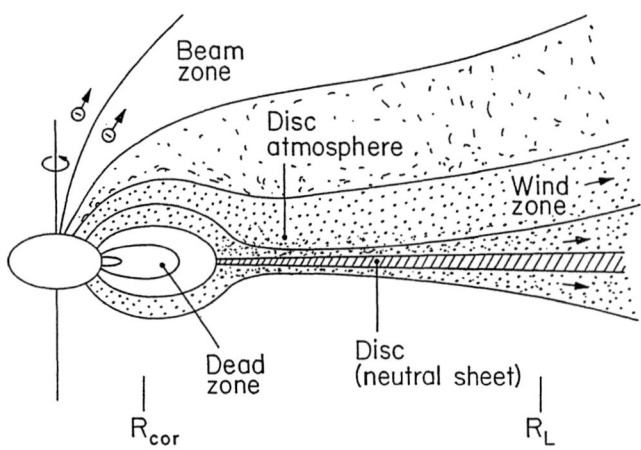

Fig. 1. The magnetospheric interaction between a rapid rotator and a disk

232

The example shown in Fig. 1 is the easiest one for numerical treatment for various reasons. Since the light cylinder is so close to the plasma injection point, Newtonian calculations are obsolete and we are forced to use special relativistic MHD. The above mentioned centrifugal instability is also at work, when the accretion disk is itself the origin for the strong magnetosphere (as proposed by BLANDFORD and PAYNE /11/). In this case, the corotation radius coincides with the plasma injection point. The strong differential rotation of such a magnetosphere implies a complicated structure for the "light cylinder", and this is a further complication for a numerical treatment of the problem. The light cylinder surface is here closest to the surface of the disk at its inner edge. In the third example, a rotating black hole sitting in the center of a thick accretion disk may be embedded into an external magnetosphere built up e. g. by surface currents of the disk (see e. g. PHINNEY /12/). Plasma from the accretion torus or a pair plasma formed by hard gamma rays can be kept up by the magnetosphere and partially accrete onto the black hole and partially escape as a wind to infinity. The plasma outflow is here driven by the rotating magnetosphere whose foot points are dragged along with the accreting plasma. The MHD boundary conditions at the surface of the hole guarantee the extraction of energy and angular momentum from the rotating black hole. This example is certainly the most intricate one from the numerical point of view, since general relativistic effects, plasma effects and curved boundaries make our life difficult.

3. The Relativistic Jet Equilibrium

In the following we describe axisymmetric equilibrium configurations for the plasma flow sketched in Fig. 1. The magnetic field is represented as

$$\underline{B} = \frac{1}{R} (\nabla\Psi \wedge \underline{e}_T) + B_T\underline{e}_T \tag{7}$$

where Ψ is the magnetic stream function which measures the magnetic flux between the rotational axis and a given field line. B_T is the toroidal magnetic field. The equations of motion for the plasma are integrable along a given flux tube (CAMENZIND /13/) and involve 5 constants of motion, $\Omega^F(\Psi)$, $\eta(\Psi)$, $E(\Psi)$, $L(\Psi)$ and $p_*(\Psi)$. These are the angular velocity of a field line which determines the light cylinder $R_L = c/\Omega^F$, the differential particle injection rate, the total energy and the total angular momentum along a flux tube, as well as the dimensionless pressure at the injection point. The poloidal velocity $u_p = \gamma v_p$ follows then from the solutions of a polynomial of degree 16 for an ion plasma with $\Gamma = 5/3$ (CAMENZIND /14/)

$$\sum_{n=0}^{16} A_n \ (X; \ E, \ L, \ \Omega^F, \ p_*; \ f/\sigma_*) \ u_p^{n/3} = 0 \qquad (8)$$

The coefficients depend on the cylindrical radius $x = R/R_L$, the constants of motion, the flux tube function f and on Michel's magnetisation parameter σ_*. In Fig. 2 we show the solutions of this wind equation in the cold limit $p_* = 0$ for a monopole flux tube (f = 1) and $\sigma_* = 0.5$. The physical wind solution starts with low velocity at the injection point, crosses the Alfven point and reaches a constant asymptotic velocity soon after the passage of the light cylinder. In the cold limit, the fast magnetosonic point moves to infinity so that the physical wind solution would cross the unphysical solution at infinity. The total energy carried by this solution is already $2.08 \ mc^2$ and about half of this energy is carried by the Poynting flux.

The geometry of the magnetosphere is in general not known, but must be determined by solving Ampère's equation

$$\nabla_\wedge \underline{B}_p = \frac{4\Pi}{c} \ \underline{j}_T. \qquad (9)$$

The toroidal current j_T now follows from the properties of the plasma flow. Its general form has been derived in CAMENZIND /14/, and it has two different components. A first part in the current is due to the rotation of the magnetosphere and well known under the name Goldreich-Julian current. For a rigidly rotating magnetosphere this part of the current can be absorbed by the left hand side of Ampère's equation. In this way we obtain the relativistic version of the Grad-Schlüter-Shafranov equation of plasma confinement

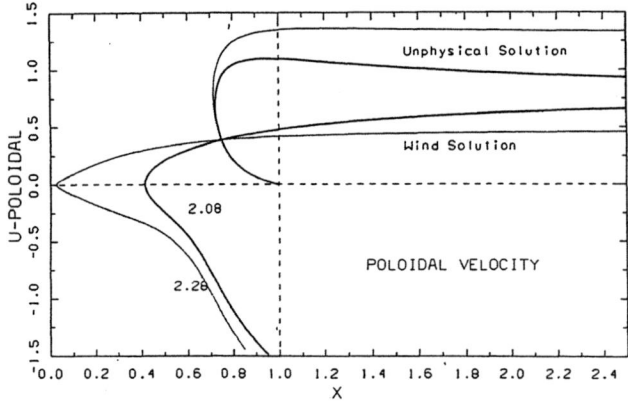

Fig. 2. The flow topology for the poloidal velocity in a monopole type flux tube with $\sigma_* = 0.5$

$$R\mathbf{\nabla} \cdot \left[\frac{1 - M^2 - x^2}{R^2} \nabla\Psi \right] = -\frac{4\Pi}{c} j^{(n)} \tag{10}$$

with a current $j^{(n)}$ now depending on the plasma content of the magnetosphere (due to drift currents). This current is now completely regular at the Alfven point defined by $M_A^2 = 1 - x_A^2$, when M is the relativistic Alfven Mach-number. This plasma current couples with $1/\sigma^2(\Psi)$, when $\sigma(\Psi)$ is defined by

$$\sigma(\Psi) = a_D / (4\Pi \, mc \, \eta(\Psi) \, R_L^2), \tag{11}$$

and a_D is the total magnetic flux involved in the problem. The parameter $\sigma(\Psi)$ is the crucial parameter which determines the asymptotic velocity of the outflow and the coupling of the toroidal plasma current in the GSS equation. When the magnetosphere is sparsely populated with plasma, we are in the high σ-limit, and the plasma current $j^{(n)}$ is of minor importance for the structure of the flux tubes in comparison to the Goldreich-Julian current. σ is also a measure for the mass-flux in the magnetosphere. For supermassive objects, we find typically

$$\dot{M} = 0.3 \, M_\odot yr^{-1} \, a_{D,34}^2 \, R_{L,16}^{-2} \, \sigma^{-1} \, \Psi \tag{12}$$

while for strongly magnetised neutron stars the mass-flux is

$$\dot{M} = 3 \cdot 10^{-5} \, M_\odot yr^{-1} \, a_{D,23}^2 \, R_{L,7}^{-2} \, \sigma^{-1} \, \Psi \tag{13}$$

Here, η is assumed to be constant and Ψ is normalised to unity.

The two equations (8) and (10) must be solved simultaneously with appropriate boundary conditions. In Fig. 3 we show the solutions for the magnetic stream lines obtained with our computer code MAGJET for an inner dipole and for η = const, σ = 100. 90 % of the magnetic flux is enclosed by the accretion disk with its inner boundary at R_{in} = 0.64 R_L (inner dashed line), and plasma is assumed to be injected at $R_* = 0.74 \, R_L$ (second dashed line). For this high value of σ, the Alfven surface practically coincides with the light cylinder. The vertical component of the magnetic field has to vanish there. As a consequence, the whole inner magnetosphere is squeezed inside the light cylinder. The last field line shown in this plot is about 10^{-5} of the total magnetic flux. The bending of the field lines towards the rotational axis starts then beyond the light cylinder, and the asymptotic structure of the magnetosphere is nicely visible at high z-values. The detailed structure does not depend on the particular upper boundary conditions, which are not yet fully relaxed in this plot for reasons of CPU time.

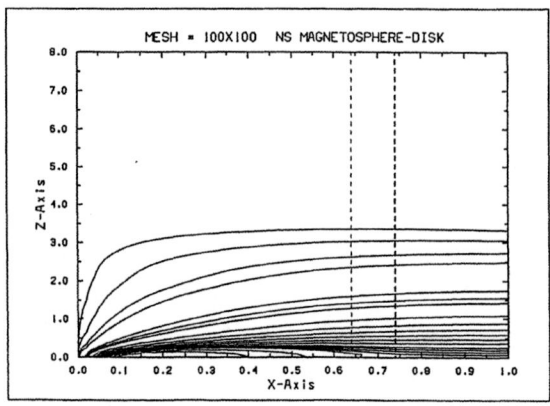

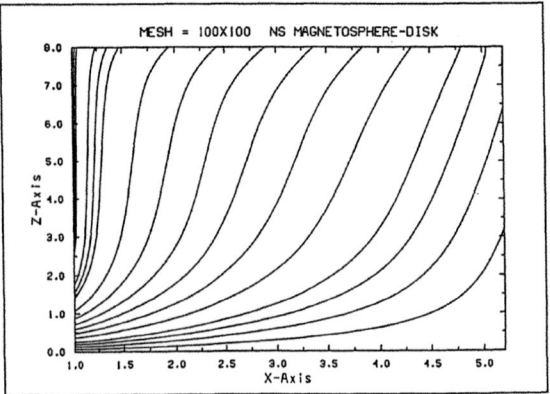

Fig. 3. The inner and outer magnetosphere for σ = 100. The cylindrical coordinates R and z are in units of R_L.

The distribution of the asymptotic Lorentz factors over the magnetosphere is shown in Fig. 4 for σ = 100, 10. The highest values are obtained in the flux tubes close to the equatorial plane (Ψ = 0.1). With moderate injection conditions, Lorentz factors of 2.5 are easily obtained, but for low values of σ (higher mass-fluxes) the outflow becomes semirelativistic. By squeezing somewhat the inner boundary conditions, even higher Lorentz factors can be achieved. This example shows therefore for the first time that relativistic jets can be produced and that collimation can be achieved on scales smaller than a light year in extragalactic sources.

In order to obtain solutions with lower values of σ, we have to include the technique of moving grids in our code, since the Alfven surface is now curved and located somewhere between the injection surface and the light cylinder. The outflow velocity observed in SS 433 is achieved for σ ≃ 0.1 with a mass-flux as observed (see equ.

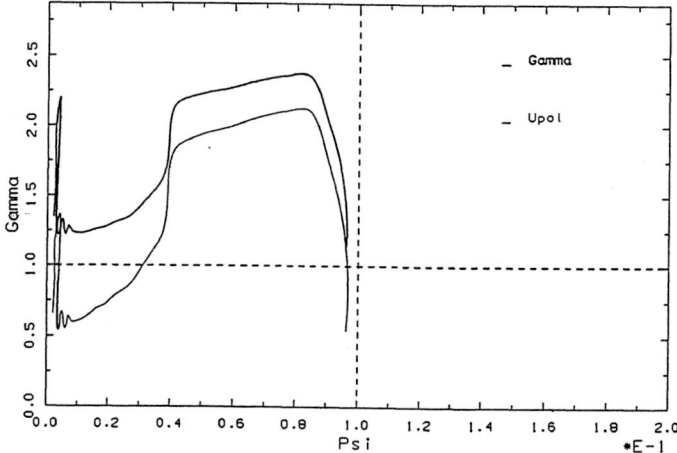

Fig. 4. The asymptotic Lorentz factor distribution

(13)). The necessary extension of our code to handle the interaction
of an accretion disk with the magnetosphere of a rapidly rotating
neutron star is in preparation.

4. Magnetized Jets

Jets produced by the interaction of a rotating magnetosphere with an
accretion disk have many interesting properties when they propagate
beyond the typical distance of 100 light cylinder radii. They consist of poloidal and toroidal magnetic fields and of a heavy ion
plasma flowing predominantly at the outer edge of the jet. When this
part of the jet interacts with the interstellar medium, the broad
and narrow emission lines can be excited. These jets also include a
current system, since the quantity RB_T is essentially only a function of the magnetic flux given by the particle injection function
$\eta(\Psi)$ and the total energy distribution $E(\Psi)$ (CAMENZIND /13/). As a
result, these jets carry a current along the axis and a return current along the surface of the jets. The current system is closed in
the head of the jet, where induced electric fields can boost electrons to relativistic energies. Such particle acceleration may also
occur when local instabilities are growing along the jet.

The development of the code MAGJET and the calculations presented
here were performed on the CRAY-XMP of the Rechenzentrum Garching
during my stay at the Max-Planck-Institut für Astrophysik in Garching.
It is a pleasure to thank F. Meyer, U. Anzer, E. Müller, H. C. Spruit
and W. Hillebrandt for strong support of this project and for many
stimulating discussions on the subject.

5. Literature

1. A. H. Bridle, R. A. Perley: Ann. Rev. Astr. Astrophys. $\underline{22}$, 319 (1984)

2. L. Rudnick, M. L. Sitko, W. A. Stein: Astron. J. $\underline{89}$, 753 (1984)

3. M. H. Ulrich, A. Altamore, A. Boksenberg, G. E. Bromage, J. Clavel, A. Elvius, M. V. Penston, G. C. Perola, M. A. J. Snijders: Nature $\underline{313}$, 747 (1985)

4. A. H. Bridle: In Physics of Energy Transport in Extragalactic Radio-Sources, ed. by A. H. Bridle, J. A. Eilek, NRAO (Virginia 1985) p. 1

5. I. W. A. Browne, R. A. Perley: Mon. Not. R. astr. Soc. $\underline{222}$, 149 (1986)

6. M. J. L. Orr, I. W. A. Browne: Mon. Not. R. astr. Soc. $\underline{200}$, 1067 (1982)

7. K. R. Lind, R. D. Blandford: Astrophys. J. $\underline{295}$, 358 (1985)

8. G. E. Eggum, F. V. Coroniti, J. I. Katz: Astrophys. J. Lett. $\underline{298}$, L41 (1985)

9. M. Camenzind: Astron. Astrophys. $\underline{156}$, 137 (1986)

10. G. Benford: Mon. Not. R. astr. Soc. $\underline{183}$, 29 (1978)

11. R. D. Blandford, D. G. Payne: Mon. Not. R. astr. Soc. $\underline{199}$, 883 (1982)

12. E. S. Phinney: In Astrophysical Jets, ed. by A. Ferrari, A. G. Pacholzcyk, Reidel (Dordrecht 1983) p. 201

13. M. Camenzind: Astron. Astrophys. $\underline{162}$, 32 (1986)

14. M. Camenzind: submitted to Astron. Astrophys. (1986)

Part V

Theoretical Aspects Related
to Cosmical Magnetic Fields

Evolving Magnetostatic Equilibria

J.J. Aly

Service d'Astrophysique, CEN Saclay, F-91191 Gif-sur-Yvette, France

We review some of the work which has been done recently on the problem of the externally driven quasi-static evolution of an ideal magnetohydrostatic configuration. In particular, we pay much attention to the possible development during such an evolution of current sheets.

1. Introduction

The theory of externally confined magnetohydrostatic (MHS) equilibria is a central topics of plasma physics, which has many important applications. First in the laboratory, where one tries to build up fusion devices able to confine a hot plasma in a stable quiescent state. And also in astrophysics, were one has very often to deal with coronae or magnetospheres in equilibrium in the external field of magnetized dense objects (cloud, star, planet ...).

In this paper we review some of the work which has been done on one particular aspect of this general theory. Essentially, we discuss how an ideal MHS equilibrium (i.e. one constituted of a perfectly conducting plasma) does evolve in a quasi-static way as a result of changes imposed on the boundary of the domain of space it occupies. Our main emphasis is on the problem of the existence – or non-existence – of MHS configurations satisfying some conditions rather than on other problems like their stability. In particular, we pay much attention to the possibility of the spontaneous development during such an evolution of current sheets (CS), i.e. of infinitesimally thin surfaces carrying a non-zero net current. This is an important question, as CS have been thought to play a key role in such poorly understood phenomena as stellar flares or stellar coronae heating.

Our plan is as follows. We first present (§2) a precise statement of our problem as well as a few general considerations on it. Then we examine in turn the evolution of configurations of the toroïdal (§ 3) and "line-tied" (§ 4) types. In each situation, we exhibit typical mechanisms leading to the creation of CS. To conclude (§ 5), we discuss briefly some possible applications, indicating in particular how the MHS theory may be used to give a partial description of the fast relaxation of a plasma from an equilibrium which is either unstable or contains CS, to a new one of lower energy, when some topological invariants are conserved during the process.

It is worth noticing that various issues concerning this problem are still highly controversial, and then that many of the conclusions in this paper have to be considered as conjectures reflecting the bias of the author. Different points of view may be found in several other recent reviews, e.g. /1/, /2/, /3/.

2. General Considerations on MHS Equilibria

2.1. Assumptions

In some region (bounded or unbounded) Ω of space, we consider a plasma which is in equilibrium in a magnetic field $\underline{B}$. We assume that : i) the magnetofluid description (discussed e.g. in /4/, /5/) may be used ; ii) the plasma is perfectly conducting, and then the frozen-in law holds ; iii) the plasma stress tensor is isotropic : $\underline{\underline{T}} = p \underline{\underline{1}}$, where p is the thermal pressure ; and iv) the gravitational forces may be neglected ; then we must assume that the system is externally confined, i.e. that it is held together by external forces applied on the boundary $\partial \Omega$. Assumption iv), which clearly restricts the possible applications of the theory to astrophysical situations, is made here for simplicity : it must be noted, however, that the very existence of the mechanisms for CS formation we would like to describe hereafter, should not be affected by the presence of gravity. With the assumptions just quoted, the equations of the MHS write

$$\nabla p = \frac{j \times B}{c} = \frac{(\nabla \times B) \times B}{4\pi} \qquad (1)$$

$$\nabla \cdot B = 0 \qquad (2)$$

It results at once from (1), that, at equilibrium, p is constant along any line of B. An important particular case of (1) corresponds to a situation in which one has $p \ll B^2/8\pi$; with that condition, one may neglect the ∇p term, and the current j flows along the lines of B : the field is force-free (then $\nabla \times B = \alpha B$, with α constant along any line).

2.2. Boundary Value Problems

To solve Eq. (1)-(2), we have to give some boundary conditions on $\partial\Omega$, which may be such that : i) they define a mathematically well-posed problem ; ii) they correspond to quantities one can physically control in some particular situation. A first indication on the type of conditions which are mathematically required may be obtained by loocking at the nature of (1)-(2), determined in particular by their "characteristics" /6/, /7/, /8/. Let us rewrite for instance (1)-(2) as

$$\nabla \times B = \nabla p \times \nabla S \qquad\qquad \nabla \cdot B = 0 \qquad (3)$$

$$B \cdot \nabla p = 0 \qquad\qquad\qquad B \cdot \nabla S = -4\pi \qquad (4)$$

where we have introduced Euler potentials compatible with (1) for j . Clearly (3) defines an elliptic problem which gives B as a function of p and S , while (4) defines an hyperbolic one – in which the (real) characteristics are the field lines themselves – which gives p and S as a function of B . Corresponding to the first problem, we need to fix for instance the value of B_n, the normal component of B , on $\partial\Omega$ (plus some "periods" if Ω is multiply connected). For the second problem, we have to give the values of p and S on some surface cutting once the lines. Then an immediate distinction arises between the configurations in which all the lines cut $\partial\Omega$ twice, and the other ones.

For the configurations of the first type, we may give the values of p and S (or j_n) on that part $\partial\Omega^+$ of $\partial\Omega$ where $B_n > 0$, say. Along with B_n, this defines a boundary value problem (BVP1) which has been abundantly considered in astrophysics. Thus a few theoretical results about the solutions of this BVP, which appears to be well-posed at least for small values of p and j_n, have been obtained for 3-D configurations (see e.g. /9/, /10/, /11/). But most of the work on it has been dealing with 2-D systems, which, according to /12/, must be invariant either by translation, or by rotation or by helical symmetry (see e.g. /13/ for a review of the various explicit solutions computed so far). In that case, one may introduce a magnetic potential ψ which is constant along the lines and integrate (4) at once by introducing two functions $f(\psi)$ and $g(\psi)$ which may be determined from the boundary conditions. Then (3) reduces to a semi-linear elliptic equation for ψ – the Grad-Shafranov equation – with a source term depending on f and g. All these results on the 2-D or 3-D BVP1 have been applied e.g. in solar physics to compute the structure of the force-free field in the solar corona from the measurements of B on the photosphere /14/, /15/ ; or to produce sequences of equilibria which could serve to modelize the quasi-static evolution of a configuration (see below) (/16/, /17/, /18/, and other references in /13/).

For configurations of the second type, which possess regions Ω_k which are not magnetically connected to $\partial\Omega$,obvious difficulties arise as : i) there is no possibilities to transmit informations about p and S from the boundary to $\partial\Omega_k$; ii) if one gives p and S on some other surface Σ_k transverse to the lines, their values are in general severely constrained as most of the lines in Ω_k cut Σ_k infinitely many times ! This does not lead to any problem for 2-D configurations, as we can again integrate (4) by introducing two arbitrary functions $F(\psi)$ and $g(\psi)$ and reduce (3) to the Grad-Shafranov equation for ψ . But severe difficulties with the compatibility of the constraints may arise in 3-D.

It should be noted, however, that the BVP we are arrived at by loocking at the characteristics, is not a physical BVP (although it may be used as a step in physical problems), because $\mathfrak{I}$ (or j_n) is not a physically controlable quantity. A more interesting BVP will actually emerge from considerations on the evolution of equilibria.

2.3. Quasi-static Evolution of MHS Equilibria

Let us now introduce the problem we want to consider in some details hereafter. Suppose that for $t > 0$ we impose on the boundary $\partial\Omega$ of some known equilibrium configuration either some motion to the conducting plasma (then deforming the shape of $\partial\Omega$ and/or moving the feet of the lines along), or some variation $\delta\,p(t)$ to the pressure given on $\partial\Omega^+$. If the time scale of these changes is much larger than the dynamical time scale naturally associated to the system (e.g. the Alfven time), which is just the characteristic time needed to reach an equilibrium, then we may assume as a first approximation that the system evolves in a quasi-static way, i.e. through a continuous sequence of equilibria. Of course, because of the frozen-in law, the topological pattern of the field lines as well as the magnetic flux through any comoving loop have to be conserved during the evolution. On the other hand the mass and entropy contents (assuming for simplifying an adiabatic evolution for the plasma) in any region bounded by field lines ("tube") not connected to $\partial\Omega$ have also to be conserved, while the pressure p inside a tube connected to has to stay equal to the imposed value at its foot on $\partial\Omega^+$.

Then computing the evolution of the field amounts to solve at each instant of time in Ω (t) a new highly non-standard BVP (BVP2) in which one imposes $B_n(t)$ on $\partial\Omega$ (t) and p(t) on $\partial\Omega^+(t)$ (boundary conditions) as well as a topological pattern for the lines, and the magnetic flux, mass and entropy contents of any "tube" not connected to (global conditions). Contrary to BVP 1, the problem so stated corresponds to a physical problem. For instance, it may be used to modelize the quasi-static evolution of the solar coronal field which is driven by the unceasing motion of the subphotospheric plasma in which it is anchored (in that situation, like in laboratory ones, the boundary conditions are effectively "given", in that sense that they are not affected by what happens in the corona owing to the high mass and energy densities of the photospheric layers ; see /5/ for a discussion of a more complex situation).

To be physically satisfying, the solution $(\underline{B}\,,\,p)(t)$ of an evolution problem must be a classical solution, i.e. a continuously differentiable function satisfying (1)-(2) in the usual sense. However, if a classical solution does not exist, it may be useful to look for generalized solutions belonging for instance to the space of finite energy functions. Solutions in that space may have some pathologies, like discontinuities ($\underline{B}$, p) accross some surfaces Σ_c , i.e. have CS (of course, Σ_c should be in equilibrium, i.e. one should have $B_N = 0$ and $[\,p + B_T^2\,/\,8\,\pi\,] = 0$ along, and $|B| < \infty$ near its boundary). The appearance of such CS - a phenomenon which could be called "Parker – Syrovatskii non-equilibrium (NE)" /2/, /19/, /20/ – although non physical (CS would be quickly destroyed by any resistive effect, however tiny it be), is a valuable indication on the actual behaviour of the plasma and on the nature of the physical effects which must be added to get a physical solution. It may also happen that a solution (in any sense) ceases to exist after some finite time T. In that case we shall say that the system has reached at T a state of global NE /3/.

2.4. An Heuristic Argument

Actually, there is a simple argument /21/, /22/, /23/, which leads to the conclusion that global NE should not occur in a closed system (i.e. one in which plasma is not allowed to flow through $\partial\Omega$: in that case, an obvious slight modification of BVP 2 has to be made). Suppose that we start from some initial equilibrium and that we change a little bit the boundary conditions. We may easily construct a field, near the original one, which satisfies the new conditions but is not in equilibrium, and use it as an initial value for the complete set of MHD equations for a non-resistive, adiabatic, but highly viscous fluid (this supposes that the heat produced by dissipation is instantaneously radiated away). Then the motion set in in the system is damped, and as $\underline{v} \rightarrow 0$, the plasma should approach a nearby equilibrium related to the initial one by the ideal MHD constraints (flux, mass, entropy) ! However, the new equilibrium may be quite singular and, for instance, contain CS /22/. Actually, this argument is far from a rigorous proof, as it takes for granted some delicate problems of convergence. However, it is quite suggestive and it could be used empirically to devise a numerical code computing a quasi-static evolution. It is worth noticing that the argument above is strongly related to the existence of a variational principle for BVP 2 [/6/, /7/, /21/] which states, in the case of a closed system, that a solution extremizes the total energy of the system considered as a functional over the set of all functions [$\underset{\sim}{B}$, p] satisfying the global and boundary conditions of the problem.

3. Quasi-Static Evolution of Toroïdal Configurations

In order to isolate specific mechanisms for CS formation, we shall restrict our attention hereafter to a few simple typical configurations. In this Section, we consider toroïdal equilibria confined in a bounded Ω [B_n / $\partial\Omega$ = 0]. Configurations of the "line-tied" type [B_n / $\partial\Omega \neq$ 0], which are expected from the analysis of the characteristics [§ 2.2.] to behave in a different way, will be studied in the next Section.

3.1. Toroidal Configurations

If a MHS equilibrium in a bounded Ω satisfies B_n / $\partial\Omega$ = 0 and ∇ p $\neq$ 0 almost every-where, the equation p = c defines, by a well-known theorem /21/, /24/, families of nested toroidal surfaces Σ to which $\underset{\sim}{B}$ and $\underset{\sim}{j}$ are tangent by [1] [then the Σ are magnetic surfaces). The singular surfaces separating the families are called separatrices, and the degenerate surface inside each family magnetic axis. The same result holds if the field is force-free in some part of Ω (p = constant), and $\nabla\alpha \neq 0$ almost everywhere. Then the only equilibria which do not require the existence of magnetic surfaces are the constant – α force-free fields ! We shall consider hereafter the simplest situation, in which Ω itself is a torus, with an external axis common to all the magnetic surfaces.

To each surface Σ of a toroidal field [equilibrium or not] are associated a poloidal flux ψ , a toroidal one $\emptyset$, and a rotational transform ι [$\emptyset$] = 2 π d ψ / d $\emptyset$ which measures, roughly speacking, the average number of times a line winds around the associated magnetic axis after one turn the long way along Σ . The lines on $\Sigma_\emptyset$ are closed if ι [$\emptyset$] / 2π is rational, and cover $\Sigma_\emptyset$ ergodically if ι[$\emptyset$] / 2 π is irrational. Of course, in the generic case [the one we consider from now on], almost all $\Sigma_\emptyset$ are ergodic [then p = p [$\emptyset$]], but the set of "rational surfaces" is dense.

A powerful method for studying the flow of the lines of an arbitrary field is obtained by expliciting its hamiltonian structure, which is strongly related to the ∇. B = 0 condition, e.g. /25/, /26/, /27/. For a field represented in a particular gauge by the vector potential $\underset{\sim}{A}$ = A$_2$ ∇x^2 + A$_3$ ∇ x^3 [x^2 and x^3 being poloidal and toroidal angles, respectively], it is for instance readily shown that the field lines equations are obtained from the hamiltonian H [p, q, t] = A$_3$, which is a function of the canonically conjugated variables q = x^2 and p = – A^2, and of t = x^3 [H is periodic in q and t]. In this formalism, to any "integrable" H [i.e. admitting a global integral of motion I [q, p, t] periodic in q, t] there corresponds in the physical space a toroidal field, and reciprocally [take in each flux cell t = x^3, q = x^2, p = $\emptyset$, where [$\emptyset$, x^2 , x^3] are now "flux coordinates" /28/, and H = ψ [$\emptyset$], obviously integrable].

From hamiltonian theory /29/, one may deduce the following important result. If $\underset{\sim}{B}_0$ is a toroidal field [H$_0$ integrable], then, for almost all perturbations ϵ $\underset{\sim}{B}_1$, $\underset{\sim}{B}_0$ + ϵ $\underset{\sim}{B}_1$ is not toroidal, i.e. there are lines not drawn on surfaces. Then, magnetic surfaces suffer from some kind of "topological instability" /27/, actually limited for small values of ϵ by the celebrated KAM theorem /29/ which asserts the conservation of the surfaces which are not "too much" rational. It is worth noticing, however, that any set of toroids filling Ω may be considered as a set of magnetic surfaces [just assign a value $\emptyset$ to each surface, and choose some arbitrary ψ [$\emptyset$] for H]. Then there is a continuum of toroidal fields, of arbitrary shapes and symmetries, which are perturbations of $\underset{\sim}{B}_0$.

3.2. 2-D Deformation of a 2-D Configuration

Explicit toroidal solutions of the MHS equations are known only in axisymmetric Ω , and then we shall take such a 2-D field as our initial state. At time t = 0 , we start deforming slowly the boundary $\partial\Omega$, making the equilibrium evolving in a quasi-static way. With our assumptions, this evolution conserves the magnetic surfaces, the fluxes and ι [$\emptyset$], as well as the mass and entropy between any two surfaces. We first assume that the deformation is also axisymmetric. Then we may introduce a magnetic potential and compute its value from the Grad-Shafranov equation if the two arbitrary functions f[ψ] and g [ψ] introduced in § 2.2. are known. Actually, f and g may be expressed as a functional of their initial values and of the geometry of the field lines in the initial and new configurations. With that expression. the equation for ψ becomes an equation of a new type /30/ – a GDE , generalized differential equation – in which the LHS contains an usual 2-D Laplace like operator acting on ψ ($\underset{\sim}{r}$), while the RHS contains a 1-D second order operator acting on ψ (V). i.e. on ψ considered as a function of the volume V inside the surface Σ_δ. The theory of GDE, which are non-linear non-local equations, is unfortunately still in its infancy /31/, and the results hereafter are based more on plausibility arguments than on complete proofs.

The simplest situation correspond to the absence of separatrix in Ω , i.e. to the existence of only one family of toroids and one magnetic axis o along which $\nabla\psi|_o = 0$. In that case. the GDE has to be solved subject to the conditions that ψ /$\partial\Omega$, ψ_o are given [ψ_o (t) = ψ_o (o) by the frozen-in law). From theoretical and numerical evidences, one may assert that this problem is well posed and does not lead to any singular behaviour /30/. The situation becomes however quite different for more complex topologies. Consider for instance a "doublet" configuration, in which there is a separatrix whose intersection with a meridian plane P is 8 - shaped. In P, there are now three points where $\nabla\psi = 0$, two elliptic ones [O - point) inside the two lobes of the 8, and an hyperbolic one c [X - point] at their contact, and, of course, the values of ψ at these points (and them on the separatrix) do not change during the evolution. Actually, one gets at once a contradiction /30/ if one assumes that ψ is well behaved near c (ψ = ψ_c + $c_{ij}x^i x^j$ + ..), as this implies when $\psi \to \psi_c$ on the one hand a bounded limit for the current j_T, but on the other hand a divergence of dV/dψ which leads in general to a divergence of j_T. A bounded j_T may be obtained only if the toroidal flux and mass contents in a flux shell tend when $\psi \to \psi_c$ towards one of the generally unphysical values 0 or ∞. Then the lines have to adjust to produce finite non-zero values of dV/dψ , p, B_φ for ψ = ψ_c. According to Grad et al. /30/, this implies the appearance of a discontinuity of $\nabla\psi$ and then of a CS all along the separatrix.

Actually, the situation is not very much better if we accept an initial configuration with a bounded j_T , as pathologies necessarily develop during the evolution, due to the natural tendency of the X-point to leave the line labelled by ψ_c , what is precluded by ideal MHD. A particular situation where the absence of a regular equilibrium may be proved without making any physical restriction on j_T, has been studied in /32/, /33/, /23/. The system considered is a cylindrical configuration [such configurations approximate large-aspect ratio toroidal equilibria when adequate periodicity conditions are imposed) confined by an external constant pressure. It may be shown that, if an equilibrium exists, the intersections of the cylinder and magnetic surfaces with a plane normal to the axis must necessarily be concentric circles, and that B_p may be zero only on the axis. Then there is only one topology compatible with equilibrium. If a different topology is initially imposed, then the system cannot relax to a regular equilibrium, and CS form.

3.3. 3-D Deformation of a 2-D Configuration

Let us now assume that the deformation imposed to the boundary of our 2-D equilibrium is 3-D. For simplifying, we use again the large aspect ratio approximation and replace, to the 0th order, our unperturbed configuration by a cylindrically symmetric one contained in r < r_0. In the perturbed state, the equation of the boundary is taken to be r_b = r_0 (1 + ϵ sin (m θ + kz]). If ϵ << 1, we may try to use a linearized theory to compute the new equilibrium. Introducing the displacement $\underset{\sim}{\xi}$ ($\underset{\sim}{r}$) needed for a plasma element to reach its new position, one may deduce from the ideal MHD constraints the variations $\delta\underset{\sim}{B}$ = $\nabla \times$ ($\underset{\sim}{\xi} \times \underset{\sim}{B}_0$) and p = $\gamma p_0 \nabla \cdot \underset{\sim}{\xi}$ of $\underset{\sim}{B}$ and p . Then, writing the new equilibrium conditions, one gets an equation for $\underset{\sim}{\xi}$ which has to be solved subject to a boundary condition expressing that the plasma moves with the boundary for r = r_b . It turns out that, if the initial field has a "resonant surface Σ_r" on which $\underset{\sim}{k} \cdot \underset{\sim}{B}_0$ = (m/r) $B_{0\theta}$ + k B_{0z} = 0, then $\underset{\sim}{\xi} \to \infty$ on Σ_r. To make $\underset{\sim}{\xi}$ bounded, it is necessary to allow for a CS on Σ_r , of intensity proportional to ϵ [see e.g. /34/, /8/, or /35/, /36/ for an analogous result in slab geometry). It is worth noticing that, if $B_{0\theta}$ is zero for r = r_1 , then the resonance condition is satisfied on this circle for k = 0 , i.e. for 2-D perturbations [this is equivalent to the situation considered in /36/). Actually, the fact that $\underset{\sim}{\xi} \to \infty$ on Σ_r may be interpreted in terms of bifurcation theory as indicating the possibility of having in a neighbourhood of the initial equilibrium a new one with a different topology /34/, /36/ (and then accessible only if the ideal MHD constraints are relaxed]. Of course, the conclusion of the calculations above may only mean the failure of the linearization method, and a more rigorous approach is needed. However, it is quite suggestive and it is likely to be confirmed by a non-linear analysis.

244

It is worth to bring together the results above and Parker's theorem /19/ on the non-existence of asymmetric equilibria analytically related to a uniform field $\underset{\sim}{B}_o$ (see also § 4.3.). Actually, Parker does not introduce explicitly the ideal MHD constraints; however, as shown in /37/, the analyticity assumption (in the parameter $\epsilon = \delta B / B_o$) precludes any change in the topology of the lines, and then provides a similar constraint. An other related argument is given by Grad /8/ who considers an hypothetical continuous sequence of 3D equilibria along which the toroidal topology is conserved but p and the toroidal current tend to zero (this sequence does not describe a quasi-static evolution as defined in this paper): this sequence should converge towards a vacuum field admitting a complete set of magnetic surfaces – a contradiction because 3-D vacuum field generally do not have such surfaces /38/. It is also worth to note that the arguments developed in /27/ are inconclusive regarding the problem considered here. The fact that the magnetic surfaces are "topologically unstable" (§ 3.1.) to most symmetry-breaking perturbations (and then that most of the magnetic fields are not compatible with equilibrium, which requires the existence of such surfaces) is certainly not sufficient to draw any conclusion on the very existence of 3-D equilibria. Indeed, as noted above, there are also 3-D perturbations conserving the flux surfaces - any perturbation constrained by ideal MHD is of this type ! - or producing new surfaces with a different topology, and nothing in the arguments of /27/ precludes continuous sequences of 3-D fields produced by some of these perturbations to be possibly equilibrium fields.

4. Quasi-static Evolution of Line-tied Configurations

In this Section, we consider the evolution of configurations for which all the magnetic lines cut twice the boundary of the domain. For simplifying the presentation, we assume that the equilibria are force-free.

4.1. Evolution of a 2-D Force-Free Field (FFF) without Critical Points

Most of the work on the evolution of line-tied configurations has been concerned with 2-D systems (invariant either by translation or by rotation) in unbounded regions, with astrophysical applications in mind. For definiteness, we shall consider here x-invariant FFF in the half-space $\{z > 0\}$. Then we may introduce a magnetic potential ψ (y, z) [$\underset{\sim}{B} = \nabla \psi \times \hat{x} + B_x (\psi) \hat{x}$] obeying a Grad - Shafranov equation. As an initial state for an evolutionary sequence of FFF, we take the potential field ψ_o corresponding to some given flux distribution on $\{z = 0\}$ (i.e. to some given ψ_o (y, 0) = g (y)). We first assume that ψ_o has no critical point in Ω or on $\partial \Omega$, and that the motion which is imposed to the plasma on $\partial \Omega$ for t > 0 is parallel to the x-axis. Then for t > 0, some lines $\mathcal{E}$ (ψ) acquire a non-zero shear X (ψ, t), defined as the difference in the x-positions of their left and right feet on $\{z = 0\}$, respectively. Clearly, X (ψ) is related to B_x (ψ) by

$$X (\psi, t) = B_x (\psi, t) \int_{\mathcal{E}_p (\psi, t)} \frac{ds_p}{|\nabla \psi|} = - B_x \frac{d\Sigma}{d\psi} \qquad (5)$$

where $\mathcal{E}_p$ is the projection of $\mathcal{E}$ onto $\{x = 0\}$, ds_p the line element along, and d Σ the area between $\mathcal{E}_p (\psi)$ and $\mathcal{E}_p (\psi + d\psi)$. Note that if we report the value of B_x given by (5) into the Grad - Shafranov equation, we get a GDE as in § 3.2. Obviously, the value of ψ (y, 0) is not changed by this type of motion.

Then the quasi-static evolution problem amounts to solve at each instant of time the equation for ψ subject to the conditions : ψ_t (y, 0) = g (y) ; topology of $\{\mathcal{E} (\psi, t)\} \sim$ topology of $\{\mathcal{E} (\psi, 0)\}$ (arcade-like) and X (ψ, t) given. We also impose to the ψ to have finite energy per unit of x - length ("asymptotic" condition).

A possible way to approach this problem analytically is to consider the well-known variational principle quoted in § 2.4. (see also /39/) which states that any function in the set $\mathcal{H} = \{ \psi \mid C (\psi) < \infty ; \psi (y, 0) = g (y) ; \text{topol.} \mathcal{E} (\psi) \sim \text{topol.} \mathcal{E} (\psi_o)\}$ which extremizes the energy functional

$$C (\psi) = \int_{\{z > 0\}} |\nabla \psi|^2 \underset{\sim}{dr} + \int_o^\infty X^2 (\psi) \left| \frac{d\psi}{d\Sigma} \right|^2 d\Sigma$$

is also a solution of the shearing problem, at least if it satisfies some regularity requirements (ψ must be a C^2 function without "flat" regions). Then to find ψ , one may first look for such an extremizing function in $\mathcal{X}$. Actually, it may be proven /23/ that there is a ψ_1 in $\mathcal{X}$ which makes C (ψ) an <u>absolute minimum</u> (then ψ_1 is absolutely stable, at least with respect to 2-D perturbations). That ψ_1 also satisfy the "regularity requirements" above is still to be completely proved, but the current developments seem to show that this must be the case, a point of view consistent with the recent numerical solutions of the problem reported in /40/.

Then it seems that with a high degree of confidence, we might admit that the shearing problem of an arcade configuration has always a stable solution, even when the shear X is large. This conclusion is in contrast with the popular idea that shearing of a FFF should lead to a global NE, i.e. that there should be a critical shear beyond which no equilibrium would be possible (/1/, /13/ and references therein). However, this idea rests not on a study of the actual problem, but of a simpler one in which one gives not X [ψ], but B_x [ψ]. In that case, general results (/11/, /41/) show that no solution exists for large values of B_x ; but one is not guaranteed to reach these critical values by an actual shearing (a point already made in /42/) - and indeed they are not reached.

Actually, an asymptotic study of the solution of the shearing problem when X [ψ, t] becomes large (taking e.g. X [ψ,t] = t Γ [ψ] and letting t $\longrightarrow \infty$) shows that B_x [ψ] tends to zero when t $\longrightarrow \infty$, the solution ψ_t converging asymptotically towards an <u>open configuration</u> in which all the currents are concentrated in an infinitesimally thin CS. Then <u>CS are also obtained in this problem</u> /43/.

Similar results hold for a large class of 2-D configurations, for instance for axisymmetric fields in $\{z > 0\}$ or around a sphere /5/. Two remarks should be in order here :

i) For axisymmetric FFF, the magnetic energy increases steadily towards the finite energy W_{op} of the totally open field corresponding to B_n /$\partial\Omega$, when the azimuthal shear, assumed to be non-zero near the inversion line (B_n = 0), increases indefinitely. Then one has $W_{FFF} < W_{op}$, and, for arbitrary ϵ , $W_{op} - W_{FFF} < \epsilon$ for a large enough shear. This result is one of the supports of the conjecture put forward in /11/, according to which in an arbitrary unbounded 3-D Ω and for a given distribution of B_n / $\partial\Omega$, W_{op} should be the <u>least upper bound</u> for the energies of all FFF. (the contradiction between this conjecture and one of the result of /17/ is analysed in /11/).

ii) For a symmetric FFF around a sphere, it has been recently shown /44/ that some particular displacements of the feet of the lines on $\partial\Omega$, however small, lead to unbounded displacements of the plasma for r $\longrightarrow \infty$, which implies a breakdown of the quasi-static approximation. Then it was conjectured that the outer lines could react to this situation by blowing open, an equatorial current sheet extending from some large distance to infinity being formed. Actually, it seems to us more likely that the field would try to approach the formal solution of the minimization problem (which, however, would be reached only for t = ∞ , because of inertial effects), as this is clearly more favourable energetically.

4.2. Evolution of a 2-D FFF with Critical Points

We keep the same geometry as in § 4.1., but we assume now that the initial potential field ψ_0 has critical points (necessarily of the X-type). Then let us first suppose that we move the feet of the lines parallel to the y-axis (no shearing, B_x = 0). At first sight, it would seem that these motions should not generate any currents and then that the field should stay potential. However, this would imply in general a change of the values of the labels ψ_k of the lines on which the X-points are located (separatrices), what is precluded by ideal MHD. To get out of this paradoxical situation, we are again leaded, as in § 3.2., to introduce C.S (with the current flowing in the x-direction). In this case, the CS (accross which there is a field reversal), stay localized in a neighbourhood of the X points, which get flattened . B is still potential almost everywhere - but not everywhere - , and the connectivity of the lines is preserved. Analytic computations of 2-D quasipotential equilibrium structures with CS have been reported by several authors /20/, /45/, /46/, /47/.

The situation is much more complex if we impose shearing motions parallel to the x-axis as in § 4.1. In that case, an argument similar to the one presented in § 3.2. /49/ show that the solution one gets cannot stay regular at the X points, unless the shear X [ψ] - that we <u>impose</u> to be finite - of the lines passing near these points is zero. Then the field has to develop some singularity, and actually, a CS appears all along the separatrices. As noted in /50/, the appearance of these CS is strongly related to the fact that <u>continuous</u> motions on the boundary generally produce functions X and $d\Sigma/d\psi$ which are <u>discontinuous</u> accross the separatrices, owing to their non-local character (their values depend on the values of physical quantities at different points), and this gives rise by (5) to discontinuous B_x. Numerical evidences for the appearance of CS along the separatrices are given in /49/. (Of course, discontinuous motions on $\partial\Omega$ produce also CS, but we do not consider this trivial case here)

4.3. Evolution of 3-D FFF

Most of the work on the evolution of 3-D FFF has been concerned with the problem of the existence of 3-D equilibria in a neighbourhood of a given simple 1-D or 2-D configuration without neutral points. For instance, Parker has considered in details the case of a uniform field $\underline{B}_0 = B_0 \hat{z}$, between two plates at $z = -h, +h$, respectively, deformed by plasma motion on $z = -h$, say. Using an expansion of the field in series of a small parameter $\epsilon = \delta B / B_0$ ($\underline{B} = \Sigma \epsilon^n \underline{B}_n$), he proves /19/ that equilibrium imposes the $\underline{B}_n$ to be all independant of z, at least <u>far from the plates</u> supposed to lie at distances much larger than the typical scale of the boundary motions. Parker interprets that result as an evidence for the necessity of the presence of a symmetry for a regular equilibrium to exist. As such a symmetry is not generally present in the system he considers (the motions on $z = -h$ may be arbitrary), he then concludes that one is in a situation of NE in which CS form.

Actually, Parker's calculation seems to us unconclusive as far as the equilibrium of a line-tied field is concerned (it may be used, however, as an argument in favor of the necessity of a symmetry for a toroidal equilibrium to be possible, see § 3.3.). Indeed, it does not prohibit the presence, near the plates, of a boundary layer through which the z-independant approximation for $\underline{B}$ valid at large distance could match arbitrary asymmetric boundary conditions.

Then taking explicitly into account the presence of the plates seem necessary and this has been done in several recent papers. In /51/, Van Ballegooijen concludes at the existence of an equilibrium for small but otherwise arbitrary deformations of the field. There is however a subtle flaw in his argument. A closed look at his calculations (in which an ordering different from Parker's is assumed, resulting in a different expansion scheme) shows indeed that the problem he actually solves is not the one in which the positions of the feet of the field lines on the boundaries are fixed (BVP 2), but the one in which the value of $\alpha = (\underline{B} . \nabla \times \underline{B}) / B^2$ is given (BVP 1). Of course, a solution of this problem gives a particular solution of the shearing problem - and then at least a particular class of 3-D solutions of BVP 2 is obtained - but, a priori, there may be displacements for which one gets no solutions or irregular ones - in which case there are not corresponding regular values of α on $\partial\Omega$ to solve BVP 1 ! However, it turns out that such displacements do not exist, and this is shown in /52/ (where BVP 2 is solved in the linear approximation ; see also /53/ for a treatment of the axisymmetric case) and in /23/ (where the successive approximations and the convergence of the expansion are studied in details). Then it seems that CS do not arise for small 3-D deformations of the field.

In a recent paper /54/, the problem has been reinvestigated without making use of perturbation arguments. Using a representation of an arbitrary field $\underline{B}$ of the given topology in terms of two functions describing the two motions (one compressible irrotational, one incompressible) which transform $\underline{B}$ into $\underline{B}_0$, Parker argues that one of these functions is completely determined by the topological pattern of the motions on $\partial\Omega$, and then that the freedhom one is left with to satisfy the equilibrium equations generally is not large enough, which seems to imply NE. However, this argument does not seem to us convincing, as it neglects in particular the possibility one has to "distribute" in an infinity of ways the incompressible part of the deformation (the one carrying on the topological pattern) along the lines.

Apart the work on the simple configuration above, there has also been some numerical calculations of fully 3-D structures (with simple topology, $B \neq 0$), which have been based either on the BVP 1 formulation /14/, or on the variational formulation of BVP 2 (§ 2.4.) /39/ (which seems a very promising way to abord the problem, but which meets difficulties much more severe than in 2-D for reasons explained in /23/). In any case, it turns out that at least large classes of regular 3-D line-tied configurations depending continuously on the boundary data do exist. The effect of the presence of critical points on the evolution of the field seems to have not yet been considered, but it looks obvious that we may apply the general argument of "non-locality" of the shear (§ 4.2.) and that CS develop along the separatrices /23/.

5. Conclusion

Quasi-static evolution of MHS equilibria subject to the ideal MHD constraints may lead in quite general conditions to the spontaneous appearance of CS and then to a local breakdown of the frozen-in-law - a phenomenon we called Parker-Syrovatskii NE. We have identified here three non-trivial elements which play an essential role in the appearance of the CS : i) <u>resonant surfaces</u>, ii) <u>neutral points and separatrices</u> ; and iii) <u>infinitely large shears</u>. The effectiveness of each of these factors was shown to depend in an essential way on the boundary conditions imposed on $B_n \big|_{\partial\Omega}$. "toroidal" configurations behaving in many respects quite differently from "line-tied" ones. Then resonant effects were found to generate CS in toroidal 2-D fields subject to symmetry-breaking perturbations, but to be absent in line-tied configurations for the existence of which no symmetry requirements appeared to be necessary. On the contrary, the presence of neutral points was proved to lead in both types of structures to the formation of CS. Actually, this last mechanism may be not very different from the previous one, as neutral points are clearly "resonant" to <u>all</u> perturbations, while a resonant surface may be considered as the loci of the points where the component of B in the direction of k vanishes. Then the third factor (large shear) was found to lead to the "asymptotic" appearance of CS in 2-D line-tied configurations in unbounded regions (but clearly the phenomenom must also exist in 3-D).

To conclude this Review, we would like to present a few remarks on the possible applications of the theory of evolution of MHS equilibria to the problems of the heating of stellar coronae and of the triggering of the stellar flares. One promising mechanism for coronal heating (e.g. /55/ and references therein) is just the dissipation of the CS continuously produced by the quasi-static evolution of the coronal field driven by photospheric motions. Although Parker's proposal for CS formation by a symmetry-breacking mechanism has been found unlikely, we have still at our disposal the possibility to produce CS along separatrices between different flux cells, or along "pseudo-separatrices", whose appearance is due to the concentration of the magnetic field on $\partial\Omega$ into flux tubes (and to the subsequent formation of neutral lines on $\partial\Omega$) /50/. As a lot of these singular surfaces are expected to be present in a magnetic corona, this mechanism should be quite efficient. An other interesting possibility is the turbulent reconnection of the sheared field, which is known to lead in several laboratory devices to a relaxation to a linear constant - α FFF , the magnetic helicity $K = \int \underline{A} \cdot \underline{B} \, d\underline{r}$ being conserved during this process /56/. Assuming that this "Taylor relaxation" also holds in coronal conditions, several authors have then computed the amount of heat which may be released /57/, /58/, /59/ (note that a redefinition of K is needed to account for $B_n \big|_{\partial\Omega} \neq 0$, /60/). It is worth noticing that this kind of theory illustrates a new field of application of the MHS theory, in which one considers the relaxation of a plasma from an unstable equilibrium (or a situation of NE) to a new one of lower energy, but with the same values of some topological quantities : MHS computations then allow to predict the final state and the energy released without having to worry about the complicated details of the transient processes (but this necessitates of course some guesses to know which quantities are conserved).

For flares, a popular idea has been that they could be due to the existence for the shear of the coronal FFF of a critical value at which global NE should set in. However, we have found this possibility quite unlikely. An other proposal /50/, which is still more a "programm" than a definitive theory, is the following one. Consider for definiteness the evolution of the simple arcade configuration of § 4.1. As long as the ideal MHD constraints are enforced, this configuration is in a state of minimal energy and nothing may happen. In actual situation, however, the plasma has always some tiny resistivity, and local changes of the topology of the lines, taking place on relatively short time scales, may be allowed. The fluxes however, are conserved, at least approximately. Then an evolution by reconnection may be possible if, for a given shear $X(\psi)$, there does exist a complex topology equilibrium configuration of <u>lower energy</u> which has the same $\psi\big|_{\partial\Omega}$, the same range of variation of ψ (conservation of the "poloidal" fluxes), and the same "global" value of $X(\psi)$, this last one being defined as the sum of the X_k (

ψ] relative to each piece $\mathcal{E}'_k(\psi)$ into which $\mathcal{E}(\psi)$ could be brocken by reconnection (conservation of the "toroidal" fluxes). Arguments showing the existence of such configurations when the shear is larger than some critical value are presented in /50/. Of course, one has just found that a <u>necessary</u> condition for the occurence of a flare may be fullfilled. That the transition between the two equilibria occurs effectively is controlled by the height of the energetic "barrier" between them. The presence of this barrier is clearly suitable initially (without it, transition would take place at the critical shear and a significant energy release would not be possible), but one would like to see its height decreasing when the shear continues increasing - a point which has still to be checked - , then allowing the flare to occur at some second critical shear.

6. References

1. B.C. Low : Solar Phys. <u>100</u>, 309 (1985)
2. E.N. Parker : in <u>"Twenty Years of Plasma Physics"</u> (ICTP, Trieste, 1984)
3. P.C. Martens : in <u>"Magnetic Reconnection and Turbulence"</u>, Cargese Workshop, 1985
4. J.P. Friedberg : Rev. Mod. Phys. <u>54</u>, 801 (1982)
5. J.J. Aly : in <u>"Magnetospheric Phenomena in Astrophysics"</u>, Taos Workshop 1984
6. H. Grad, H. Rubin : in "2nd Conf. on Peaceful Uses of At. En." Geneva, <u>31</u>, 190 (1958)
7. H. Grad : Phys. Fluids <u>7</u>, 1283 (1964)
8. H. Grad : Int. J. Fusion En. <u>3</u>, 33 (1985)
9. M.M. Molodensky : Sov. Astr. AJ <u>10</u>, 578 (1967)
10. M. Bineau : Comm. Pure. App. Math. <u>25</u>, 77 (1972)
11. J.J. Aly : Ap. J. <u>283</u>, 349 (1984)
12. J.W. Edenstrasser : J. Plasma Phys. <u>24</u>, 515 (1980)
13. J. Birn, K. Schindler : in <u>"Solar Flare MHD"</u> (Gordon § Breach 1981)
14. T. Sakurai : Solar Phys. <u>69</u>, 43 (1981)
15. T. Sakurai, M. Makita, K. Shibasaki : MPA/LPARL Workshop, München 1985
16. P.A. Sturrock, E.T. Woodbury : in Intern. School of Phys. Enrico Fermi, Course 39
17. C.W. Barnes, P.A. Sturrock : Ap. J. <u>174</u>, 659 (1972)
18. B.C. Low : Ap. J. <u>212</u>, 234 (1977)
19. E.N. Parker : <u>"Cosmical Magnetic Fields"</u> (Oxford Univ. Press 1979)
20. S.I. Syrovatskii : Ann. Rev. Astr. Ap. <u>19</u>, 163 (1981)
21. M.D. Kruskal, R.M. Kulsrud : Phys. Fluids <u>1</u>, 265 (1958)
22. H.K. Moffat : J. Fluid Mech. <u>159</u>, 359 (1985)
23. J.J. Aly : "Shearing of line-tied force-free fields", Preprint (1986)
24. V.I. Arnold : Ann. Inst. Fourier <u>16</u>, 319 (1966)
25. K.J. Whiteman : Rep. Prog. Phys. <u>40</u>, 1033 (1977)
26. J.R. Cary, R.G. Littlejohn : Ann. Phys. <u>151</u>, 1 (1983)
27. K.C. Tsinganos, J. Distler, R. Rosner : Ap. J. <u>278</u>, 409 (1984)
28. R.D. Hazeltine, J.D. Meiss : Phys. Rep. <u>121</u>, 1 (1985)
29. V.I. Arnold : <u>"Mathematical Methods of Classical Mechanics"</u> (Springer-Verlag 1978)
30. H. Grad, P.N. Hu, D.C. Stevens : Proc. Nat. Ac. Sci USA, <u>72</u>, 3789 (1975)
31. J. Mossino : <u>"Inégalités Isopérimétriques et Applications en Physique"</u> (Hermann 1984)
32. S.I. Vainshtein : Sov. Phys. JETP <u>38</u>, 270 (1974)
33. S.I. Vainshtein, E.N. Parker : Ap. J. <u>304</u>, 821 (1986)
34. G. Laval, R. Pellat, P.H. Rebut : Nucl. Fus. <u>3</u>, 99 (1963)
35. N.A. Bobrova, S.I. Syrovatskii : Sol. Phys. <u>61</u>, 379 (1979)
36. T.S. Hahm, R.M. Kulsrud : Phys. Fluids <u>28</u>, 2412
37. R. Rosner, E. Knobloch : Ap. J. <u>262</u>, 349 (1982)
38. J.R. Cary : Phys. Fluids <u>27</u>, 119 (1984)
39. T. Sakurai : Pub. Astr. Soc. Japan <u>31</u>, 209 (1979)
40. W. Zwingman : PhD. Thesis, Bochum (1985)
41. J. Heyvaerts, J.M. Lasry, M. Schatzman, P. Witomsky : Astr. Ap. <u>111</u>, 104 (1982)
42. K. Jockers : Sol. Phys. <u>50</u>, 405 (1976)
43. J.J. Aly : Astr. Ap. <u>143</u>, 19 (1985)
44. B.C. Low : Ap. J. <u>307</u>, 205 (1986)
45. E.R. Priest, M.A. Raadu : Sol. Phys. <u>43</u>, 177 (1975)
46. T.J. Tur, E.R. Priest : Sol. Phys. <u>48</u>, 89 (1976)
47. Y.Q. Hu, B.C. Low : Sol. Phys. <u>81</u>, 107 (1982)
48. T. Amari, J.J. Aly : "Current sheets in 2-D potential fields", Preprint (1986)
49. W. Zwingman, K. Schindler, J. Birn : Sol. Phys. <u>99</u>, 133 (1985)
50. J.J. Aly : "Constraints on magnetic energy release in the solar corona", Preprint (1986)

51. A. Van Ballegooijen : Ap. J. $\underline{298}$, 421 (1985)
52. T. Sakurai, R.H. Levine : Ap. J. $\underline{248}$, 217 (1981)
53. E.G. Zweibel, A.H. Boozer : Ap. J. $\underline{295}$, 642 (1985)
54. E.N. Parker : Geoph. Ap. Fluid. Dyn. $\underline{34}$, 243 (1986)
55. E.N. Parker : Ap. J. $\underline{264}$, 642 (1983)
56. J.B. Taylor : Rev. Mod. Phys. $\underline{58}$, 741 (1986)
57. J. Heyvaerts, E.R. Priest : Astr. Ap. $\underline{137}$, 63 (1984)
58. P.K. Browning, E.R. Priest : Astr. Ap. $\underline{159}$, 129 (1986)
59. P.K. Browning, T. Sakurai, E.R. Priest : Astr. Ap. $\underline{158}$, 217 (1986)
60. M.A. Berger, G.B. Field : J. Fluid Mech. $\underline{147}$, 133 (1984)

Avoiding Magnetic Monopoles in Numerical MHD Calculations

M. Schmidt-Voigt

Max-Planck-Institut für Physik und Astrophysik, Institut für Astrophysik, Karl-Schwarzschild-Str. 1, D-8046 Garching, Fed. Rep. of Germany

Beside the equations for mass, momentum and energy conservation, a numerical magnetohydrodynamical (MHD) code (assuming infinite conductivity) has to solve the magnetic field equation:

$$\frac{\partial}{\partial t}\vec{B} = [\vec{\nabla} \times [\vec{v} \times \vec{B}]] \ , \tag{1}$$

which automatically fulfills $(\vec{\nabla} \cdot \vec{B}) = 0$ if used as an initial condition. However, due to discretization errors usually artificial magnetic monopoles develop and grow during the computation, what causes an artificial force parallel to the field, energy and momentum no longer being conserved (BRACKBILL and BARNES [1]). Stationarity may not be achieved. In the present study we discuss the effects of the monopoles, and we get rid of them in treating $(\vec{\nabla} \cdot \vec{B}) = 0$ as a dynamical condition: RAMSHAW [2] shows that field equation (1) and solenoidal condition together are equivalent with the two equations (for $(\vec{\nabla} \cdot \vec{B}) = 0$)

$$\frac{\partial}{\partial t}\vec{B} = [\vec{\nabla} \times [\vec{v} \times \vec{B}]] + \vec{\nabla}\Psi(\vec{x},t) \ , \tag{2}$$

$$\vec{\nabla}^2\Psi = \frac{\partial}{\partial t}(\vec{\nabla} \cdot \vec{B}) \ . \tag{3}$$

To implement (2) and (3) to a MHD code, it is more convenient to define the potential Φ with (Δt is the time step)

$$\Phi(\vec{x}) = \int_{t-\Delta t}^{t} \Psi(\vec{x},t')\,dt' \ . \tag{4}$$

After having calculated $\vec{B}(\vec{x},t)$ from (1), we solve the Poisson equation

$$\vec{\nabla}^2\Phi = -(\vec{\nabla} \cdot \vec{B}) \ . \tag{5}$$

Finally $\vec{B}(\vec{x},t)$ from (1) has to be replaced by

$$\vec{B}'(\vec{x},t) = \vec{B}(\vec{x},t) + \vec{\nabla}\Phi \ , \tag{6}$$

what is equivalent to solving (2) and (3), and yields the ordinary field equation (1) with $(\vec{\nabla} \cdot \vec{B}) = 0$ and $\tilde{\Phi} = const.$ on the boundary surface.

To solve the MHD equations, a three dimensional implicit code was developed, based on the implicit factored scheme ("ADI" differences) by BEAM and WARMING [3],[4]. After each time step the monopoles are eliminated in repeating (5) and (6) until

$$(\vec{\nabla} \cdot \vec{B})_{rel} = \frac{(\vec{\nabla} \cdot \vec{B})}{|\vec{B}|} \, |\, \Delta \vec{x} \,| < \omega \, , \tag{7}$$

where $|\, \Delta \vec{x} \,|$ is the local mean distance between the (not necessarily equidistant) gridpoints, and $\omega \simeq 10^{-4}$.

To study the effects of the artificial monopoles (and to check the accuracy of the code), we calculated a stellar explosion as it can be treated analytically by the self similarity solution of LOW [5], [6]: A star of a given mass ($\simeq 1 \, M_\odot$) blows off an outer layer, thereby dragging out an intrinsic stellar magnetic dipole field ($\simeq 10^3 G$) to a radial configuration. Since during the evolution the Lorentz forces play an important role ($\beta = \vec{B}^2 / 8\pi p \sim 1$), this problem seems to be an excellent test of a MHD code. We used 512 gridpoints (8 in each direction), the time step being around 1.2 times the Courant step. One octant of the azimuthally symmetric problem was calculated in a cartesian mesh. To measure the differences between analytical and numerical solution, we compute the relative standard deviation of mass, momentum and total energy density as well as magnetic field. Within the "vector error" we consider all components of the vectors $\vec{v}$ and $\vec{B}$ separately, in the "scalar error" just their norms, thereby avoiding large errors that occur if one component is small compared to the others.

In Fig.1 the standard deviation of two models (just differing in whether $(\vec{\nabla} \cdot \vec{B})$ is corrected or not) is shown as a function of evolutionary time. It can be seen that for $t > 700 \, s$ the error increases almost linearly with time, if the monopoles are not eliminated, whereas it reaches a constant level around 0.9% for $(\vec{\nabla} \cdot \vec{B}) = 0$.

In Fig.2 both vector and scalar error are shown logarithmically. At $t \simeq 800 \, s$ one component of $\vec{B}$ changes sign, thus being small against the others and introducing a large vector error. Again we can reduce the error significantly if we correct for $(\vec{\nabla} \cdot \vec{B}) = 0$: The vector error of about 100% for $(\vec{\nabla} \cdot \vec{B}) \neq 0$ is just around a few percent for $(\vec{\nabla} \cdot \vec{B}) = 0$. Hence the field structure is much better resolved in the latter case.

The line denoted with "$(\vec{\nabla} \cdot \vec{B})$ - Transport" takes into account a simple correction by SCHMIDT and WEGMANN [7]. If we allow $(\vec{\nabla} \cdot \vec{B}) \neq 0$, the error in the field equation will be about $\vec{v}(\vec{\nabla} \cdot \vec{B})$, what can easily be seen in writing

$$[\vec{\nabla} \times [\vec{v} \times \vec{B}]] = \vec{v}(\vec{\nabla} \cdot \vec{B}) - \vec{B}(\vec{\nabla} \cdot \vec{v}) + (\vec{B} \cdot \vec{\nabla})\vec{v} - (\vec{v} \cdot \vec{\nabla})\vec{B} \, . \tag{8}$$

To avoid this error, we simply add $-\vec{v}(\vec{\nabla} \cdot \vec{B})$ as a source to the field equation (physically the monopoles are transported out of the region where they locally develop). This greatly reduces the "final error" ($t > 1000 \, s$) to about 0.2%. Even if the $(\vec{\nabla} \cdot \vec{B})$ correction is applied, the $(\vec{\nabla} \cdot \vec{B})$ transport works since we correct for centered differences, whereas in the code upwind or centered differences are used in super or sub(magneto)sonic regions. In Figs.1 and 2 the flow is always supersonic. If the flow gets subsonic ($t > 2500 \, s$ as the velocity in Low's solution is $\sim t^{-1}$), the correction (8) is not as effective: The velocities are too small to support a sufficient transport of the monopoles. How this affects the overall solution of the flow can be seen from Fig.3, where the scalar error is shown as a function of $(\vec{\nabla} \cdot \vec{B})_{rel}$ (the

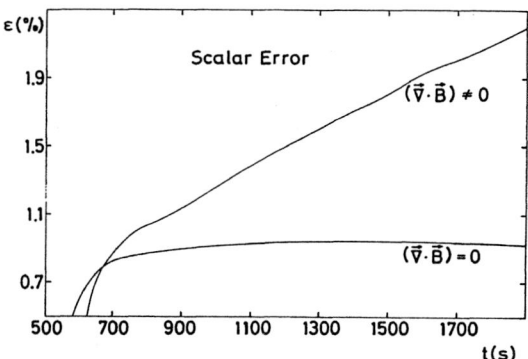

Fig. 1: *Scalar error ε versus evolutionary time for two models with* $(\vec{\nabla} \cdot \vec{B}) = 0$ *(corrected) and* $(\vec{\nabla} \cdot \vec{B}) \neq 0$ *(not corrected).*

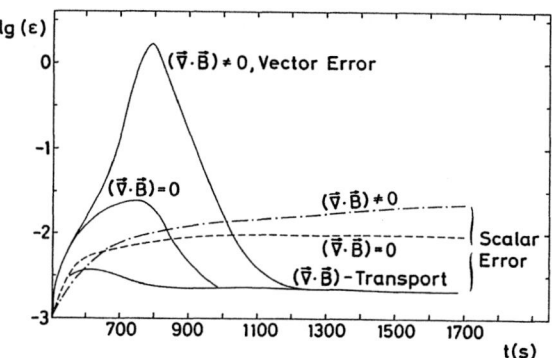

Fig. 2: *Vector and Scalar error ε plotted logarithmically as a function of evolutionary time as in Fig.1.* "$(\vec{\nabla} \cdot \vec{B})$ - Transport" *takes into account a simple correction by SCHMIDT and WEGMANN [7].*

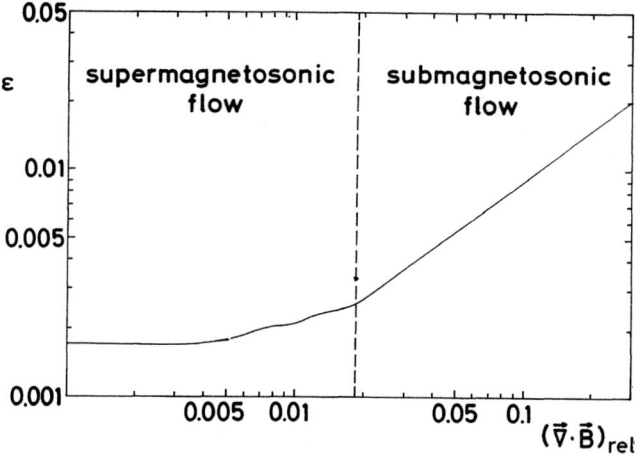

Fig. 3: *Scalar error as a function of* $(\vec{\nabla} \cdot \vec{B})_{rel}$ *from (7).*

maximum value at each time step). The figure is a result of a variety of test calculations. Just the correction (8) is taken into account. In the supersonic region $(\vec{\nabla} \cdot \vec{B})_{rel}$ is always below 1%, the scalar error about 0.2%. Hence there is no influence on the dynamics (if $(\vec{\nabla} \cdot \vec{B})$ transport is used). In the subsonic region $(\vec{\nabla} \cdot \vec{B})_{rel}$ exceeds about 1% and the error increases almost linearly with $(\vec{\nabla} \cdot \vec{B})_{rel}$. Finally unphysical solutions, such as negative pressure or density, were found. Hence for subsonic flow the $(\vec{\nabla} \cdot \vec{B})$ correction is necessary to avoid large errors due to the monopoles.

To summarize: We always have to avoid errors due to artificial monopoles (Fig.1). If we consider supersonic flow, just adding the source $-\vec{v}(\vec{\nabla} \cdot \vec{B})$ to the field equation suffices, whereas in subsonic regions the monopoles have to be eliminated by other methods (like the one used). The monopoles influence the general flow structure if $(\vec{\nabla} \cdot \vec{B})_{rel} > 1\%$.

REFERENCES

1. Brackbill, J.U., Barnes, D.C.: 1980, J. Comp. Phys. **35**, 426
2. Ramshaw, J.D.: 1983, J. Comput. Phys. **52**, 592
3. Beam, R.M., Warming, R.F.: 1976, J. Comp. Phys. **22**, 87
4. Beam, R.M., Warming, R.F.: 1978, AIAA J. **16**, 393
5. Low, B.C.: 1982a, Astrophys. J. **254**, 796
6. Low, B.C.: 1982b, Astrophys. J. **261**, 351
7. Schmidt, H.U., Wegmann, R.: 1980, Comp. Phys. Comm. **19**, 309

Stochastic Particle Acceleration
at Magnetohydrodynamic Shock Waves

W. Dröge

Max-Planck-Institut für Radioastronomie, Auf dem Hügel 69,
D-5300 Bonn 1, Fed. Rep. of Germany

Abstract

The acceleration of energetic particles at a single shock wave in the test–particle
approximation is considered. Using the concept of a microscopic treatment a steady–
state solution of the cosmic ray transport equation in momentum space describing
first–order Fermi acceleration of energetic charged particles at a plane parallel
shock and second–order Fermi acceleration in the downstream region or in both the
downstream and upstream regions of the shock is derived. The solution depends on
the shock compression ratio, the momentum dependence of the spatial diffusion coefficient and the alfvenic Mach number. In the limit of no second–order Fermi acceleration and a constant spatial diffusion coefficient the power law characteristic
of first–order Fermi acceleration depending only on the compression ratio obtained
previously is recovered. The second–order effects lead to a flattening of the spectrum for any shock other than a strong adiabatic shock and make the spectral index
more independent of the shock's compression ratio.

The basic features of the distribution function in momentum space f(p) of energetic
particles in the vicinity of a plane parallel shock undergoing first–order Fermi acceleration as they scatter back and forth across the shock and second–order Fermi
acceleration by MHD waves or turbulence can be studied in terms of a
Fokker–Planck– or "Leaky–box"–equation of the form (DRÖGE [1])

$$\frac{\partial f}{\partial t} - \frac{1}{p^2} \frac{\partial}{\partial p} \left[p^2 \left[\langle D(p) \rangle \frac{\partial f}{\partial p} - \langle \frac{\Delta p}{\Delta t} \rangle \, f \right] \right] + \frac{f}{T} = Q(p) \tag{1}$$

where T is an escape time describing the advection of particles from the shock and
Q(p) is a source term representing the injected particles.

Here we will follow the "microscopic" or "random walk" approach of shock acceleration given by BELL [2] (see also LAGAGE and CESARSKY [3], DRURY [4]). At a
microscopic level diffusive shock acceleration describes how an energetic particle
gains energy as it scatters back and forth across a shock being reflected by resonant Alfven waves or other forms of plasma turbulence on both sides of the shock
(Fig. 1).

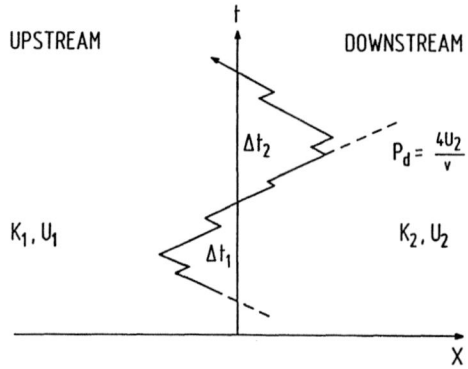

UPSTREAM   DOWNSTREAM

K_1, U_1 K_2, U_2

$P_d = \dfrac{4U_2}{v}$

X

Figure 1: Schematic diagramme of a particle's random walk in the vicinity of the shock in the course of which it cycles through the shock many times (details see text)

A particle crossing the shock from downstream to upstream, which subsequently recrosses the shock has a mean residence time $\Delta t_1 = 4K_1/(vU_2)$ in the upstream region. Similarly a particle crossing the shock from downstream to upstream, returns to the shock after a mean residence time of $\Delta t_2 = 4K_2/(vU_2)$. While the probability not to return to the shock is zero in the upstream region, there exists a finite escape probability downstream given by

$$P_d = \frac{4U_2}{v} \tag{2}$$

Thus the time for a complete cycle is

$$\Delta t = \frac{4}{v}\left[\frac{K_1}{U_1} + \frac{K_2}{U_2}\right] \tag{3}$$

and from (A2) and (A3) the escape time becomes

$$T = (P_d/\Delta t)^{-1} = \frac{1}{U_2}\left[\frac{K_1}{U_1} + \frac{K_2}{U_2}\right] \tag{4}$$

Defining an averaged spatial diffusion coefficient parallel to the magnetic field $K_\| = K_1 + rK_2$ and with $r = U_1/U_2$, the compression ratio of the shock, we obtain

$$T = \frac{K_\|}{r\,U_2^2} \tag{5}$$

The fractional momentum gain at each completed loop is

$$\Delta p = \frac{4}{3}\frac{U_1 - U_2}{v}\,p \tag{6}$$

so that the momentum change rate of particles that complete a cycle is

$$\left\langle \frac{\Delta p}{\Delta t} \right\rangle = \frac{U_1 - U_2}{3(K_1/U_1 + K_2/U_2)}\,p = \frac{r-1}{3r}\frac{U_1^2}{K_\|}\,p \tag{7}$$

256

The effects of second-order Fermi acceleration can be described by a momentum diffusion coefficient $D(p)$. In the case of Alfven waves $D(p)$ is given by (SKILLING [5])

$$D(p) = \frac{V_A^2}{9K} p^2 \qquad (8)$$

(V_A: Alfven velocity). Taking into account the residence times of the particles up- and downstream we are able to define an averaged momentum diffusion coefficient

$$\langle D(p)\rangle = \alpha \frac{\Delta t_1}{\Delta t} \frac{V_{A1}^2}{9K_1} + \frac{\Delta t_2}{\Delta t} \frac{V_{A2}^2}{9K_2} \qquad (9)$$

where the parameter α ($0 \le a \le 1$) describes the efficiency of second-order Fermi acceleration in the upstream region. On letting $V_{A1}^2/U_1 = V_{A2}^2/U_2$ and with (3) and (8) we obtain

$$\langle D(p)\rangle = \frac{\alpha+1}{9} \frac{V_{A1}^2}{K_\parallel} p^2 \qquad (10)$$

From (5), (7), (9) and with $\chi = \alpha+1$ equation (1) becomes

$$\frac{\partial f}{\partial t} - \frac{1}{p^2} \frac{\partial}{\partial p} \left[\frac{\chi V_{A1}^2}{9 \, K_\parallel} p^4 \frac{\partial f}{\partial p} - \frac{r-1}{3r} \frac{U_1^2}{K_\parallel} p^3 \, f \right] + \frac{rU_2^2}{K_\parallel} f = Q(p) \qquad (11)$$

The steady-state solution to this equation in the absence ($\chi=0$) of the second-order Fermi acceleration term for monoenergetic injection at the shock, $Q = q_0 \, \delta(p-p_0)$, is ($p > p_0$)

$$f(p) \propto K_\parallel(p) \, p^{-\frac{3r}{r-1}} \qquad (12)$$

which is consistent with the volume-integrated spectrum given by BELL [2]. Including the second-order Fermi acceleration term again for a constant diffusion coefficient $K_\parallel$ we obtain

$$f(p) \propto p^{-\sigma} \qquad (13)$$

with

$$\sigma = \frac{3}{2} \left(1 - \frac{r-1}{\chi r} M_A^2 \right) + \frac{3}{2} \left(1 + \frac{r-1}{\chi r} M_A^2 \right) \left[1 + \frac{4 \, M_A^2}{\chi r \left(1 + \frac{r-1}{\chi r} M_A^2 \right)^2} \right]^{1/2} \qquad (14)$$

where we introduced the Alfvenic Mach number $M_A = U_1/V_{A1}$. For an adiabatic shock wave propagating in a non-relativistic medium parallel to the magnetic field,

$$M_A^2 = \frac{3r\beta}{4-r} \qquad (15)$$

where $\beta = C_s^2/V_{A1}^2$ is the square of the ratio of the upstream sound to Alfven speed. Using (15) in (14) we find

$$\sigma = \frac{3}{2} \left[1 - \frac{3\beta(r-1)}{\chi(4-r)} \right] + \frac{3}{2} \left[1 + \frac{3\beta(r-1)}{\chi(4-r)} \right] \left[1 + \frac{12\chi\beta(4-r)}{\left\{\chi(4-r) + 3\beta(r-1)\right\}^2} \right]^{1/2} \qquad (16)$$

which is plotted in Figure 2 for the case $\beta = 1$. In the limit of $\chi \to 0$ (that is second-order Fermi acceleration is negligible) equation (16) reduces to the familiar first-order result (eq. 12). For $\chi = 1$ or 2, corresponding to second-order effects from downstream only ($\chi = 1$) or from upstream and downstream ($\chi = 2$), respectively, there is a strong effect on the particle spectral index from second-order Fermi acceleration, and this effect is most noticeable for moderate to weak shocks with compression ratios somewhat less than about 3, as can be seen from Figure 2.

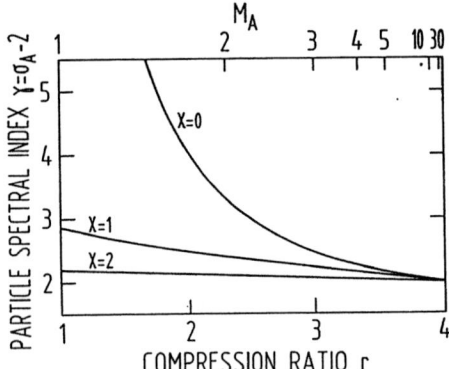

Figure 2: Dependence of particle spectral index on the compression ratio r (assuming $\beta = 1$) for a plane parallel adiabatic shock for the cases of second-order Fermi acceleration which is (a) absent ($\chi=0$), (b) downstream only ($\chi=1$), (c) downstream and upstream ($\chi=2$)

We suggest that the combination of first-order Fermi acceleration at a shock wave with second-order Fermi acceleration behind or on both sides of the shock front is the mechanism responsible for particle acceleration in shocks with compression ratios less than the strong adiabatic value of 4. This therefore appears to be a likely scenario for the acceleration of energetic particles in the solar corona, where due to a strong magnetic field both effects can be of the same strength, and possibly in the shocks of supernova remnants, if, following the work of SOLINGER et al. [6] and LERCHE and VASYLIUNAS [7], due to thermal conduction from the hot interior of the remnant the compression ratio is reduced to $r \approx 2.4$.

Acknowledgements

A number of stimulating discussions with Drs. L. Drury, I. Lerche, M. Ostrowski and R. Schlickeiser are gratefully acknowledged.

References

1. W. Dröge: Ph.D. Thesis, University of Bonn (1986)
2. A.R. Bell: Monthly Notices Roy. Astron. Soc. _182_, 147 and 443 (1978)
3. P.O. Lagage, C.J. Cesarsky: Astron. Astrophys. _118_, 223 (1983)
4. L.O.C. Drury: Rept. Progr. Phys. _46_, 973 (1983)
5. J. Skilling: Monthly Notices Roy. Astron. Soc. _172_, 557 (1975)
6. A. Solinger, S. Rappaport, J. Buff: Astrophys. J. _201_, 381 (1975)
7. I. Lerche, V.M. Vasyliunas: Astrophys. J. _210_, 85 (1976)

Synchrotron Radiation in a Plasma
with Random Magnetic Fields

A. Crusius

Max-Planck-Institut für Radioastronomie, Auf dem Hügel 69,
D-5300 Bonn 1, Fed. Rep. of Germany

1. Introduction

In most of the powerful cosmic synchrotron sources, as radio galaxies and active
galactic nuclei, the measured degree of linear polarization is smaller than the one
expected theoretically for a completely unidirectional magnetic field (KELLERMANN
and PAULINY-TOTH [1], MILEY [2], ANGEL and STOCKMAN [3]). Therefore a field
configuration can be assumed which consists of a homogeneous and a random component. The spectral power radiated into a particular linear polarization mode σ
($\sigma = 1,2$) at frequency ν then is expressed as the sum of the homogeneous (P_h),
with percentage q, and the random (P_r), percentage 1−q, contribution:

$$P^{\sigma}(\nu,\theta) = q\, P_h^{\sigma}(\nu,\theta) + \frac{1}{2}\,(1-q)\,P_r(\nu) \tag{1}.$$

q = 0 and q = 1 refer to the cases of emitted power in 0% and 100% ordered magnetic field, respectively. θ denotes the angle between the field lines and the direction of emission. The factor 1/2 is due to the fact that only half of the power $P_r(\nu)$
is measured in each of the two polarization modes. In the following we will show
how to get an analytical expression for the random part of the spectrum. Two integrals appearing there are solved exactly. The result obtained that way is easily
transferred to synchrotron radiation by relativistic electrons embedded in a thermal
plasma which affects the spectrum at low frequencies (Razin-Tsytovich cutoff).

2. Synchrotron emissivity in random magnetic fields

We assume that the magnetic field B has stochastic directions on scales small compared to the size of the cosmic source but large compared to the Larmor radii of the
radiating particles so that the synchrotron formulas are still applicable. In that case
$P_r(\nu)$ can be calculated by averaging the total spontaneously emitted power for a
homogeneous field over the whole range of solid angles:

$$P_r(\nu) = \frac{1}{4\pi}\int d\Omega \sum_{\sigma=1}^{2} P_h^{\sigma}(\nu,\theta) = \frac{1}{2}\int d\theta \, \sin\theta \, \frac{1}{4\pi} \int_{o}^{\infty} dE \, N(E) \, Q_h(\nu,\theta,E) \tag{2}.$$

In eq. (2) N(E) denotes the energy distribution of the radiating particles and
$Q_h(\nu,\theta,E)$ is the spontaneously emitted spectral power of a single electron in vacuum
which is given by (see e.g. MOFFET [4]):

$$Q_h(\nu,\theta,E) = 4\pi\ c_2\ B\ \sin\theta\ \frac{\nu}{\nu_c}\ \int_{\nu/\nu_c}^{\infty} dt\ K_{5/3}(t) \tag{3}$$

with

$$c_2 = 3^{1/2}\ e^3/(4\pi mc^2)\ ,$$

$$\nu_c = \frac{3eB}{4\pi mc}\ \left(\frac{E}{mc^2}\right)^2\ \sin\theta = c_1\ B\ E^2\ \sin\theta \tag{4}.$$

$K_{5/3}$ denotes a modified Bessel function of order 5/3. If we define

$$x \equiv \nu/(c_1\ B\ E^2) \tag{5}$$

we may write

$$P_r(\nu) = c_2\ B\ \int_0^{\infty} dE\ N(E)\ R(x) \tag{6}$$

where

$$R(x) = \frac{x}{2}\ \int_0^{\pi} d\theta\ \sin\theta\ \int_{x/\sin\theta}^{\infty} dt\ K_{5/3}(t) \tag{7}.$$

In the following we prove that R(x) can be integrated to give

$$R(x) = \frac{1}{2}\ \pi x\ \left[W_{0,\frac{4}{3}}(x)\ W_{0,\frac{1}{3}}(x)\ -\ W_{\frac{1}{2},\frac{5}{6}}(x)\ W_{-\frac{1}{2},\frac{5}{6}}(x)\right] \tag{8}$$

where $W_{\lambda,\mu}(x)$ denotes Whittaker's function. By noting that $\sin\theta$ is symmetric around $\theta = \pi/2$ and partially integrating (7) with respect to θ we get

$$R(x) = x^2\ \int_0^{\pi/2} d\theta\ \cos^2\theta\ \sin^{-2}\theta\ K_{5/3}\left(\frac{x}{\sin\theta}\right) \tag{9}.$$

Substituting $\theta = \pi/2 - \beta$ yields

$$R(x) = I_1(x)\ -\ I_2(x) \tag{10}$$

with

$$I_1(x) = x^2\ \int_0^{\pi/2} d\beta\ \cos^{-2}\beta\ K_{5/3}(x/\cos\beta) \tag{11}$$

$$I_2(x) = x^2\ \int_0^{\pi/2} d\beta\ K_{5/3}(x/\cos\beta) \tag{12}.$$

I_2 can be readily solved with (GRADSHTEYN and RYZHIK [5], p. 741)

$$\int_0^{\pi/2} d\beta\ \frac{\cos(2\lambda\beta)}{\cos\beta}\ K_{2\mu}\left(\frac{x}{\cos\beta}\right) = \frac{\pi}{2x}\ W_{\lambda,\mu}(x)\ W_{-\lambda,\mu}(x) \tag{13}$$

to

$$I_2(x) = \frac{\pi}{2} x \; W_{\frac{1}{2},\frac{5}{6}}(x) \; W_{-\frac{1}{2},\frac{5}{6}}(x) \tag{14}.$$

Integral (11) is rewritten by substituting

$$\cosh t = \cos^{-1}ß \tag{15},$$

which implies

$$dß \cos^{-2}ß = dt \cosh t \tag{16}.$$

We obtain

$$I_1(x) = x^2 \int_0^\infty dt \cosh t \; K_{5/3}(x \cosh t) \tag{17}$$

which reduces to

$$I_1(x) = \frac{x^2}{2} K_{4/3}\left(\frac{x}{2}\right) \; K_{1/3}\left(\frac{x}{2}\right) \tag{18}$$

since (GRADSHTEYN and RYZHIK [5], p. 727)

$$\int_0^\infty dt \cosh (2\mu t) \; K_{2\nu}(2a \cosh t) = \frac{1}{2} K_{\mu+\nu}(a) \; K_{\mu-\nu}(a) \tag{19}.$$

The combination of (14) and (18) yields for (10)

$$R(x) = \frac{x^2}{2} K_{4/3}\left(\frac{x}{2}\right) \; K_{1/3}\left(\frac{x}{2}\right) - \frac{\pi x}{2} \; W_{\frac{1}{2},\frac{5}{6}}(x) \; W_{-\frac{1}{2},\frac{5}{6}}(x) \tag{20}$$

Modified Bessel functions are related to Whittaker functions as (ABRAMOWITZ and STEGUN [6], p. 377)

$$K_\nu(z) = \left(\frac{\pi}{2z}\right)^{1/2} W_{0,\nu}(2z) \tag{21}.$$

By inserting (21) into (20) we obtain the emissivity function R(x) in the form of eq. (8) which completes the proof.

Using the asymptotic expansions of Whittaker functions for small and large arguments we derive

$$R(x) \simeq \begin{cases} \dfrac{2^{1/3}}{5} \; \Gamma^2\left(\frac{1}{3}\right) x^{1/3} = 1.80842 \; x^{1/3} & \text{for } x \ll 1 \\[2ex] \dfrac{\pi}{2} e^{-x} \left(1 - \dfrac{99}{162} x^{-1}\right) & \text{for } x \gg 1 \end{cases} \tag{22}$$

A plot and a table of R(x) for values $10^{-2} \leqslant x \leqslant 10$ can be found in CRUSIUS and SCHLICKEISER [7]. There it is also shown that in case of a power-law distribution $N(E) = N_0 E^{-s}$ of the relativistic electrons one gets a result that has been known for a long time. The interesting point is that our averaging procedure is independent of the energy distribution of the radiating particles.

3. Synchrotron emissivity in a plasma with random magnetic fields

If the relativistic electrons are embedded in a thermal plasma the total spontaneously emitted spectral synchrotron power of a single electron in a homogeneous field is given by (ZHELEZNYAKOV [8], MELROSE [9])

$$Q_h(\nu,\theta,\gamma) = \frac{q_0 \nu}{\gamma^2}\left[1 + \left(\frac{\gamma \nu_p}{\nu}\right)^2\right] \int_{\frac{x}{\sin\theta}\left[1 + \left(\frac{\gamma \nu_p}{\nu}\right)^2\right]^{3/2}}^{\infty} dy\, K_{5/3}(y) \qquad (23)$$

with

$$q_0 = \frac{4e^2\pi}{3^{1/2}c} \quad , \qquad \gamma = \frac{E}{mc^2} \quad ,$$

and ν_p is the plasma frequency. x is defined in eq. (5). To calculate that part of the power which is due to the random magnetic field component we can apply the same procedure as in part 2. We obtain

$$Q_r(\nu,\gamma) = \frac{q_0 \pi \nu}{2\gamma^2}\left[1 + \left(\frac{\gamma \nu_p}{\nu}\right)^2\right] CS\left[x\left\{1 + \left(\frac{\gamma \nu_p}{\nu}\right)^2\right\}^{3/2}\right] \qquad (24)$$

for the angle-average of (23), where we have written $R(x) \equiv 1/2\, \pi x\, CS(x)$, i.e.

$$CS(x) = W_{0,\frac{4}{3}}(x)\, W_{0,\frac{1}{3}}(x) - W_{\frac{1}{2},\frac{5}{6}}(x)\, W_{-\frac{1}{2},\frac{5}{6}}(x) \qquad (25).$$

Equation (24) allows a detailed analysis of the influence of a thermal plasma on the synchrotron spectrum which is discussed in CRUSIUS and SCHLICKEISER [10].

References

1. K.I. Kellermann, I.I.K. Pauliny-Toth: Ann. Rev. Astron. Astrophys. 19, 373 (1981)
2. G. Miley: Ann. Rev. Astron. Astrophys. 18, 165 (1980)
3. J.R.P. Angel, H.S. Stockman: Ann. Rev. Astron. Astrophys. 18, 321 (1980)
4. A.T. Moffet: In Stars and Stellar Systems, Vol. 9, ed. by G.P. Kuiper and B.M. Middlehurst (Chicago University Press, Chicago 1965), p. 211
5. I.S. Gradshteyn, I.M. Ryzhik: Tables of Integrals, Series, and Products (Academic Press, New York 1965)
6. M. Abramowitz, I.A. Stegun: Handbook of Mathematical Functions (National Bureau of Standards, Washington 1970)
7. A. Crusius, R. Schlickeiser: Astron. Astrophys. 164, L16 (1986)
8. V.V. Zheleznyakov: Soviet Phys. JETP 24, 381 (1966)
9. D.B. Melrose: Plasma Astrophysics, Vol. 1, (Gordon and Breach Science Publishers, New York, London, Paris 1980)
10. A. Crusius, R. Schlickeiser: in preparation

A Few Summarizing Remarks

W. Hillebrandt

Max-Planck-Institut für Physik und Astrophysik, Institut für Astrophysik, Karl-Schwarzschild-Str. 1, D-8046 Garching, Fed. Rep. of Germany

In order to keep their models simple theorists usually tend to neglect the influence of magnetic fields on the dynamics of astrophysical systems, and only after observations revealed the importance of magnetic fields were they willing to take them into consideration. The physics of the outer layers of our sun is a first and typical example, and the dynamics of the interstellar matter has turned out to be another clear case.

The present workshop was devoted to interstellar magnetic fields, and both theories and observations were discussed extensively. Theoretical models mostly dealt with the questions how the observed large-scale structures of the magnetic fields can be explained and how they can be sustained over the life of a galaxy. Dynamo models borrowed from solar physics were presented by various groups (see figure 1) which could successfully reproduce observed ring-like and spiral structures as well, and there was general agreement that dynamos are necessary to sustain the magnetic fields. But, as usual, answers to certain questions raise

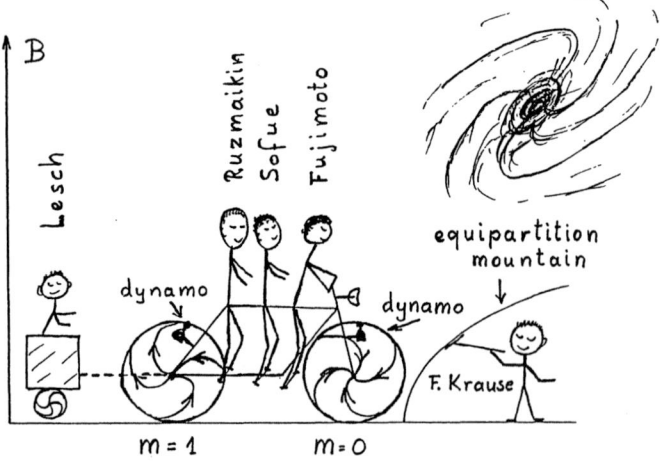

Figure 1: An international collaboration exciting dynamos and approaching the equipartition mountain. It is not yet clear whether or not they can reach the top. Note that the back wheel (m=1) is not exactly bi-symmetric but shows a more stable structure.

new ones. The origin of the seed fields, for example, is still subject to considerable controversies. Moreover, classical dynamo models neglect the feedback of the magnetic fields on the motion of the plasma. Since, however, the observed fields don't seem to be weak but are rather close to equipartition, nonlinear effects may indeed be important and should be taken into consideration.

In a series of talks presented by members of the Bonn group exciting new results of radio observations of polarized emission from various galaxies were shown (see figure 2). Evidence was shown for ring-like structures in some cases and bisymmetric fields in others, supporting the interpretation that they are due to dynamo action. However, high resolution observations such as have been performed for M31 are highly desirable for other galaxies in order to strengthen this conclusion.

Very often magnetic fields are aligned with the spiral arms of galaxies. This has led to suggestions that magnetic fields may play an important role in

Figure 2: Members of the Bonn group observing ring-like and spiral structures of magnetic fields with radio telescopes. One telescope, however, is pointing towards dwarf galaxies rather than a giant spiral. An intruder from Tel Aviv obviously prefers an optical instrument.

the formation process of spiral arms challenging the canonical density wave theory (figure 3). Although these unconventional ideas have not been worked out as well as the standard models they certainly deserve further attention.

Observations of intense linearly polarized emission from a region near the center of our own galaxy were also reported (figure 4). The emission region contains a central source and jet-like structures extending out to about 100 pc. The structure of the magnetic field in the extended features seem to be rather complicated but can be interpreted by means of a twisted poloidal field which may be due to the interaction of a jet with a poloidal field. This explanation is so far a purely phenomenological model only and does not give reasons for the existence of such a field and an emitting plasma, but it adds new evidence to the picture that considerable activity is going on in the center of our galaxy.

The acceleration and/or collimation of gas flows by magnetic fields were discussed in several theoretical talks and were related to the formation of galactic winds and jets from active galactic nuclei (see figures 5 and 6). Again the necessity of including magnetic fields in the theoretical models seem to be unavoidable and, with the new generation of supercomputers, even 3-dimensional simulations are becoming feasible. Problems of similar nature arise also in protostars as was pointed out on the basis of optical polarization measurements during the workshop.

Figure 3: Spiral Arms as Ejection Phenomena?

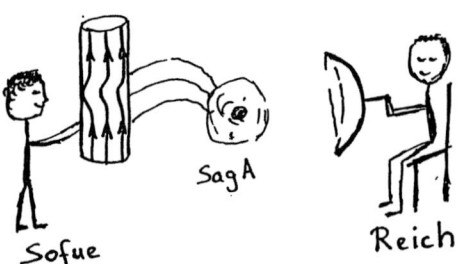

Figure 4: Observations of polarized emission from the galactic center region and a simple theoretical explanation.

Magnetic fields, finally, may also influence the velocities of matter in outer regions of galaxies thereby creating rotation curves which usually are interpreted as being evidence for large amounts of dark matter (figure 7). Although this idea is certainly appealing, the model that has been presented is not yet conclusive, in particular since the motion of matter was assumed to be confined in the galactic plane and strong field concentrations will also push it in vertical direction.

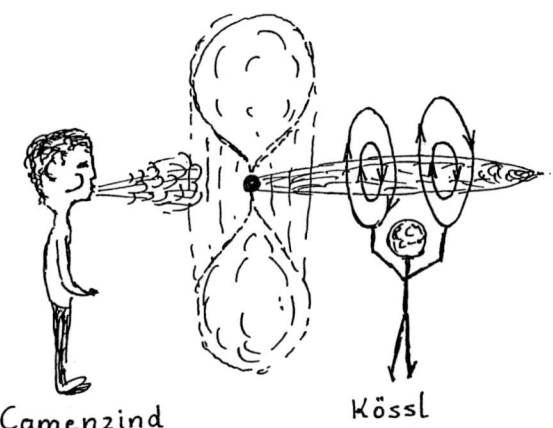

Camenzind   Kössl

Figure 5: The formation of jets in active galactic nuclei by centrifugally driven MHD winds and their collimation by toroidal magnetic fields.

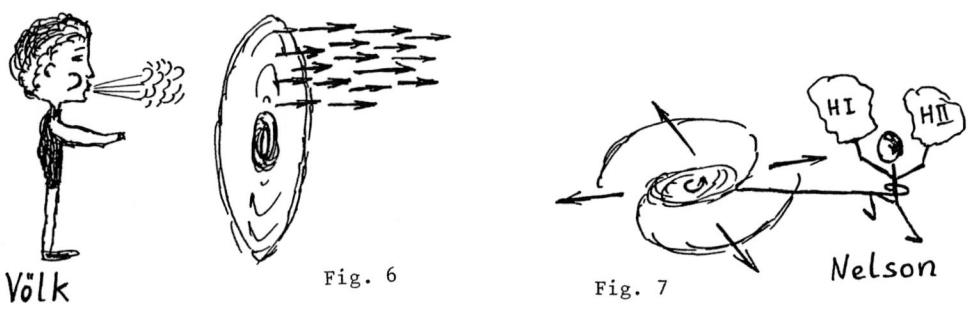

Völk   Fig. 6   Fig. 7   Nelson

Figure 6: The formation of galactic winds. Note that in contrast to figure 5 the disk rather than the galactic nucleus is blown at leading to an uncollimated outflow of particles.

Figure 7: A new explanation of high rotation velocities of outer regions of galaxies. It can be seen that matter is forced into rapid rotation because it is coupled to a strong radial component of a magnetic field.

The summary given here is definitely by no means complete and reflects a biased view of some of the highlights only. But we all left the castle with the impression that interstellar magnetic fields are an exciting and new area of research which deserves considerable effort in the years to come.

Conference Summary

R. Wielebinski

Max-Planck-Institut für Radioastronomie, Auf dem Hügel 69,
D-5300 Bonn 1, Fed. Rep. of Germany

In the last five days we saw many new observational results and heard many new theories about magnetic fields. The new radio data, obtained both at Effelsberg and more recently with the VLA, gave us a new insight into the magnetic field structure in galaxies. Certainly the one dominating impression is the large–scale alignment of the field lines over kpc scales. We saw VLA maps of the field structure in M31 down to 150 pc scale. For me the most amazing result was the good correlation between the field direction derived from optical photopolarimetry and the radio vectors obtained with the VLA for M51. This result means that in face–on galaxies the same magnetic fields that align the dust particles (Davis–Greenstein effect) are also generating radio waves by the synchrotron process. The quality of the optical data and the radio results show that indeed serious studies of magnetic fields in galaxies are possible. Also the use of the Zeeman effect, at least in our Galaxy, offers new possibilities of direct measurement.

Two different theoretical schools have come to word in this conference. The classical dynamo concept was shown to be extendable to give not only axisymmetric fields but also bisymmetrical structures. These developments are, of course, encouraging since we still have no proof of the existence of intergalactic magnetic fields. The alternative approach, that of numerical modelling, is now developed so far that three–dimensional computations seem feasible. Magnetic fields can also be included in some of these simulations. Since magnetic fields are seen in pulsars, supernova remnants, accretion disks etc., their use in numerical simulations seems unavoidable. Certainly I expect great progress in some of the detailed studies in the near future.

The question of the magnetic fields in the universe at large deserves also greater attention. The search for magnetic fields between galaxies using quasar rotation measures is one possibility. Various new and interesting experiments have been proposed. We should hear of further new results at the next conference on magnetic fields in galaxies.

The Sun does not need dynamos to shine.

270

List of Participants

ALY, J.J.	Service d'Astrophysique – CEN Saclay, Gif–sur–Yvette, France
ANZER, U.	Max–Planck–Institut für Physik und Astrophysik, Garching, F.R.G.
ARP, H.	Max–Planck–Institut für Physik und Astrophysik, Garching, F.R.G.
ASSEO, E.	Centre de Physique Théorique, Ecole Polytechnique, Palaiseau, France
BATTANER, E.	Cátedra de Astrofísica, Universidad de Granada, Spain
BECK, R.	Max–Planck–Institut für Radioastronomie, Bonn, F.R.G.
BERKHUIJSEN, E.M.	Max–Planck–Institut für Radioastronomie, Bonn, F.R.G.
BROWNE, P.F.	Dept. of Pure and Applied Physics, University of Manchester, U.K.
BUCZILOWSKI, U.R.	Max–Planck–Institut für Radioastronomie, Bonn, F.R.G.
CAMENZIND, M.	Landessternwarte, Heidelberg–Königstuhl, F.R.G.
CRUSIUS, A.	Max–Planck–Institut für Radioastronomie, Bonn, F.R.G.
CUGNON, P.	Observatoire Royal de Belgique, Bruxelles, Belgium
DRÖGE, W.	Max–Planck–Institut für Radioastronomie, Bonn, F.R.G.
FEITZINGER, J.V.	Astronomisches Institut, Ruhr–Universität Bochum, F.R.G.
FÜRST, E.	Max–Planck–Institut für Radioastronomie, Bonn, F.R.G.
FUJIMOTO, M.	Dept. of Physics, Nagoya University, Nagoya, Japan
GRÄVE, R.	Max–Planck–Institut für Radioastronomie, Bonn, F.R.G.
HILLEBRANDT, W.	Max–Planck–Institut für Physik und Astrophysik, Garching, F.R.G.
HUMMEL, E.	Max–Planck–Institut für Radioastronomie, Bonn, F.R.G.
JUNKES, N.	Max–Planck–Institut für Radioastronomie, Bonn, F.R.G.
KLEIN, U.	Radioastronomisches Institut der Universität Bonn, F.R.G.
KÖSSL, D.	Max–Planck–Institut für Physik und Astrophysik, Garching, F.R.G.
KRAUSE, F.	Zentralinstitut für Astrophysik der Akademie der Wissenschaften der DDR, Potsdam–Babelsberg, G.D.R.
KRAUSE, M.	Max–Planck–Institut für Radioastronomie, Bonn, F.R.G.
KRONBERG, P.P.	Dept. of Astronomy, University of Toronto, Canada
KUNDT, W.	Institut für Astrophysik und Extraterrestrische Forschung der Universität Bonn, F.R.G.
LESCH, H.	Max–Planck–Institut für Radioastronomie, Bonn, F.R.G.
LOISEAU, N.	Max–Planck–Institut für Radioastronomie, Bonn, F.R.G.
MEYER, F.	Max–Planck–Institut für Physik und Astrophysik, Garching, F.R.G.
MEYER–HOFMEISTER, E.	Max–Planck–Institut für Physik und Astrophysik, Garching, F.R.G.
NELSON, A.H.	Dept. of Applied Mathematics and Astronomy, University College, Cardiff, U.K.
REICH, W.	Max–Planck–Institut für Radioastronomie, Bonn, F.R.G.
REIF, K.	Radioastronomisches Institut der Universität Bonn, F.R.G.
RUZMAIKIN, A.	Keldysh Institute of Applied Mathematics, Moscow, U.S.S.R.
SANCHEZ–SAAVEDRA, M.L.	Cátedra de Astrofísica, Universidad de Granada, Spain.
SCARROTT, S.M.	Dept. of Physics, University of Durham, U.K.

SCHMIDT-VOIGT, M. Max-Planck-Institut für Physik und Astrophysik, Garching, F.R.G.
SIEBER, W.   Max-Planck-Institut für Radioastronomie, Bonn, F.R.G.
SOFUE, Y.   Dept. of Astronomy, University of Tokyo, Japan
SPICKER, J.   Astronomisches Institut, Ruhr-Universität Bochum, F.R.G.
THIEMANN, H.   Max-Planck-Institut für Radioastronomie, Bonn, F.R.G.
TOSA, M.   Astronomical Institute, Tohoku University, Sendai, Japan
VERSCHUUR, G.L.   Arecibo Observatory, Puerto Rico, USA
VÖLK, H.   Max-Planck-Institut für Kernphysik, Heidelberg, F.R.G.
WIELEBINSKI, R.   Max-Planck-Institut für Radioastronomie, Bonn, F.R.G.

BREUER, G.   Max-Planck-Institut für Radioastronomie, Bonn, F.R.G.
(Conference Secretary)

Index of Contributors

K. Rohlfs

Tools of Radio Astronomy

1986. 127 figures. XII, 319 pages. (Astronomy and Astrophysics Library). ISBN 3-540-16188-0

Contents: Introduction. – Concordance Relation: ISCU-AB-AAA. – Abbreviations. – Periodicals, Proceedings, Books, Activities. – Applied Mathematics, Physics. – Astronomical Instruments and Techniques. – Positional Astronomy, Celestial Mechanics. – Space Research. – Theoretical Astrophysics. – Sun. – Earth. – Planetary System. – Stars. – Interstellar Matter, Nebulae. – Radio Sources, X-ray Sources, Cosmic Rays, Stellar Systems, Galaxy, Extragalactic Objects, Cosmology. – Author Index. – Subject Index. – Object Index.

Physical Processes in Comets, Stars and Active Galaxies

Proceedings of a Workshop, Held at Ringberg Castle, Tegernsee, May 26–27, 1986

Editors: W. Hillebrandt, E. Meyer-Hofmeister, H.-C. Thomas

1987. 46 figures. Approx. 200 pages. ISBN 3-540-17766-3

This volume contains the lectures presented at a workshop at Ringberg Castle in 1986 on the occasion of the 60th birthdays of Rudolf Kippenhahn and Hermann Ulrich Schmidt. It covers a wide span of seemingly disparate topics ranging from comets, the sun, single and binary stars to active galactic nuclei. Although the astrophysical objects addressed are quite different, the physical processes involved are very much the same. The emphasis on the common physics underlying this wide variety of celestial phenomena makes this a unique book for researchers and students of astrophysics and astronomy.

Springer-Verlag
Berlin Heidelberg New York
London Paris Tokyo

Springer

Weak and Electromagnetic Interactions in Nuclei

Proceedings of the International Symposium on Weak and Electromagnetic Interactions in Nuclei, Heidelberg, 1–5 July 1986

Editor: H. V. Klapdor

1986. 556 figures. Approx. 1100 pages.
ISBN 3-540-17255-6

With contributions by: *B. H. Wildenthal:* Shell-Model Analyses of Weak and Electromagnetic Data. – *G. A. Leander:* Electromagnetic Properties of Nuclei at High Spins. – *C. Gaarde:* Δ-Excitations in Nuclei. – *K. Heyde:* Mixed-Symmetry States in Proton-Neutron Systems. – *F. E. Close:* QCD and Fermi Gas Model Interpretations of the E. M. C. Effect. – *E. L. Berger:* Massive Lepton Pair Production – the Drell-Yan Process – with Nuclear Targets. – *L.-L. Chau:* Status of Electroweak Theory for Heavy Quark Decays and CP Violation. – *R. N. Mohapatra:* Constrains on the Left-Right Symmetric Models of Weak Interactions. – *J. Dubach:* Theoretical Aspects of the Weak Decay of Hypernuclei. – *D. O. Caldwell:* New Limits on Neutrino Masses and Right-Handed Currents from Double Beta Decay. – *J. N. Bahcall:* Solar Neutrinos. – *A. Yu. Smirnov:* Neutrino Oscillations in Matter. – *W. Kündig:* An Upper Limit for the Electron Antineutrino Mass. – *R. Engfer:* Study of Rare and Forbidden μ and π- Decays. – *H. Ohtsubo:* Muon Capture in Nuclei and Determination of Weak Coupling Constants. – *V. M. Lobashev:* An Experimental Search for the Neutron Electric Dipole Moment. – *S. J. Freedman:* Search for Short-Lived Axions. – *P. Langacker:* The Present Status of Proton Decay and Baryon Number Nonconservation. – *R. D. Peccei:* Quasi Standard Model Physics. – *J. P. Derendinger:* Superstrings. – *E. W. Kolb:* Baryon and Lepton Number Violation in Astrophysics. – *R. H. Brandenberger:* The Inflationary Universe: Progress and Problems. – *D. N. Schramm:* Weak Interaction and the Large Scale Structure of the Universe.

Concepts and Trends in Particle Physics

Proceedings of the XXV Internationale Universitätswochen für Kernphysik 1986 der Karl-Franzens-Universität Graz at Schladming (Steiermark, Austria) February 19–27, 1986

Editor: H. Latal, H. Mitter

1987. Approx. 350 pages. ISBN 3-540-17372-2

The twenty-fifth Winter School in Schladming was devoted to important concepts in theoretical particle physics. These include: the standard model of QCD supersymmetric models, superstrings and unification. The authors present their material in considerable detail while at the same time developing the background material necessary for lectures held for advanced students.

The Kaluza-Klein Theory is presented together with a no-go-theorem for certain nonabelian extensions and its consequences for the compactification of extra space dimensions. Two lectures on supersymmetry deal with Yang-Mills fields, supergravity, anomalies and various aspects of gauge covariance. Two long and critical articles present in historical context the mathematical structure of superstring theories and explain why it may help to solve open problems in particle physics and its unification with gravity. Mass issues in QCD and electroweak interactions and a survey on the transition from hadronic matter to the quark-gluon plasma are among the topics treated in QCD. The only experimental contribution is a lecture on experiments beyond the standard model also pointing to the overlap between accelerator physics and modern cosmology.

Springer-Verlag
Berlin Heidelberg New York
London Paris Tokyo